GENETIC MAPS

Locus Maps of Complex Genomes

FIFTH EDITION

Stephen J. O'Brien, Editor
Laboratory of Viral Carcinogenesis, National Cancer Institute

BOOK 3

Lower Eukaryotes

COLD SPRING HARBOR LABORATORY PRESS 1990

GENETIC MAPS

First edition, March 1980
Second edition, June 1982
Third edition, June 1984
Fourth edition, July 1987
Fifth edition, February 1990

GENETIC MAPS
Locus Maps of Complex Genomes
Fifth Edition
BOOK 3 Lower Eukaryotes

Printed in the United States of America

Library of Congress Catalog Card Number 84-644938
ISBN 0-87969-344-4
ISSN 0738-5269

Cover design by Leon Bolognese

All Cold Spring Harbor Laboratory Press publications may be ordered directly from Cold Spring Harbor Laboratory, Box 100, Cold Spring Harbor, New York 11724. Phone 1-800-843-4388. In New York (516) 367-8423.

PREFACE

"The map is a sensitive indicator of the changing thought of man, and few of his works seem to be such an excellent mirror of culture and civilization."

Norman J.W. Thrower
Maps and Men, 1972

The early geographic explorers are revered by their countrymen and descendants throughout the world for pioneering the discovery and charting of foreign lands. The precision and attention to detail accorded to their charts and maps are almost always taken for granted today. Astronauts, and then those who saw their photographs from space, have marveled at the resemblance of these real images of the Florida peninsula, the boot of Italy, and the British Isles to the maps we had grown up with. John Noble Wilford, in his fascinating monograph entitled *The Mapmakers,* reminds us that it is only within the last quarter of the twentieth century that it could be said that the earth has been mapped. It is humbling to consider how important these maps have been to our history, our culture, and our way of life today; in this context, the early cartographers were indeed intrepid.

The new explorers of the eukaryote and prokaryote genome will certainly one day be considered with equivalent veneration, since the topography of their charts and maps may prove as daunting as the earth's surface. The animal and plant genomes are products of several hundred million years of biological evolution, and there are clearly far more secrets to be deciphered in the genomes that can be anticipated in future generations. However, genetic detectives have begun to unravel some of the mysteries of genetic organization, and as we enter the final decade of the century, the scientific community (as well as the enlightened public) has begun to grasp the enormous value of generating and expanding genetic maps of man, of agriculturally significant plants and animals, and of model systems that allow us a glimpse of how genes are organized, replicated, and regulated.

The first genetic map was formulated by A.H. Sturtevant in 1913 and consisted of five genes arranged in a linear fashion along the X chromosome of the fruit fly, *Drosophila melanogaster.* In the ensuing decades, gene mapping of numerous species has proceeded deliberately and cumulatively in organisms as diverse as flies, corn, wheat, mink, apes, man, and bacteria. All of these maps, whether based on recombination, restriction, physical or DNA sequence, are predicated on Sturtevant's logical notion that gene order on a chromosome could be displayed as a linear array of genetic markers. The results of these efforts in more than 100 genetically studied organisms are the basis of *GENETIC MAPS: Locus Maps of Complex Genomes.*

During the preparation of the previous edition, it occurred to me that the rate of growth of the gene mapping effort was so rapid that we should prepare to publish future editions in multiple volumes. The fifth edition is a realization of that idea. *Genetic Maps* now consists of six smaller books, each based on an arbitrary subdivision of biological organisms. These are BOOK 1 Viruses; BOOK 2 Bacteria, Algae, and Protozoa; BOOK 3 Lower Eukaryotes; BOOK 4 Nonhuman Vertebrates; BOOK 5 Human Maps; BOOK 6 Plants. Each of these is available in paperback at a modest cost from Cold Spring Harbor Laboratory Press. The entire compendium can be purchased as a hardback volume of 1098 map pages suitable for libraries and research institutions.

Our original intent was to publish in one volume complete, referenced genetic maps of every organism with a substantive group of assigned loci. We intentionally excluded DNA sequences, since these are easily available in several computer databanks (Genbank, EMBL, and others). Text was to be kept at a minimum, and the maps would be both comprehensive and concise. The collection was to be updated every 2–3 years, and each new volume would contain a complete revision, rendering the previous volume obsolete.

The original publication of *Genetic Maps* was an experiment, which I believe now can be judged a success. The execution of such a venture depended heavily on the support and cooperation of thousands of geneticists throughout the worldwide scientific community. This cooperation was graciously extended, and the result was an enormously valuable and accessible collection. On behalf of my colleagues in the many fields that we call genetic biology, I gratefully acknowledge the numerous geneticists, scientists, and readers who have contributed to or corrected these maps. To ensure the continuation of these heroic efforts in the future, readers are encouraged to send to me any suggestions for improvement, particularly suggestions for new maps to be included, as well as constructive criticisms of the present maps. When new organisms are recommended, I would also appreciate names and addresses of prospective authors. In addition, readers are encouraged to supply corrections, reprints, and new mapping data to the appropriate authors who may be included in future editions.

The compilation of each genetic map, like the drafting of the first geographic maps, is an extremely tedious, yet important, assignment. All of the authors deserve special thanks for the large efforts they have expended in contributing their maps. Even in this computer age, we often feel like we are proofreading the telephone book, but I, for one, believe that the final product makes the effort worthwhile. Finally, I acknowledge specifically my editorial assistants, Patricia Johnson and Virginia Frye, who have cheerfully and expeditiously carried the bulk of the editcrial activities for the present volume, and Annette Kirk, Lee Martin, and John Inglis of Cold Spring Harbor Laboratory Press for support and advice on the preparation of this edition.

Stephen J. O'Brien, *Editor*

CONTENTS

COMPLETE CONTENTS OF THE FIFTH EDITION

PROTOZOA

Paramecium tetraurelia	K.J. Aufderheide and D. Nyberg
Tetrahymena thermophila	P.J. Bruns

BOOK 3 Lower Eukaryotes

FUNGI

Dictyostelium discoideum	P.C. Newell
Neurospora crassa	
nuclear genes	D.D. Perkins
mitochondrial genes	R.A. Collins
restriction polymorphism	R.L. Metzenberg and J. Grotelueschen
Saccharomyces cerevisiae	
nuclear genes	*R.K. Mortimer, D. Schild, C.R. Contopoulou, and J.A. Kans*
mitochondrial DNA	*L.A. Grivell*
Podospora anserina	D. Marcou, M. Picard-Bennoun, and J.-M. Simonet
Sordaria macrospora	G. Leblon, V. Haedens, A.D. Huynh, L. Le Chevanton, P.J.F. Moreau, and D. Zickler
Coprinus cinereus	J. North
Magnaporthe grisea (blast fungus)	D. Z. Skinner, H. Leung, and S.A. Leong
Phycomyces blakesleeanus	M. Orejas, J.M. Diaz-Minguez, M.I. Alvarez, and A.P. Eslava
Schizophyllum commune	C.A. Raper
Ustilago maydis	A.D. Budde and S.A. Leong
Ustilago violacea	A.W. Day and E.D. Garber
Aspergillus nidulans	
nuclear genes	A.J. Clutterbuck
mitochondrial genome	T.A. Brown

NEMATODE

Caenorhabditis elegans	M.L. Edgley and D.L. Riddle

INSECTS

Drosophila melanogaster	
biochemical loci	G.E. Collier
in situ hybridization data	E. Kubli and S. Schwendener
cloned genes	J. Merriam, S. Adams, G. Lee, and D. Krieger
Aedes (Stegomyia) aegypti	L.E. Munstermann
Aedes (Protomacleaya) triseriatus	L.E. Munstermann
Drosophila pseudoobscura	W.W. Anderson
Anopheles albimanus	S.K. Narang and J.A. Seawright
Anopheles quadrimaculatus	S.K. Narang, J.A. Seawright, and S.E. Mitchell
Nasonia vitripennis	G.B. Saul, 2nd

BOOK 4 Nonhuman Vertebrates

RODENTS

Mus musculus (mouse)	
nuclear genes	M.T. Davisson, T.H. Roderick, D.P. Doolittle, A.L. Hillyard, and J.N. Guidi
DNA clones and probes, RFLPs	J.T. Eppig
retroviral and cancer-related genes	C.A. Kozak
Rattus norvegicus (rat)	G. Levan, K. Klinga, C. Szpirer, and J. Szpirer
Cricetulus griseus (Chinese hamster)	
nuclear genes	R.L. Stallings, G.M. Adair, and M.J. Siciliano
CHO cells	G.M. Adair, R.L. Stallings, and M.J. Siciliano
Peromyscus maniculatus (deermouse)	W.D. Dawson
Mesocricetus auratus (Syrian hamster)	R. Robinson
Meriones unquiculatus (Mongolian gerbil)	R. Robinson

OTHER MAMMALS

Felis catus (cat)	S.J. O'Brien
Canis familiaris (dog)	P. Meera Khan, C Brahe, and L.M.M. Wijnen
Equus caballus (horse)	L.R. Weitkamp and K. Sandberg
Sus scrofa domestica L. (pig)	G. Echard
Oryctolagus cuniculus (rabbit)	R.R. Fox
Ovis aries (sheep)	G. Echard
Bos taurus (cow)	J.E.Womack
Mustela vison (American mink)	O.L. Serov and S.D. Pack
Marsupials and Monotremes	J.A. Marshall Graves

PRIMATES

Primate Genetic Maps	N. Creau-Goldberg, C. Cochet, C. Turleau, and J. de Grouchy
Pan troglodytes (chimpanzee)	
Gorilla gorilla (gorilla)	
Pongo pygmaeus (orangutan)	
Hylobates (Nomascus) concolor (gibbon)	
Macaca mulatta (rhesus monkey)	
Papio papio, hamadryas, cynocephalus (baboon)	
Cercopithecus aethiops (African green monkey)	
Cebus capucinus (capuchin monkey)	
Microcebus murinus (mouse lemur)	
Saquinus oedipus (cotton-topped marmoset)	P.A. Lalley
Aotus trivirgatus (owl monkey)	N.S.-F. Ma

FISH

Salmonid fishes *Salvelinus, Salmo, Oncorhynchus*	B. May and K.R. Johnson
Non-Salmonid fishes *Xiphophorus, Poeciliopsis, Fundulus, Lepomis*	D.C. Morizot

AMPHIBIAN

Rana pipiens (leopard frog)	D.A. Wright and C.M. Richards

BIRD

Gallus gallus (chicken)	R.G. Somes, Jr., and J.J. Bitgood

BOOK 5 Human Maps

BOOK 6 Plants

BOOK 3

LOWER EUKARYOTES

Genetic map of Dictyostelium discoideum (Cellular Slime Mould)

June 1989

PETER C. NEWELL
Department of Biochemistry
University of Oxford
South Parks Road
Oxford OX1 3QU, U.K.

NOTE: All markers have been mapped by parasexual genetic techniques. Square brackets [] denote genes mapped by restriction fragment length polymorphisms in combination with parasexual analysis.

Linkage Group	Gene Symbol	Mutant Phenotype/[Gene function]	Reference
I	abpA	Lacks 95 Kd actin-binding protein α-actinin	46
	act-8	[Actin 8 gene]	51
	acrB	Growth in the presence of acriflavin (100ug/ml)	33
	aggB	Aggregation deficient	8
	benA	Growth in the presence of benlate (600 ug/ml)	29
	cadA	Growth in the presence of cadmium sulfate (150 ug/ml)	9
	couB	Sensitive to coumarin (1.3 mM)	36
	couI	Sensitive to coumarin (1.3 mM)	43
	cycA	Growth in the presence of cycloheximide (500 ug/ml)	1
	devA	Developmental mutation	4
	matA	Mating-type locus (also mata)	30
	modA	Alteration of glycosidase protein modification	7
	radA	Sensitive to U.V. and γ-irradiation	6
	sprA	Round spores	1
	$tRNA^{Val}$(GUU)C	[Valine tRNA (GUU) gene]	53
	$tRNA^{Val}$(GUU)E	[Valine tRNA (GUU) gene]	53
	$tRNA^{Val}$(GUU)F	[Valine tRNA (GUU) gene]	53
	tsgE	Temperature sensitive for frowth	1
	tsgI	Temperature sensitive for growth	7
	tsgL	Temperature sensitive for growth	18
	tsgQ	Temperature sensitive for growth	36
II	acrA	Growth with acriflavin (100 ug/ml) or methanol (2%)	3
	actM6	[Actin M6 gene]	51
	act-12	[Actin 12 gene]	51
	aggA	Aggregation deficient	8
	aggF	Aggregation deficient	11
	aggI	Aggregation deficient (redesignated fgdA, see ref: 42)	11
	arsA	Growth in the presence of Na arsenate (1.5 mg/ml)	33

Linkage Group	Gene Symbol	Mutant Phenotype/[Gene function]	Reference
II continued			
	arsB	Growth in the presence of Na arsenate (1.5 mg/ml)	33
	axeA	Axenic growth (provided that axeB is present)	3
	axeC	Axenic growth (provided axeA and axeB are present)	17
	casA	Able to fruit in the presence of cAMP-S (2 uM)	46
	cbpA	Altered discoidin (CBP) protein	22
	clyA	Wide band of unaggregated amoebae in colony	28
	couC	Sensitive to coumarin (1.3 mM)	36
	couE	Sensitive to coumarin (1.3 mM)	36
	couF	Sensitive to coumarin (1.3 mM)	36
	daxA	Failure to accumulate discoidin I during axenic growth	56
	devB	Developmental mutation	4
	disA	Unable to form discoidin	49
	fgdA	Unresponsive (frigid) to cAMP	42
	motA	Temperature sensitive motility	56
	nysA	Resistance to nystatin (100ug/ml)	52
	oaaA	Development sensitive to ω-amino acids	17
	radA	Sensitive to U.V. and γ-irradiation	18
	slgE	Unable to develop beyond slug stage	39
	sprB	Long spores	5
	stkA	Differentiation only into stalk cells	25
	stmB	Formation of large aggregation streams	19
	stmD	Formation of large aggregation streams	19
	stmF	Large aggregation streams: inactive cGMP pde	19,47
	$tRNA^{Val}$(GUU)D	[Valine tRNA (GUU) gene]	53
	$tRNA^{Val}$(GUU)I	[Valine tRNA (GUU) gene]	53
	tsgD	Temperature sensitive for growth	1
	tsgF	Temperature sensitive for growth	4
	tsgH	Temperature sensitive for growth	7
	tsgP	Temperature sensitive for growth	25
	whiA	White spore mass	1
III	acrC	Growth in the presence of acriflavin (100 ug/ml)	4
	alpA	Altered alkaline phosphatase activity	21
	axeB	Axenic growth (provided axeA is present)	3
	bsgA	Unable to grow on Bacillus subtilis	12
	couD	Sensitive to coumarin (1.3 mM)	36
	couG	Sensitive to coumarin (1.3 mM)	43
	couJ	Sensitive to coumarin (1.3 mM)	43
	disB	Unable to form discoidin	49
	radB	Sensitive to U.V. and γ-irradiation	6
	radC	Sensitive to U.V. irradiation	6
	radE	Sensitive to U.V. and γ-irradiation	20
	radG	Sensitive to U.V. and γ-irradiation	18
	rdeC	Rapid development	40
	slgB	Unable to develop beyond the slug stage	39

Linkage Group	Gene Symbol	Mutant Phenotype/[Gene function]	Reference
III continued			
	slgD	Unable to develop beyond the slug stage	39
	slgG	Unable to develop beyond the slug stage	39
	stmC	Formation of large aggregation streams	19
	stmE	Formation of large aggregation streams	19
	tsgA	Temperature sensitive for growth	2
	tsgC	Temperature sensitive for growth	2
	tsgJ	Temperature sensitive for growth	42
	tsgN	Temperature sensitive for growth	30
	tsgR	Temperature sensitive for growth	36
	whiB	White spore mass	25
IV	acrD	Growth in the presence of acriflavin (100 ug/ml)	32
	aggJ	Aggregation deficient	10
	aggL	Aggregation deficient	34
	bwnA	Formation of brown pigment	1
	bsgC	Unable to grow on Bacillus subtilis	24
	couH	Sensitive to coumarin (1.3 mM)	43
	drsA	Discoidin restoring (suppressor of disB)	49
	ebrA	Growth in the presence of ethidium bromide (35 ug/ml)	14
	ebrB	Growth in the presence of ethidium bromide (35 ug/ml)	27
	folA	Unable to chemotax to folate	55
	frtA	Spore mass tends to slide to base of fruiting body	31
	minA	Colonies grow very slowly on SM agar	28
	nagA	Altered N-acetylglucosaminidase activity	16
	nysC	Resistance to nystatin (100 ug/ml)	52
	pdsA	Deficient in phosphodiesterase	26
	radD	Sensitive to U.V. and γ-irradiation	18
	radF	Sensitive to U.V. and γ-irradiation	18
	rdeA	Rapid development	13
	slgC	Long delay at slug stage	39
	slgI	Unable to develop beyond slug stage	39
	sprH	Round spores	28
	sprJ	Spores fail to mature	29
	$tRNA^{Val}$(GUU)G	[Valine tRNA (GUU) gene]	53
	tsgB	Temperature sensitive for growth	2
	whiC	White spore mass	25
	phgB	Altered phagocytic ability	45
V		No established markers	
VI	gluA	Altered β-glucosidase activity	16
	manA	α-mannosidase-1 deficient	7
	nysB	Resistance to nystatin (100 ug/ml)	52
	modB	Altered post-translational modification of proteins	44
	stlA	Deficient in stalk formation	25
	tsgV	Temperature sensitive for growth	48

Linkage Group	Gene Symbol	Mutant Phenotype/[Gene function]	Reference
VII	bsgB	Unable to grow on Bacillus subtilis	25
	cobA	Growth with cobaltous chloride (300-360 ug/ml)	15,31
	couA	Sensitive to coumarin (1.3 mM)	31
	fgdB	Aggregation deficient: unresponsive (frigid) to cAMP	42
	frtB	Formation of aggregates in rings in clones on agar	31
	phgA	Altered phagocytic ability	45
	relA	Recessive lethal (causes cell death in haploids)	54
	relB	Recessive lethal (causes cell death in haploids)	54
	slgJ	Unable to develop beyond slug stage	39
	stmA	Formation of large aggregation streams	19
	tsgG	Temperature sensitive for growth	4,31
	tsgK	Temperature sensitive for growth	19
	tsgM	Temperature sensitive for growth	31

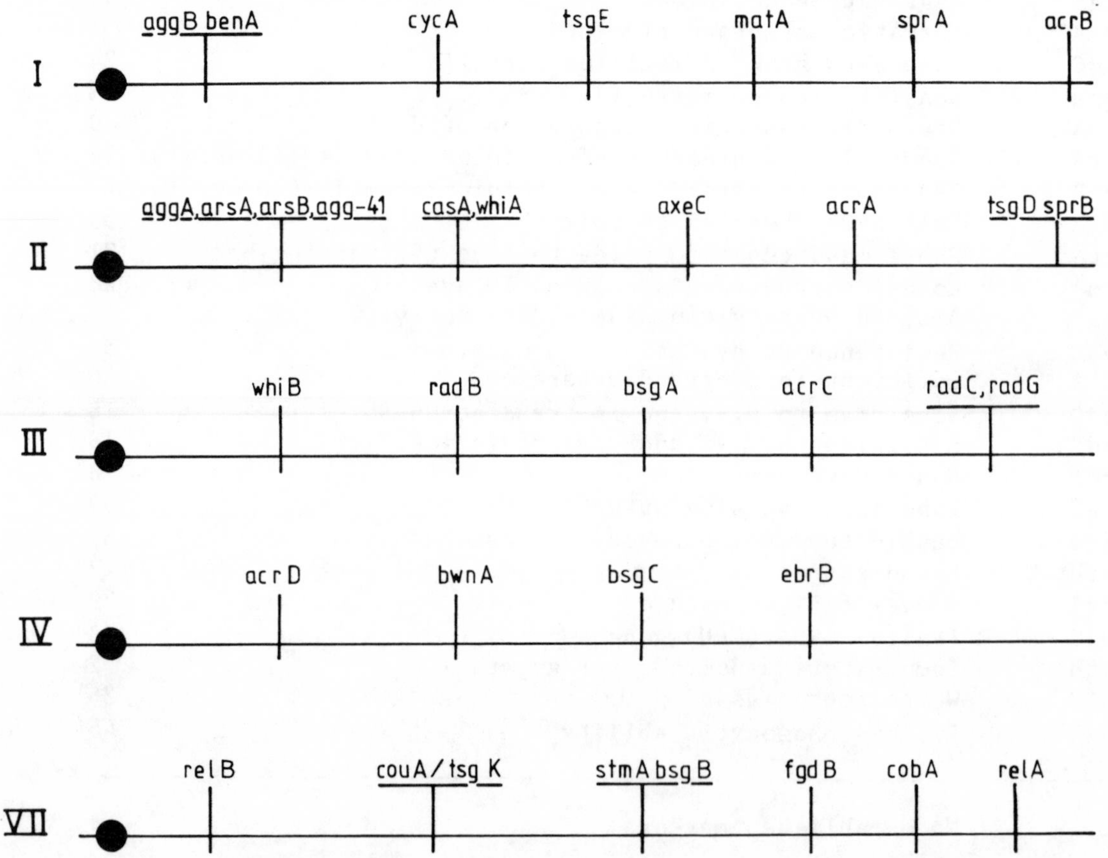

Figure 1 Map of genetic markers on Dictyostelium discoideum linkage groups I, II, III, IV, and VII. The figure indicates the relative order of the markers as found from mitotic recombination studies (Refs: 3,8,5,38,36,and 35).

REFERENCES

1. Katz, E. R., and Sussman M. (1972) Proc. Natl. Acad. Sci. USA 69, 495-498.
2. Kessin, R. H., Williams, K. L., and Newell P. C. (1974) J. Bacteriol. 119, 776-783-mutants.
3. Williams, K. L., Kessin, R. H., and Newell P. C. (1974) J.Gen. Microbiol. 84, 59-69.
4 Rothman, F. G., and Alexander E. T. (1975) Genetics 80, 715-731.
5. Mosses, D. M., Williams, K. L., and Newell P. C. (1975) J.Gen. Microbiol. 90, 247-259.
6. Welker, D. L., and Deering R. A. (1976) J. Gen. Microbiol. 97, 1-10.
7. Free, S. J., Schimke, R. T., and Loomis W. F. (1976) Genetics 84, 159-174.
8. Williams, K. L., and Newell P. C. (1976) Genetics 82, 287-307.
9. Williams, K. L., and Newell P. C. (1976) Unpublished.
10. Coukell, M. B. (1977) Mol. Gen. Genet. 151, 269-273.
11. Coukell, M. B., and Roxby N. M. (1977) Mol. Gen. Genet. 151, 275-288.
12. Newell, P. C., Henderson, R. F., Mosses, D., and Ratner D. I. (1977) J. Gen. Microbiol. 100, 207-211.
13. Kessin, R. H. (1977) Cell 10, 703-708.
14. Wright, M. D., Williams, K. L., and Newell P. C. (1977) J. Gen. Microbiol. 102, 423-426.
15. Ratner, D. I., and Newell P. C. (1978) J. Gen. Microbiol. 109, 225-236.
16. Loomis, W. F. (1978) Genetics 88, 277-284.
17. North, M. J., and Williams K. L. (1978) J. Gen. Microbiol.107, 223-230.
18. Welker, D. L., and Deering R. A. (1978) J. Gen. Microbiol. 109, 11-23.
19. Ross, F. M., and Newell P. C. (1979) J. Gen. Microbiol. 115, 289-300.
20. Coukell, M. B., and Cameron A. M. (1979) J. Gen. Microbiol. 114, 247-256.
21. MacLeod, C. L., and Loomis W. F. (1979) Dev. Genet. 1, 109-121.
22. Ray, J., Shinnick, T. M., and Lerner R. A. (1979) Nature 279, 215-221.
23. Welker, D. L., and Williams K. L. (1980) J. Gen. Microbiol. 116, 397-407.
24. Welker, D. L., and Williams K. L. (1980) FEMS Microbiol. Lett. 9, 179-183.
25. Morrissey, J. H., Wheeler, S., and Loomis W. F. (1980) Genetics 96, 115-123.
26. Barra, J., Barrand, P., Blondelet, M. H., and Brachet P. (1980) Mol. Gen. Genet. 177, 607-613.
27. Welker, D. L., and Williams K. L. (1980) Unpublished.
28. Williams, K. L., Robson, G. E., and Welker D. L. (1980) Genetics 95, 289-304.
29. Williams, K. L., and Welker D. L. (1980) Develop. Genet. 1, 355-362.
30. Williams, K. L., and Welker D. L. (1980) Unpublished.
31. Welker, D. L., and Williams K. L. (1980) J. Gen. Microbiol. 120, 149-159.
32. Welker, D. L., and Williams K. L. (1981) Chromosoma 82, 321-332.
33. Williams, K. L. (1981) FEMS Microbiol. Lett. 11, 317-320.

REFERENCES (cont'd)

34. Glazer, P. M., and Newell P. C. (1981) J. Gen. Microbiol. 125, 221-232.
35. Welker, D. L., and Williams K. L. (1982) Genetics 102, 691-710.
36. Welker, D. L., and Williams K. L. (1982) J. Gen. Microbiol. 128, 1329-1343.
37. Van Haastert, P. J. M., Van Lookeren Campagne, M. M., and Ross F. M. (1982) FEBS Lett. 147, 149-152.
38. Wallace, J. S., and Newell P. C. (1982) J. Gen. Microbiol. 128, 953-964.
39. Ross, F. M., and Newell P. C. (1982) J. Gen. Microbiol. 128, 1639-1652.
40. Abe, K., Okada, Y., Wada, M., and Yanagisawa K. (1983) J. Gen. Microbiol. 129, 1623-1628.
41. Kasbekar, D. P., Madigan, S., and Katz E. R. (1983) Genetics 104, 271-277.
42. Coukell, M. B., Lappano, S., and Cameron A. M. (1983) Devel. Genet. 3, 283-297.
43. Welker, D. L., and Williams K. L. (1983) J. Gen. Microbiol. 129, 2207-2216.
44. Murray, B. A., Wheeler, S., Jongens, T., and Loomis W. F. (1984) Mol. Cell. Biol. 4, 514-519.
45. Duffy, K. T. I., and Vogel G. (1984) J. Gen. Microbiol. 130, 2071-2077.
46. Wallraff, E., Welker, D. L., Williams, K. L., and Gerisch G. (1984) J. Gen. Microbiol. 130, 2103-2114.
47. Coukell, M. B., and Cameron A. M. (1985) J. Bact. 162, 427-429.
48. Welker, D. L., and Williams K. L. (1985) Genetics 109, 341-364.
49. Alexander, S., Cibulsky, A. M., and Cuneo S. D. (1986) Mol. Cell. Biol. 6, 4353-4361.
50. Wallraff, E. M., Schleicher, M., Modersitzki, D., Reiger, G., Isenberg, G., and Gerisch G. (1986) EMBO J. 5, 61-67.
51. Welker, D. L., Hirth, K. P., Romans, P., Noegel, A., Firtel, R. A., and Williams K. L. (1986) Genetics 112, 27-42.
52. Welker, D. L. (1986) Genetics 113, 53-62.
53. Dingermann, T., Amon, E., Williams, K. L., and Welker, D. L. (1987) Mol. Gen. Genet. 207, 176-187.
54. Welker, D.L., and Williams, K. L. (1987) Genetics 115, 101-106.
55. Segall, J. E., Bominaar, A. A., Wallraff, E., and De Wit R. J. W. (1988) J. Cell Sci. 91, 479-489.
56. Kayman, S. C., Birchman, R., and Clarke M. (1988) Genetics 118, 425-436.

NEUROSPORA CRASSA GENETIC MAPS, June, 1989

David D. Perkins

Department of Biological Sciences, Stanford University, Stanford, California 94305-5020

The maps are based largely on information compiled in a 1982 compendium [reference 64]. Newly mapped loci and new information on linkage have now been incorporated in the maps or added under each figure. The figures show gene loci whose order is established with reasonable certainty. Gene order is based either on meiotic crossing-over data from 3-point crosses or on duplication coverage. Rearrangement breakpoints are shown when they have been used for seriation or for mapping chromosome tips. Order of markers in parentheses on the maps is uncertain. The centromere is represented as a circle. Loci are listed below the maps if their order is uncertain relative to genes shown on the map. Percentages and fractions indicate recombination among random ascospore progeny, unless stated otherwise. Where no reference is given, either the data are documented in [64] or are from this laboratory, unpublished. — — *Distances on the maps are only rough approximations. rec* genes with local effects differ in many of the strains used for mapping, and recombination values for the same interval can vary tenfold or more in crosses of different parentage [see reference 21]. For this reason no map scale is indicated. Linkage group I is estimated to be at least 200 map units long, and the total for all seven groups probably exceeds 1000 [65]. One centimeter thus represents about 7% recombination. — — Mapping data have come from many workers. For loci mapped before 1982 (both on the maps and listed below them) references are given in reference 64, which also gives information on phenotype and scoring. References for loci added to the map figures since 1982 are as follows: *ace-8* [46], *Bml* [64], *cot-5* [31a], *cpc-1* [69b], *cum* [9], *cya-8* [68], *cya-7* [68], *dgr-1* [1], *lah-1* [34], *mus-16* [37, 74], *mus-23* [34], *mus-26* [35], *oak* [68], *oli* [48], *pmb* [64], *prd-4* [79, 18a], *Sk-2*, *Sk-3* [9], *T(NM183)* [9], *Tel-IIL, -IIR, -IIIR, -IVL, -VR* [73], *Tel-VIIL* [73, 53], *tRNA^{LEU}* [32a]. References for loci mapped since 1982 but not included in the map figures are given in square brackets in the lists below the maps; where no reference is given the data are from this laboratory, unpublished. For background and references on mapping methodology, use of chromosome rearrangements, linkage group-chromosome correlations, genetic nomenclature, see [64, 65, 63]. Stocks of most mutants are available from the Fungal Genetics Stock Center, Department of Microbiology, University of Kansas Medical Center, Kansas City KS 66103. Stock lists are published by the Stock Center. — — Numerous workers have generously provided information prior to publication as indicated in the references. Support has come from N.I.H. Grants AI-01462 and K6-GM-4899.

Gene symbols and names of mapped loci are listed below, followed by a list of unmapped genes. The list of mapped loci includes some that are not shown on or beneath the maps because they have not been tested for allelism with nearby genes that have similar phenotypes. All these symbols should be considered preempted and should be avoided when new loci are designated. The word "requirement" is implied but not stated in naming auxotrophs.

SYMBOLS OF MAPPED LOCI: *ace*--acetate, *acon*--aconidiate, *acr*--acriflavine resistant, *acu*--acetate utilization, *ad*--adenine, *adh*--adherent, *ads*--adenine sensitive, *aga*--arginase, *age*--ageing of conidia, *al*--albino, *alc*--allantoicase, *aln*--allantoinase, *am*--amination deficient (glutamate dehydrogenase), *amy*--amylase (see *exo, sor)*, *amyc*--amycelial, *aod*--alternate oxidase deficient, *arg*--arginine, *arg^R*--arginine resistant, *aro*--aromatic biosynthesis, *ars*--aryl sulfatase, *asc*--defective ascospore formation, *asco*--ascospore color (see *lys-5)*, *asn*--asparagine, *asp*--aspartic acid, *at*--attenuated, *atr*--aminotriazole resistant, *aza*--azapurine resistant. — *B^m*--mauve colony, *bal*--balloon, *Ban*--Banana asci, *bat*--basic amino acid transport (see *pmb*), *bd*--band, *bis*--biscuit (see *pk*--peak), *Bml*--Benomyl resistant, *bn*--button, *bs*--brown ascospore. — *caf*--caffeine resistant, *can*--(see *cnr*), *car*--glucose transport, *cat*--catalase, *ccg*--clock-controlled gene, *cdc*--cell division cycle, *cel*--chain elongation (saturated fatty acid), centromere--open circle on map, *chol*--choline, *chr*--chrono (rhythm), *cl*--clock, *cnr*--canavanine resistant, *cog*--recognition (recombination control), *coil*--coiled hyphae, *chol*--choline, *cmt*--copper metallothionein (the symbol *cum*, proposed by [25], is preempted), *col*--colonial, *con*--sequences transcribed differentially during conidiation, *cot*--colonial temperature-sensitive, *cox*--cytochrome oxidase subunit, *cpc*--crosspathway control, *cpd*--cyclic phosphodiesterase (orthophosphate regulated), *cpk*--cAMP-dependent protein kinase, *cpl*--chloramphenicol sensitive, *cpt*--carpet, *cr*--crisp, *crib*--cytoplasmic ribosome, *crp*--cytoplasmic ribosomal protein, *csh*--cushion, *csp*--conidial separation, *csr*--cyclosporin resistant, *cum*--cumulus, *cut*--cut, *cwl*--crosswall, *cy*--curly, *cya*--cytochrome *a*, *cyb*--cytochrome *b*, *cyc*--cytochrome *c*, *cyh*--cycloheximide resistant, *cys*--cysteine, *cyt*--cytochrome. — *da*--dapple, *del*--delicate, *dgr*--deoxyglucose resistant, *dir*--dirty, *dn*--dingy, *do*--doily, *dot*--dot, *dow*--downy, *dr*--drift. — *eas*--easily wettable, *edr*--edeine resistant, *en*( )--enhancer of ( ), *erg*--ergosterol (nystatin resistant), *eth*--ethionine resistant, *exo*--exoamylase. — *f*--fast (see *su([mi-1])*), *fdu*--fluorodeoxyuridine resistant, *ff*--female fertility, *fi*--fissure, *fl*--fluffy, *fld*--fluffyoid, *flm*--flame (see *os*), *fls*--fluffyish, *fmf*--female and male fertility, *for*--formate, *fpr*--*p*-fluorophenylalanine resistant, *fr*--frost, *frq*--frequency (rhythm), *fs*--female sterile, *Fsp*--Four-spore ascus, *Fsr*--5S RNA. — *gln*--glutamine, *glp*--glycerol phosphate utilization, *gluc*--aryl beta-glucosidase, *gly*--glycerol utilization (see *glp*), *gpi*--glucose phosphate isomerase, *gran*--granular, *gsp*--giant ascospore, *gua*--guanine, *gul*--gulliver (suppressor of *cot-1*). — *H3*, *H4*--histones H3, H4, *hbs*--homebase (anonymous DNA fragment), *het*--heterokaryon (vegetative) incompatibility, *hgu*--histidyl-glycine uptake, *his*--histidine, *hlp*--histidinol permeability, *hom*--homoserine, *hss*--histidine sensitive. — *Iasc*--Indurated ascus, *ile*--isoleucine, *ilv*--isoleucine + valine, *In*( )--inversion, *inl*--inositol, *int*--intense (pigment), *inv*--invertase, *ipa*--it pokes along. — *kyn*--kynureninase. — *lah*--laccase halo, *le*--lethal (ascospore), *leu*--leucine, *lox*--L-amino acid oxidase, lp--lump, *lys*--lysine, lys^R--lysine resistant. — *mac*--methionine + adenine, *mat*--mat, *mb*--male barren, *md*--mad, *med*--medusa, *mei*--meiotic, *mel*--melon, *mep*--6-methylpurine resistant, *met*--methionine, [*mi*-]--maternal inheritance (cytoplasmic), *mig*--migration of trehalase, *mo*--morphological, *mod*( )--modifier of ( ), *moe*--morphological environment sensitive, *mt*--mating type (*A*) or (*a*), *mtr*--methyl tryptophan resistant, *mts*--methyltryptophan sensitive, *mus*--mutagen sensitive. — *nada*--NAD(P) glycohydrase, *nap*--neutral and acidic amino acid permeability, *nd*--natural death, *ndc*--nuclear division cycle, *nic*--nicotinic acid, *nit*--nitrate nonutilizer, *nmr*--nitrogen metabolite regulation, *nt*--nicotinic acid or tryptophan, *NO*--nucleolus organizer, *nuc*--nuclease, *nuh*--nuclease acid halo. — *oak*--oak, *oli*--oligomycin resistant, *opi*--overproduction of inositol, *os*--osmotic sensitive, *ota*--ornithine transminase, *oxD*--D-amino acid oxidase. — *pab*--*p*-aminobenzoic acid, *pan*--pantothenic acid, *pat*--patch, *pcon*--phosphatase control, *pde*--phosphodiesterase, cyclic, GTP-regulated, *pdx*--pyridoxine, *pe*--peach, *pen*--perithecial neck, *per*--perithecial color, *pf*--puff, *pgov*--phosphorus governance, *phe*--phenylalanine, *pho*--phosphatase, *pi*--pile, *pk*--peak (=*bis*), *pl*--plug, *pma*--plasma membrane H^+ATPase, *pmb*--permease basic amino acid, *pmg*--permease general amino acid, *pmn*--permease neutral amino acid, *prd*--period (rhythm), *preg*--phosphatase regulation, *Prf*--Perforated, *pro*--proline, *psi*--protein synthesis initiation, *pt*--phenylalanine plus tyrosine, *put*--putrescine (see *spe*), *pyr*--pyrimidine. — *qa*--quinate catabolism. — *R*--Round ascospore, *r*(*Sk*-)-resistant to *Sk*- , *rec*--recombination affector, *rg*--ragged, *rib*--riboflavin, *rip*--ribosome production, *ro*--ropy, *rol*--ropy-like. — *sar*--surfactant resistant, *sat*--satelliteless, *sc*--scumbo, *scon*--sulfur metabolism control, *scot*--spreading colonial temperature sensitive, *scr*--scruffy, *sdh*--succinic dehydrogenase, *ser*--serine, *sf*--slow fine hyphae, *sfo*--sulfonamide dependent, *sh*--shallow, *shg*--shaggy, *sk*--skin, *Sk*--Spore killer, *slo*--slow growth, *smco*--semicolonial, *sn*--snowflake, *so*--soft, *sod*--superoxide dismutase, *sor*--sorbose resistant, *sp*--spray, *spco*--spreading colonial, *spe*--spermidine, *spg*--sponge, *SPIAE*--anonymous DNA fragment, *ss*--synaptic sequence, *ssu*--supersuppressor, *st*--sticky, *su( )*--suppressor of ( ), *suc*--succinic acid. — *T*--tyrosinase (structural gene), *T( )*--translocation, *ta*--tufted aerial, *Tel*--telomere, *tet*--tetrazolium reduction, *thi*--thiamine, *thr*--threonine, *ti*--tiny, *tng*--tangerine, *tol*--tolerant, *Tp( )*--transposition, *tre*--trehalase, *trk*--transport of potassium, *trp*--tryptophan, *ts*--tan ascospore, *tu*--tuft, (*tub-2*--tubulin-2. Used by [61]. *Bml-1* has priority), *ty*--tyrosinase regulation, *tyr*--tyrosine, *tys*--tyrosine sensitive. — *uc*--uracil salvage/uracil transport, *ud*--uridine (pyrimidine nucleoside) transport, *udk*--uridine kinase, *un*--unknown heat-sensitive defect, *ufa*--unsaturated fatty acids, *upr*--UV photoreactivation/UV sensitive, *ure*--urease, *uvs*--ultraviolet sensitive. — *val*--valine, (*van*--vanadate resistant; renamed *pho-4* [51]), *vel*--velvet, *vma*--vacuolar membrane ATPase. — *wa*--washed, *wc*--white collar, *ws*--white ascospore. — *xdh*--xanthine dehydrogenase. — *ylo*--yellow.

UNMAPPED GENES: *acp*--ATP/ADP carrier protein [see 25], *acp*^{i}--acetate permease (inducible), *act*--actin [see 25], *agr-1*--altered glucose repression [43], *an*--anaerobic, *apu*--accumulation of purines, *azs*--azide sensitive, *cell-1*-cellobiase/cellulase, *cni-1*--cyanide insensitive, *crp-2* [82], *cyb-2*, *cyc-1* [78], *cyh-3*, *fs-1* (I or II), *des*--delta subunit of ATP synthetase [see 25], *fes*--iron-sulfur subunit of ubiquinol-cytochrome c reductase [see 25], *fz*-- fuzzy, *glt*--glycyl-leucyl-tyrosine resistant, *grg-1*--glucose repressible gene [52], *gul-2*, *gul-6*, *has*--hydroxamic acid sensitive, *het-i*, *ipm-1*, *ipm-2*--isopropylmalate permeation, *ma-1*, *ma-2*--malate utilization, *mea-1*--methylammonium resistant, *mod(os-5)* [24], *mus(FK125)*, *mus(FK128)* [38], *nuh-7*, *nuh(19)*, *nuh(22)* [39], *opi-1*--(IVR or VR: linked to *cot-1* (9%) in *alcoy* [75]), *pde-1* [31], *pho-1* (II?), *prl-1*--mitochondrial proteolipid ATP synthetase [see 25] (the symbol pl, proposed by [25], is preempted), *pts-1*-protease-1, *sar-3*, *sit-1*, *-2*, *-3*, *-4*, *-5*,--siderophore transport, *sg*--spontaneous germination, *sor-6*, *ssu-10*, *ssu(WRU79)*, *su(arg-1)-1*, *su(inl)*, *su(met-7)-2*, *su(pan-2 - Y153M66)*, *su(pe)*, *su(rg-2)*, *su(trp-3*^{td3}*)*, *su(trp-3*^{td6}*)*, *su(trp-3*^{td201}*)-3*, *tub*--tubercidin resistant [50], *var-1*--variant-1 (self-lysis).

REFERENCES: [1] Allen, K. E., *et al.* 1989 J. Bacteriol. *171*:53-58. — [2] Barthelmess, I. B., personal communication. — [3] Barthelmess, I. B., and D. Krüger, personal communication. — [4] Benarous, R. *et al.* 1988 Genetics *119*:805-814. — [5] Berlin, V., and C. Yanofsky 1985 Mol. Cell. Biol. *5*:849-855. — [6] Bowman, B. J., personal communication. — [7] Bowman, B. J., *et al.* 1988 J. Biol. Chem. *263*:14002-14007. — [8] Bowman, E. J., *et al.* 1988 J. Biol. Chem. *263*:13994-14001. — [9] Campbell, J. L., and B. C. Turner 1987 Genome *29*:129-135. — [10] Chan, W. L., 1977 Ph.D. thesis, U. of Malaya, Kuala Lumpur. — [11] Chang, T., R. L. Metzenberg and B. Weisblum, unpublished. — [12] Chary, P., 1989 Ph.D. dissertation, U. New Mexico. — [13] Chary, P., and D. O. Natvig 1989 J. Bacteriol. *171*:2646-2652. — [14] Chow, C. Ming, personal communication. — [15] David, M., M. Saurez, M. S. Sachs and U. L. RajBhandary, personal communication. — [16] DeBusk, R. M., and S. Ogilvie, personal communication. — [17] De Busk, R. M., and S. Ogilvie 1984 J. Bacteriol. *160*:656-661. — [18] DeLange, A. M., and N. C. Mishra 1981 Genetics *97*:247-259; 1982 Mutat. Res. *96*:187-199. — [18a] Dunlap, J. C., personal communication. — [19] Dunn, L. T., 1981 M.A. Thesis, University of N. Carolina, Greensboro. — [20] Eversole, P. E., *et al.* 1985 Mol. Cell. Biol. *5*:1301-1306. — [21] Fincham, J. R. S., *et al.* 1979 *Fungal Genetics*, 4th Ed. pp. 248ff. — [22] Forsthoefel, A. M., and N. C. Mishra 1983 Genet. Res. *41*:271-286. — [23] Furukawa, K., and K. Hasunuma 1984 Jpn. J. Genet. *59*:181-194; Hasunuma, personal communication. — [24] Grindle, M., and G. H. Dolderson 1986 Trans. Brit. Mycol. Soc. *87*:457-460. — [25] Gurr, S. J., and J. R. Kinghorn 1988 Fungal Genet.

Newsl. *35*:10, or Chapter 5 in "Gene Structure in Eukaryotic Microbes" (J. R. Kinghorn, ed.), 1987, J. R. L. Press. — [26] Hager, K. M., *et al.* 1986 Proc. Natl. Acad. Sci. U.S.A. *83*:7693-7697. — [27] Harding, R. W., *et al.* 1984 Neurospora Newsl. *31*:23-25. — [28] Harding, R. W., personal communication. — [29] Hasunuma, K., and Y. Shinohara 1985 Curr. Genet. *10*:197-203. — [30] Hasunuma, K., and Y. Shinohara 1986 Curr. Genet. *10*:893-901. — [31] Hasunuma, K. 1988 Jpn. J. Genet. *63*:556, and personal communication. — [31a] Howlett, B. J., personal communication. — [32] Huiet, L., *et al.* 1984 Proc. Natl. Acad. Sci. U.S.A. *81*:1174-1178; — [32a] Huiet, L., *et al.* 1984 Nucl. Acids Res. *12*:5757-5765. — [33] Inoue, H., and C. Ishii 1984 Mutat. Res. *125*:185-194, and personal communication. — [34] Inoue, H., personal communication. — [35] Inoue, H., and C. Ishii 1985 Mutat. Res. *152*:161-168. — [36] Inoue, H., and C. Ishii, personal communication. — [37] Inoue, H., and A. L. Schroeder 1988 Mutat. Res. *194*:9-16. — [38] Käfer, E., and D. Luk 1988 Fungal Genet. Newsl. *35*:11-13; 1989 Mutat. Res. *217*:75-81. — [39] Käfer, E., and G. R. Witchell 1984 Biochem. Genet. *22*:403-417. — [40] Käfer, E., personal communication. — [41] Kinsey, J. A., personal communication. — [42] Koch, J., and I. B. Barthelmess 1986 Fungal Genet. Newsl. *33*:30-33. — [43] Kreader, C. A., and J. E. Heckman 1989 Nucl. Acids Res. *15*:9027- 9042. — [44] Kubelik, A. R., personal communication via A. M. Lambowitz. — [45] Kuiper, M. T. R., 1987 Ph.D. Thesis, University of Groningen; Kuiper, M. T. R., *et al.* 1987 J. Biol. Chem. *263*:2840-2847. — [46] Kuwana, H., and M. Kubata 1983 Jpn. J. Genet. *58*:579-589. — [47] Lerch, K., personal communication. — [48] Loros, J., *et al.* 1986 Genetics *114*:1095-1110. — [49] Loros, J., *et al.* 1989 Science *243*:385-388. — [50] Magill, J. M., *et al.* 1982 J. Bacteriol. *152*:1292-1294. — [51] Mann, B. J., *et al.* 1989 Gene (in press). — [52] McNally, M. T., and S. J. Free 1988 Curr. Genet. *14*:545-551. — [53] Metzenberg, R. L., personal communication. — [55] Metzenberg, R. L. and J. Grote-leuschen 1989 Fungal Genet. Newsl. *36*. — [56] Metzenberg, R. L., *et al.* 1985. Proc. Natl. Acad. Sci. U.S.A. *82*:2067-2071. — [57] Morgan, J. L., S. Brown and S. N. Bennett 1985 Genetics *110*:s79, and personal communication. — [58] Münger, K., *et al.* 1985 EMBO J. 4:2665-2668. — [59] Munkres, K. D., and C. A. Furtek 1984 Mech. Age. Devel. *25*:63-77. — [60] Murayama, T., *et al.* 1985 Arch. Microbiol. *142*:109-112. — [61] Orbach, M. J., *et al.* 1986 Mol. Cell. Biol. *6*:2452-2461. — [62] Perkins, D. D., 1986 Fungal Genet. Newsl. *33*:34. — [63] Perkins, D. D., 1986 J. Genet. *65*:121-144. — [64] Perkins, D. D., *et al.* 1982 Chromosomal loci of *Neurospora crassa*. Microbiol. Rev. *46*:426-570. — [65] Perkins, D. D., and E. G. Barry 1977 Advan. Genet. *19*:133-285. — [66] Perkins, D. D., *et al.* 1986 Can. J. Genet. Cytol. *28*:971-981. — [67] Perkins, D. D., 1986 Fungal Genet. Newsl. *33*:33-34. — [68] Perkins, D. D., and V. C. Pollard 1987 Fungal Genet. Newsl. *34*:52-53. — [69a] Pitkin, J. 1989, personal communication. — [69b] Plamann, M. D. 1989, personal communication. — [70] Raju, N. B., 1987 Mycologia *79*:697-706.— [71] Sachs, M. S., *et al.* 1986 J. Biol. Chem. *261*:869-873. — [72] Sachs, M. S., *et al.* 1989 J. Mol. Cell. Biol. *9*:566-577. — [73] Schechtman, M. 1989 Fungal Genet. Newsl. *36*: . — [74] Schroeder, A. L., 1988 Pp. 77-98 in *DNA repair. A laboratory manual of research procedures*. Vol. 3 (Friedberg, E. C., and P. C. Hanawalt, eds.). — [75] Shablik, M., *et al.* 1988 Mol. Gen. Genet. *213*:140-143. — [76] Sigmund, R. D., *et al.* 1985 Bioch. Genet. *23*:89-103. — [77] Simmons, J., *et al.* 1987 Fungal Genet. Newsl. *34*:55-56. — [78] Stuart, R. A., *et al.* 1987 EMBO J. *6*:2131-2137. — [79] Taylor, W., and J. F. Feldman, personal communication. — [80] Tropschug, M., *et al.* 1988 J. Biol. Chem. *263*:14433-14440. — [81] Turner, B. C., personal communication. — [82] Tyler, B. 1989, personal communication. — [83] Vidiera, A., *et al.* 1988 Genome *30*:802-807. — [84] White, C., *et al.* 1985 Genetics *110*:217-227.

I

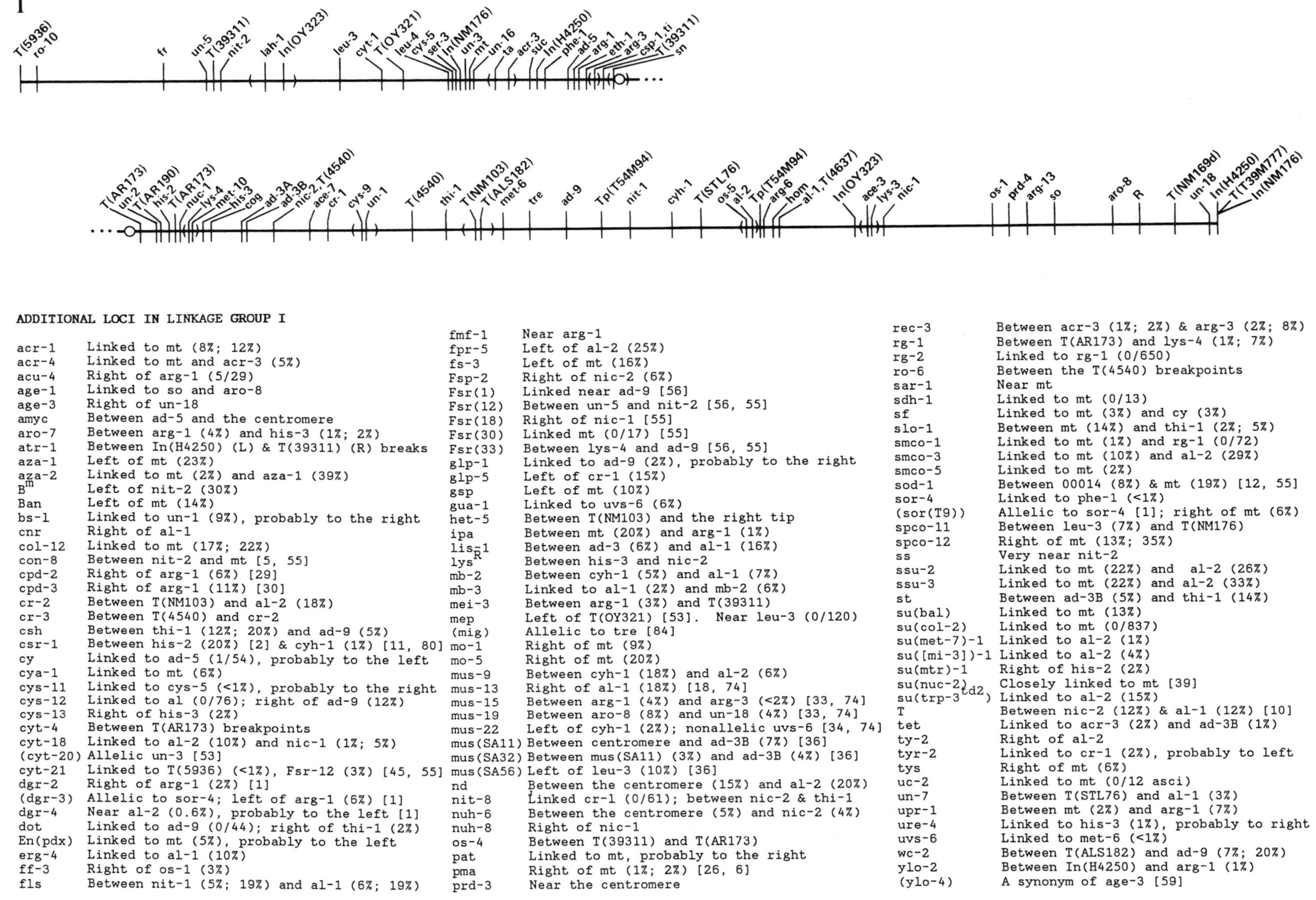

ADDITIONAL LOCI IN LINKAGE GROUP I

acr-1	Linked to mt (8%; 12%)
acr-4	Linked to mt and acr-3 (5%)
acu-4	Right of arg-1 (5/29)
age-1	Linked to so and aro-8
age-3	Right of un-18
amyc	Between ad-5 and the centromere
aro-7	Between arg-1 (4%) and his-3 (1%; 2%)
atr-1	Between In(H4250) (L) & T(39311) (R) breaks
aza-1	Left of mt (23%)
aza-2	Linked to mt (2%) and aza-1 (39%)
B^m	Left of nit-2 (30%)
Ban	Left of mt (14%)
bs-1	Linked to un-1 (9%), probably to the right
cnr	Right of al-1
col-12	Linked to mt (17%; 22%)
con-8	Between nit-2 and mt [5, 55]
cpd-2	Right of arg-1 (6%) [29]
cpd-3	Right of arg-1 (11%) [30]
cr-2	Between T(NM103) and al-2 (18%)
cr-3	Between T(4540) and cr-2
csh	Between thi-1 (12%; 20%) and ad-9 (5%)
csr-1	Between his-2 (20%) [2] & cyh-1 (1%) [11, 80]
cy	Linked to ad-5 (1/54), probably to the left
cya-1	Linked to mt (6%)
cys-11	Linked to cys-5 (<1%), probably to the right
cys-12	Linked to al (0/76); right of ad-9 (12%)
cys-13	Right of his-3 (2%)
cyt-4	Between T(AR173) breakpoints
cyt-18	Linked to al-2 (10%) and nic-1 (1%; 5%)
(cyt-20)	Allelic un-3 [53]
cyt-21	Linked to T(5936) (<1%), Fsr-12 (3%) [45, 55]
dgr-2	Right of arg-1 (2%) [1]
(dgr-3)	Allelic to sor-4; left of arg-1 (6%) [1]
dgr-4	Near al-2 (0.6%), probably to the left [1]
dot	Linked to ad-9 (0/44); right of thi-1 (2%)
En(pdx)	Linked to mt (5%), probably to the left
erg-4	Linked to al-1 (10%)
ff-3	Right of os-1 (3%)
fls	Between nit-1 (5%; 19%) and al-1 (6%; 19%)
fmf-1	Near arg-1
fpr-5	Left of al-2 (25%)
fs-3	Left of mt (16%)
Fsp-2	Right of nic-2 (6%)
Fsr(1)	Linked near ad-9 [56]
Fsr(12)	Between un-5 and nit-2 [56, 55]
Fsr(18)	Right of nic-1 [55]
Fsr(30)	Linked mt (0/17) [55]
Fsr(33)	Between lys-4 and ad-9 [56, 55]
glp-1	Linked to ad-9 (2%), probably to the right
glp-5	Left of cr-1 (15%)
gsp	Left of mt (10%)
gua-1	Linked to uvs-6 (6%)
het-5	Between T(NM103) and the right tip
ipa	Between mt (20%) and arg-1 (1%)
lis-1	Between ad-3 (6%) and al-1 (16%)
lys^R	Between his-3 and nic-2
mb-2	Between cyh-1 (5%) and al-1 (7%)
mb-3	Linked to al-1 (2%) and mb-2 (6%)
mei-3	Between arg-1 (3%) and T(39311)
mep	Left of T(OY321) [53]. Near leu-3 (0/120)
(mig)	Allelic to tre [84]
mo-1	Right of mt (9%)
mo-5	Right of mt (20%)
mus-9	Between cyh-1 (18%) and al-2 (6%)
mus-13	Right of al-1 (18%) [18, 74]
mus-15	Between arg-1 (4%) and arg-3 (<2%) [33, 74]
mus-19	Between aro-8 (8%) and un-18 (4%) [33, 74]
mus-22	Left of cyh-1 (2%); nonallelic uvs-6 [34, 74]
mus(SA11)	Between centromere and ad-3B (7%) [36]
mus(SA32)	Between mus(SA11) (3%) and ad-3B (4%) [36]
mus(SA56)	Left of leu-3 (10%) [36]
nd	Between the centromere (15%) and al-2 (20%)
nit-8	Linked cr-1 (0/61); between nic-2 & thi-1
nuh-6	Between the centromere (5%) and nic-2 (4%)
nuh-8	Right of nic-1
os-4	Between T(39311) and T(AR173)
pat	Linked to mt, probably to the right
pma	Right of mt (1%; 2%) [26, 6]
prd-3	Near the centromere
rec-3	Between acr-3 (1%; 2%) & arg-3 (2%; 8%)
rg-1	Between T(AR173) and lys-4 (1%; 7%)
rg-2	Linked to rg-1 (0/650)
ro-6	Between the T(4540) breakpoints
sar-1	Near mt
sdh-1	Linked to mt (0/13)
sf	Linked to mt (3%) and cy (3%)
slo-1	Between mt (14%) and thi-1 (2%; 5%)
smco-1	Linked to mt (1%) and rg-1 (0/72)
smco-3	Linked to mt (10%) and al-2 (29%)
smco-5	Linked to mt (2%)
sod-1	Between 00014 (8%) & mt (19%) [12, 55]
sor-4	Linked to phe-1 (<1%)
(sor(T9))	Allelic to sor-4 [1]; right of mt (6%)
spco-11	Between leu-3 (7%) and T(NM176)
spco-12	Right of mt (13%; 35%)
ss	Very near nit-2
ssu-2	Linked to mt (22%) and al-2 (26%)
ssu-3	Linked to mt (22%) and al-2 (33%)
st	Between ad-3B (5%) and thi-1 (14%)
su(bal)	Linked to mt (13%)
su(col-2)	Linked to mt (0/837)
su(met-7)-1	Linked to al-2 (1%)
su([mi-3])-1	Linked to al-2 (4%)
su(mtr)-1	Right of his-2 (2%)
su(nuc-2)	Closely linked to mt [39]
$su(trp-3^{td2})$	Linked to al-2 (15%)
T	Between nic-2 (12%) & al-1 (12%) [10]
tet	Linked to acr-3 (2%) and ad-3B (1%)
ty-2	Right of al-2
tyr-2	Linked to cr-1 (2%), probably to left
tys	Right of mt (6%)
uc-2	Linked to mt (0/12 asci)
un-7	Between T(STL76) and al-1 (3%)
upr-1	Between mt (2%) and arg-1 (7%)
ure-4	Linked to his-3 (1%), probably to right
uvs-6	Linked to met-6 (<1%)
wc-2	Between T(ALS182) and ad-9 (7%; 20%)
ylo-2	Between In(H4250) and arg-1 (1%)
(ylo-4)	A synonym of age-3 [59]

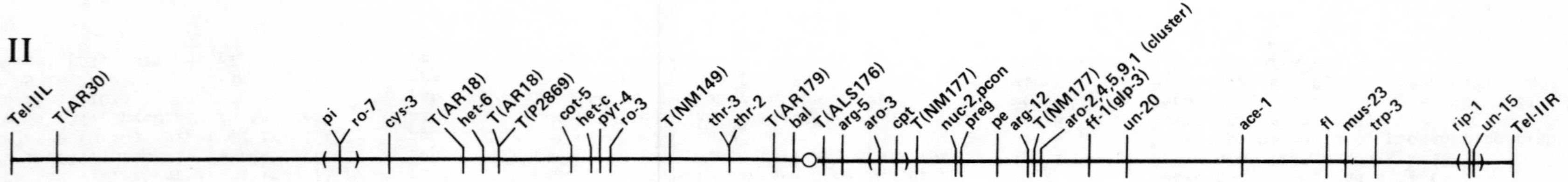

II

Tel-IIL T(AR30) pi ro-7 cys-3 T(AR18) het-6 T(AR18) T(P2869) cot-5 het-c pyr-4 ro-3 T(NM149) thr-3 thr-2 T(AR179) bal T(ALS176) arg-5 aro-3 cpt T(NM177) nuc-2,pcon preg pe arg-12 T(NM177) aro-2,4,5,9,1 (cluster) ff-1(glp-3) un-20 ace-1 fl mus-23 trp-3 rip-1 un-15 Tel-IIR

ADDITIONAL LOCI IN LINKAGE GROUP II

acr-5	Between arg-5 (6%) and pe (9%) [68]
acu-5	Linked to arg-5 (6%) and aro-3 (7%)
alc-1	Linked to pe (10%)
aod-2	Linked to arg-5 (7%) and thr-3 (16%)
ccg-2	Between Fsr(3) and Fsr(34) [49]
col-10	At or near pi
con-6	Between hbs and Fsr(55) [5, 55]
(cox-5)	Between arg-5 and Fsr(55). Allelic to cya-4 [72]
cpd-4	Right of arg-12 (7%) [30]
cwl-1	Linked to arg-5 (3%)
cya-4	Between arg-5 and Fsr(55). Allelic to cox-5 [72]
cyb-3	Left of ro-3 (9%)
cyt-12	Between thr-3 (38%) and trp-3 (18%)
da	Linked to thr-3 (3%); right of T(NM149)
eas	Linked to rip-1 (1/151), trp-3 (0/71) and fl (1/52)
en(am)-2	Near pe
(fs-2)	Tentative linkage to cot-5 (14/48)
Fsp-1	Right of pe (4%)
Fsr(3)	Between H3H4 and ccg-2 [55]
Fsr(17)	Right of trp-3 [56]
Fsr(21)	Between arg-12 and trp-3 [55]
Fsr(32)	Between Fsr(52) and arg-5 [56]
Fsr(34)	Between arg-12 and trp-3. Right of ccg-2 [56, 55]
Fsr(52)	Between pyr-4 and Fsr(32) [56, 55]
Fsr(54)	Between Fsr(32) and arg-5 [56, 55]
Fsr(55)	Between arg-5 and preg [56]
glp-2	Between T(ALS176) and T(NM177)
H3, H4	Between Fsr(55) and Fsr(3) [55]
hbs	Between Fsr(54) and con-6 [55]
het-d	Right of fl (25%)
lp	Right of thr-3 (10%)
leu-6	Right of Fsr(17) [4, 55]
mus-7	Between arg-5 (8%; 12%) and nuc-2 (11%)
mus-24	Left of pyr-4 (8%) [34, 74]
mus-27	Right of nuc-2 (16%) [40, 38]
nuc	One or more loci right of arg-12 [22]
nuh-5	Linked to trp-3 (30%); near T(4637)
nuh-9	Linked to arg-5 [39]
(pcon)	Probably allelic to nuc-2 (0/854)
pmg	Linked to pyr-4 (0/207) [16]
(ro-9)	Probably allelic to da [62]
scr	Linked to arg-12 (2%), probably to the left
spco-14	Linked to arg-5 (7%)
su([mi-1])-3	Linked to fl (31%)
su([mi-1])-4	Linked to fl (22%) and arg-5 (40%)
su([mi-1])-10	Linked to arg-5 (29%) and fl (24%)
tng	Between T(NM149) and arg-5 (6%; 14%)
tu	Between pe (8%) and fl (19%)
uc-1	Linked to pyr-4 (31%)
ure-3	Between arg-12 (7%; 12%) and ace-1 (14%)
vma-2	Linked near Fsr(3) [8]
(ylo-3)	Allelic to fl. Renamed fl^Y
xdh-1	Linked to pe (14%)

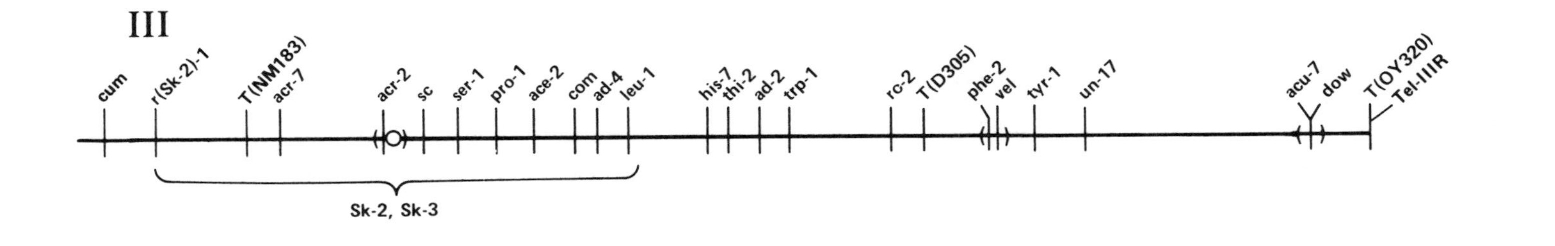

ADDITIONAL LOCI IN LINKAGE GROUP III

acon-2	Linked to vel (6%) and tyr-1 (9%; 14%)
acr-6	Linked to shg (0/368)
aza-3	Linked to trp-1 (14%)
cat-1	Left of trp-1 (17%) [77]
cat-3	Right of trp-1 and Fsr(45) [12, 55]
col-16	Between acr-2 (6%) and leu-1 (1%)
con-1	Between thi-4 and con-7 [5, 55]
con-7	Linked to trp-1 (0/18) [5, 55]
(cox-4)	See cyt-8
cpk	Linked to un-17 (5%) [60]
cya-7	Linked to ad-4 (25%)
cyt-8	Linked near ad-4. Allelic to cox-4 [15]
cyt-22	Linked to cum (6%), acr-7 (19%) [44]; left of con-1 [55]
erg-3	Linked to dow (10%), probably to the left
ff-5	Between pro-1 (2%) and met-8 (1%)
ff-6	Linked near ty-1
fpr-3	Linked to trp-1 (<1%) and thi-2 (5%)
Fsr(45)	Right of tyr-1 [56]. Between trp-1 and cat-3 [55]
gluc-1	Linked to dow (10%)
het-7	Between T(D305) and the right tip
lox	Linked to un-17 (5%) [16]
mei-4	Linked to leu-1 (12%), probably to the left
met-8	Between ff-5 (1%; 4%) and ad-4 (4%)
mo-4	Right of leu-1 (8%)
mus-20	Between trp-1 (22%) and phe-2 (14%) [33, 74]
mus-21	Right of trp-1 (20%) [40, 38]
nit-7	Between trp-1 (26%; 32%) and dow (45%)
nuh-1	Between leu-1 (4%) and nuh-2 (<1%)
nuh-2	Between nuh-1 (<1%) and trp-1 (11%)
os-8	Close to ad-2 and trp-1 [57]
ota	Linked to pro-4 (4%); between ad-4 (15%) and tyr-1 (14%)
pgov	Linked to tyr-1 (1 to 4%), probably to the right
prd-1	Linked to acr-2 (5%) and pro-1 (20%)
pro-4	Linked to thi-2 (0/78) and ota (4%)
r(Sk-3)	Between cum (4%) and acr-7 (17%). Left of T(NM183) [9]
ser-5	Linked to trp-1 (1%) and ser-1 (12%)
shg	Linked to trp-1 (7%)
(Sk-2)	A haplotype. Blocks recombination r(Sk-2) to leu-1 [9]
(Sk-3)	A haplotype. Blocks recombination r(Sk-2) to leu-1 [9]
sor-3	Linked to ad-4 (7%)
spg	Linked to sc (<1%) and thi-4 (0/103)
su([mi-1])-1	Linked to ad-4 (14%)
su(trp-3^{td2})-2	Linked to leu-1 (22%)
thi-4	Linked to acr-2, spg, and sc (<1%)
trk	Linked to leu-1 (0/92)
ty-1	Linked to tyr-1 (0%; 6%) and dow (21%)
un-6	Right of sc (21%) and acr-2 (6-20%)
un-14	Right of acr-2 and thi-4 (8%-20%); left of leu-1 (5%)
un-21	Between acr-2 and met-8, un-6
uvs-4	Left of ad-4 (4%)
uvs-5	Linked to vel (1%)

IV

Tel-IVL T(AR33) uvs-3 cys-10 acon-3 ace-4 cut fi psi T(ALS159) pyr-1 pdx-1 T(S1229) cys-15 oxD met-1 mus-26 col-4 arg-2 T(S4342) arg-14 T(NM152) pyr-3 his-5 trp-4 leu-2 ilv-3 ad-6 ro-1 pan-1 cot-1 his-4 met-5 nit-3 pyr-2 mat T(NM152) cys-4 uvs-2 T(S4342) pmb T(AR209) T(T54M50)

ADDITIONAL LOCI IN LINKAGE GROUP IV

acu-2	Between leu-2 (11%) and pan-1 (6%)
ads	Linked to col-4
aod-1	Left of trp-4 (23%)
bd	Right of pan-1 (2%)
car	Linked to cys-10 (1%)
cel	Linked to pan-1 (1%)
chol-1	Linked to ad-6 (1%), probably to the right
chol-4	Linked to cot-1 (32%) [68]
coil	Between arg-2 (2%) and leu-2 (12%)
col-1	Linked to pan-1 (0/47 asci) and cot-1 (3%)
col-5	Linked to cot-1 (1%), probably to the right
col-6	Linked to centromere (0/28 asci) and pan-1 (21%)
col-8	Linked to pan-1 (4%; 13%)
con-5	Linked cot-1 (0/18) [5, 53]
con-9	Linked cot-1 (0/18) [5, 53]
con-10	Linked cot-1 (0/18) [5, 53]
cot-3	Left of pan-1 (16%; 25%); probably right of arg-2
cpd-1	Right of pyr-2 (19%) [29]
cr-4	Linked to cot-1 (22%) [68]
crib-1	Linked to met-1 (6%)
cya-5	Right of pan-1 (2%)
cya-6	Right of pan-1 (2%)
cys-14	Linked to cot-1 (21%)
cyt-5	Left of trp-4 (9%)
cyt-19	Linked to cyt-5 (1/201)
dn	Linked to mat (1%); right of pyr-2 (4%)
fdu-2	Right of cys-4 (2%)
fld	Left of his-5 (2%)
Fsr(4)	Right of met-5 [55]
Fsr(13)	Between Fsr(51) and cot-1 [56]
Fsr(51)	Between Fsr(62) and Fsr(13) [56]
Fsr(62)	Between pyr-1 and Fsr(51) [56]
Fsr(63)	Left of pyr-1 [56]
gpi-1	Linked to ad-6 (10%)
gua-2	Linked to cot-1 (5%)
gul-3	Linked to cot-1 (10%) and pyr-2 (7%)
hss-1	Linked to cys-4 (2%) [68]
int	Linked to pan-1 (0/50)
le-1	Linked to pan-1 (1%; 2%); right of cot-1 (14%)
med	Linked to met-5 (5%) and pan-1 (8%)
mei-1	Linked to arg-2 (<1%), probably to the right
met-2	Linked to ilv-3 (0/129); between trp-4 (6%) and pan-1 (4%)
mod(sc)	Linked to pan-1 (17%)
moe-3	Left of pan-1 (17%; 25%) and bd (20%)
mtr	Between pdx-1 (2%) and col-4 (1%)
mus-8	Linked to mtr (1%) and pdx-1 (6%)
mus-17	Between pan-1 (2%) and cys-4 (25%) [34, 74]
mus-30	Left of trp-4 (4%) [38]
mus(SA51)	Between pan-1 and uvs-2 [36]
nada	Left of ad-6 (18%)
nit-4	Linked col-4 (2%), probably right; left of pan-1 (6%; 27%)
nit-9	Right of nit-4 (9%)
os-2	Right of cot-1 (4%)
ovc	Between col-4 (10%) & met-5 (14%) [27]; nonallelic int [28]
pf	Linked to mat (3%). Right of pyr-2 (2%)
pho-3	Linked to pan-1 (<1%)
pho-5	Linked to pyr-1 (2%) [53]
pt	Right of T(S1229) and pdx-1 (2%); left of col-4 (2%)
rib-2	Between T(S4342) and chol-1
rol-1	Linked to pdx-1 (0/88)
ser-4	Right of arg-2 (<1%)
smco-4	Linked to pan-1 (8%)
smco-8	Linked to pan-1 (1-7%) and smco-4 (30%)
smco-9	Linked to pan-1 (5%) and smco-4 (2%) and smco-8 (13%)
spco-8	Linked to pan-1 (23%)
SPIAE	Linked to IVL telomere [73]
su([mi-1])-14	Linked to arg-2 (14%)
thi-5	Linked to pan-1 (1%)
tol	Linked to trp-4 (1%), probably to the left
uc-5	Right of cot-1 (32%)
ufa	Linked to met-5 (7%) [68]
un-8	Linked to pyr-1 (0/47); between T(ALS159) and col-4 (5%)
un-12	Linked to col-4 (0/73) and pdx-1 (5%)

V

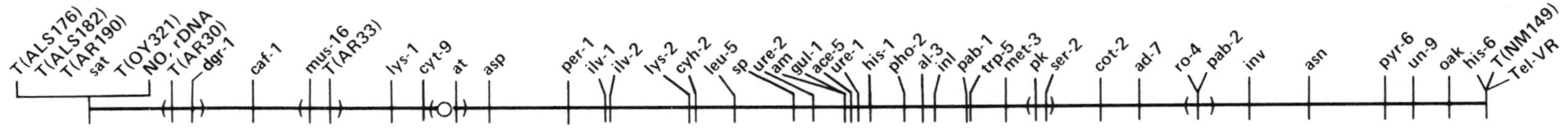

T(ALS176) T(ALS182) T(AR190) sat T(OY321) NO, rDNA T(AR30) dgr-1 caf-1 mus-16 T(AR33) lys-1 cyt-9 at asp per-1 ilv-1 ilv-2 lys-2 cyh-2 leu-5 sp ure-2 am gul-1 ace-5 ure-1 his-1 pho-2 al-3 inl pab-1 trp-5 met-3 pk ser-2 cot-2 ad-7 ro-4 pab-2 inv asn pyr-6 un-9 oak his-6 T(NM149) Tel-VR

ADDITIONAL LOCI IN LINKAGE GROUP V

acu-1	Right of asn (21%)
acu-3	Between inl (7%) and asn (20%)
arg-4	Between sp (1%; 11%) and inl (2%; 4%)
Asc(KH2A83)	Left of al-3; probably right of his-1 [23]
ccg-1	Between lys-1 (18%) and cyh-2 (11%) [49]
chol-3	Between at (18%) and al-3 (19%; 47%) [68]
cl	Right of pk (<1-2%)
cmt	In V by RFLP mapping [47]
col-9	Between inl (16%) and asn (5%)
con-2	Linked to rDNA (2/18) [5, 55]
con-4a	Between con-2 (2/18) and lys-1 (1/18) [5, 55]
cot-4	Between rol-3 (5%) and inl (10%)
cox-6	(See cya-2)
cya-2	Between al-3 and inl. Allelic to cox-6 [15]
cyb-1	Between al-3 (24%) and his-6 (10%)
en(am)-1	Between am-1 (8%) and inl (1%)
erg-1	Between pk (2%) and asn (9%)
erg-2	Left of inl (6%)
fpr-1	Linked to cyh-2 (<1%)
fpr-4	Right of inl (11%)
Fsr(9)	Between Fsr(16) and lys-1 [56]
Fsr(16)	Between con-4a and Fsr(9) [56]
Fsr(20)	Right of inl [56]
gln-1	Linked to inl (2%), probably to the right
glp-6	Left of inl (30%)
gran	Linked pl (0/75); between pab-2 (1%; 8%) & his-6 (8%; 27%)
hgu-4	Between cyh-2 (7%) and ure-2 (10%)
Iasc	Right arm
lis-3	Right of inl (4%)
md	Between sh (3%) and sp (9%)
Mei-2	Linked near inl
mus-11	Linked to his-6; right of pab-2 (23%) [38, 74]
mus-12	Left of inl (10%) [18, 74]
mus-18	Linked to T(AR190); left of caf-1 (20%) [34, 74]
mus-28	Left of lys-1 (10%) [38]
mus(SA18)	Between ure-2 and am; between leu-5 and his-1 (3%) [36]
mus(SA53)	Between lys-1 and cyh-2 [36]
mus(SA60)	Between inl and his-6 [36]
mus(SA66)	Between mus-8 and leu-5 [36]
mus(SC17)	Left of inl (27%)
nap	Linked to inl (15%); right of ure-2 (32%)
ndc-1	Left of arg-4 (2%)
nmr-1	Between am (3%; 7%), gln-1 (4%; 10%); probably allelic MS5 [17]
nuh-3	Between cyh-2 (4%) and al-3 (17%)
nuh(23)	Linked to nuh-3 (6%)
pl	Linked to gran (0/75); between asn (1%; 9%) & his-6 (16%)
prd-2	Right arm
Prf	Right of al-3 (3%) [70]
pro-3	Linked inl, pab-1 (0/74); between his-1 (4%) & pk (2%; 6%)
rec-1	Between ro-4 (7%) and asn (5%)
rec-2	Between sp and am
rol-3	Between ilv-1 (2%) and cot-4 (5%)
scon	Probably linked to VR
scot	Between al-3 (7%) and his-6 (11%)
sh	Between ilv (4%; 7%) and md (3%)
smco-6	Linked to asn (6%) near pyr-6
smco-7	Linked rol-3 (0/154), probably L of sp (12%); R of ilv-1 (2%)
spco-9	Linked to asn (6%); right of met-3 (18%)
spe-1	Between am and his-1 (1%) [20]
spe-2	Left of am and spe-1 (5%) [69a]
spe-3	Right of inl (4%) [69a]
su(ile-1)	Right of met-3 (4%)
su([mi-1])-f	Left of inl (10%)
ts	Linked to inl (4%)
uc-4	Between inl (12%) and his-6 (11%)
udk	Left of uc-4 (29%)
un-11	Linked to al-3 (0/48)
un-19	Linked to al-3 (9%) and un-11 (14%)
val	Linked to ilv-2 (0/135)
vma-1	Linked near cyh-2, inl [6, 8]
wa	Linked to inl (6%)

ADDITIONAL LOCI IN LINKAGE GROUP VI

acu-6	Left of cys-1 (3%)
age-2	Right of ws-1
chr	Between chol-2 (10%) and pan-2
cmt	In VI by RFLP mapping [47]
con-3	Linked to Bml [5, 55]
con-11	Linked to trp-2 [5, 55]
cya-3	Between chol-2 (10%) and cyt-2 (10%)
cyt-2	Between cya-3 (10%) and lys-5 (6%)
cyt-15	No data [83]
cyt(U-28)	No data. Not allelic with cyt-15 [83]
edr-1	Linked to ad-1 and pan-2 (0/125)
edr-2	Left of ad-1 (19%)
fpr-6	Between pan-2 and trp-2
Fsr(50)	Between Bml and trp-2 [56, 55]
glp-4	Between ad-1 (0%; 2%) and rib-1 (3%; 4%)
gul-5	Linked to trp-2 (10%)
het-8	Between chol-2 (19%) and ad-8 (12%)
het-9	Between centromere and trp-2
lis-2	Between chol-2 (11%) and trp-2 (25%)
mod-5	Near the centromere (3%); linked to trp-2
moe-2	Linked to trp-2 (14%), probably to the left
mts	Linked to ylo-1 (<1%). Probably allelic to cpc-1 [42]
mus-14	Near lys-5 [18, 74]
mus-29	Right of chol-2 (14%) [38]
mus(FK133)	Right of ylo-1 (2.5% - 8%). Not allelic to mus-9 (10%) [36]
nuh-10	Linked to trp-2 [39]
sor-1	Left of ylo-1 (3%)
spco-7	Linked to ad-1 (0/65); between ylo-1 (4%) and T(AR209)
spco-13	Linked to trp-2 (16%) and the centromere (1/10 asci)
ssu-7	Between ad-8 (8%) and ylo-1 (14%)
su(mtr^{26})	Linked to trp-2; in the right arm
su(trp-5)	Near aro-6 (allelic?)
un-13	Linked to ylo-1 (2%) and lys-5 (4%)
un-23	Right of trp-2 (5%; 27%); left of ws-1
ws-2	Linked to trp-2 (2%) and ylo-1 (16%)

ADDITIONAL LOCI IN LINKAGE GROUP VII

adh	Linked do (0/53), between cyt-7 (9%) & nic-3 (4%; 11%)
aga	Between wc-1 (2%) and arg-10 (24%; 27%)
aln-1	No data
ars	Right of thi-3 (2%; 5%); left of ile-1 (1%), met-7 (<1%)
bn	Linked met-7 (0/93), wc-1 (1%); right of T(T54M50)
cat-2	Right of ars-1 (27%) [77]; between for & cox-8 [55]
col-2	Linked to met-7 (1%)
(col-3)	Allelic bn [67]
col-17	Linked to nt (14%) and spco-5 (6%)
cox-8	Linked to nt [15]. Right of un-10 [55]
cpc-2	Linked to arg-10 (16-21%) [3].
cyt-6	Linked to wc-1 (2%) (indicated left)
gul-4	Linked to nic-3 (17%)
het-10	Between T(5936) and right tip
hlp-1	Between sfo (1%; 9%) and nt (28%; 37%)
hlp-2	Between hlp-1 (8%; 25%) and nt (29%)
ile-1	Between ars (1%) and wc-1 (<1%; 2%)
kyn-1	Linked to nic-3 (30%) and wc-1 (20%)
le-2	Linked to met-7 (7%)
mb-1	Linked to nic-3 and wc-1 (23%)
mel-1	Left of thi-3 (27%)
mo-2	Linked to nt (29%)
mus-10	Right of met-7 (7%)
mus-25	Right of met-7 (4%; 6%); not allelic to mus-10 [33, 74]
mus(SA33)	Between met-7 (4%), un-10 (8%); nonallelic mus-10, mus-25 [36]
mus(SA50)	Right of nic-3 [36]
pen-1	Linked to csp-2 (4%) [68]
pho-4	Left of nic-3 (4%) [51]
qa gene cluster	Order [32, 32a]: tRNA^{LEU} qa-1F qa-1S qa-y qa-3 qa-4 qa-2 qa-x . . met-7
rol-2	Linked to met-7 (0/298)
sfo	Between thi-3 (6%) and hlp-1 (1%; 9%)
slo-2	Left of met-7 (2%)
sor-2	Linked to nt (31%)
spco-5	Linked to col-17 (6%) and nt (20%)
spco-6	Linked to spco-5 (8%); right of thi-3 (4%)
ssu-1	Between met-7 (14%) and su(trp-3^{td201})-1 (10%)
ssu-4	Between nic-3 (28%) and met-7 (20%)
su([mi-1])-5	Left of nic-3 (23%)
su(trp-3^{td201})-1	Right of ssu-1 (10%; 13%); left of arg-10 (7%)
su(trp-3^{td201})-2	Linked to su(trp-3^{td201})-1, probably to the left
ud-1	Between met-7 (27%) and arg-10 (10%)
un-22	Linked to met-7 (1%)
(van)	Renamed pho-4 [51]

NEUROSPORA CRASSA Laboratory strain 74-OR23-1A: Mitochondrial Genes

September, 1989.

Richard A. Collins.
Department of Botany
University of Toronto
Toronto, Ontario, Canada M5S 3B2

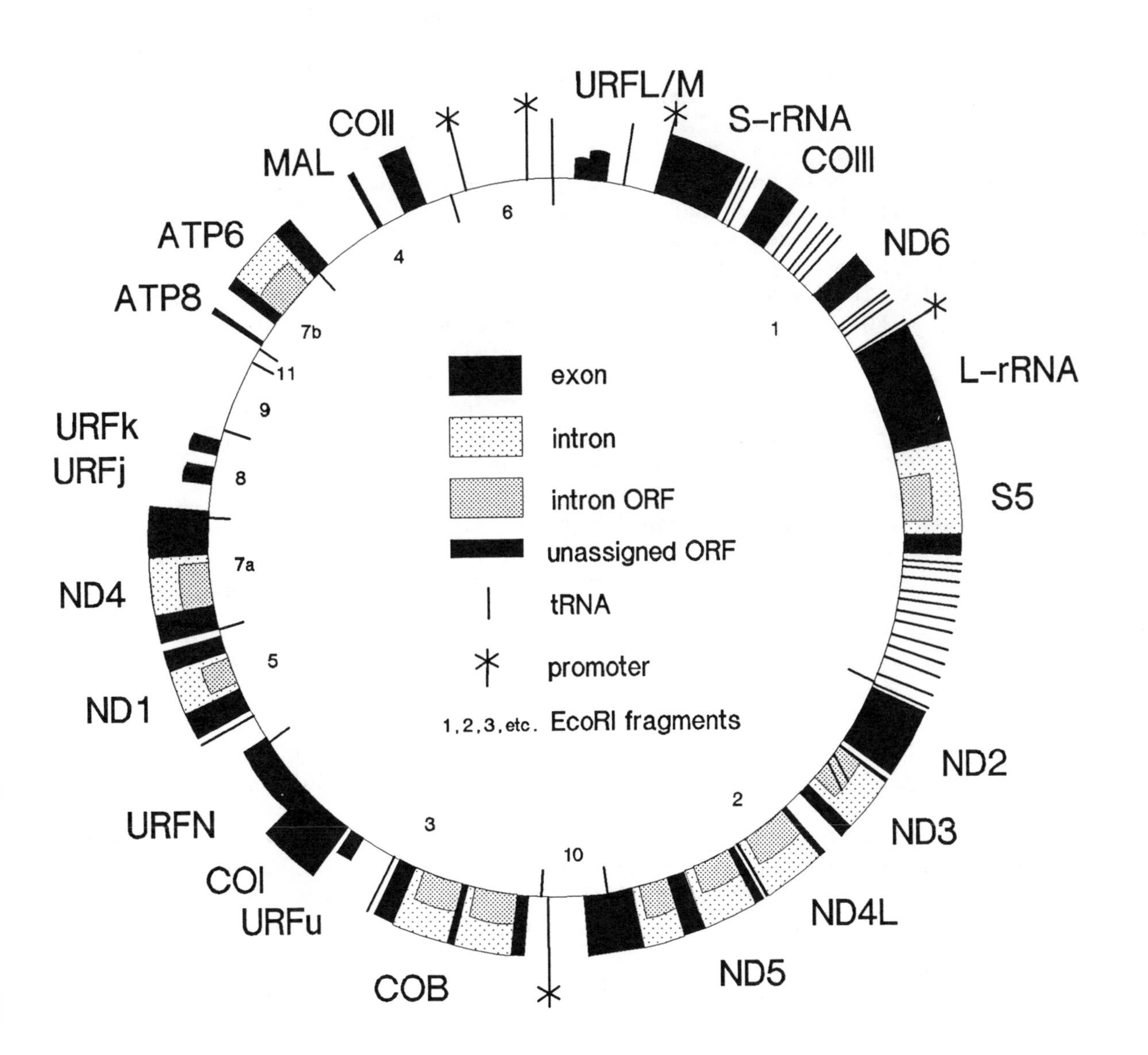

Mitochondrial DNA in the standard laboratory strain of Neurospora crassa is a circular molecule of about 62 kb (1,2). The nucleotide sequence of 94% of the genome has been determined. Unsequenced regions include exon 1 of L-rRNA, the 3' half of S-rRNA, a small region upstream of ND2, and part of EcoRI-11. The EcoRI sites and fragment numbers are indicated on the inside of the map. A detailed restriction map for 15 enzymes can be found in ref. 3. In other wild-type N. crassa strains a large number of polymorphisms, predominantly insertions and deletions, have been found. Many of the larger insertions are optional introns (4,5).

The following genes have been identified:

large and small rRNAs and 27 tRNAs (1, 6-12).

COI (13,14); COII (15); COIII (16): these encode subunits I, II, and III of cytochrome c oxidase, respectively.

COB (17,18): apocytochrome b

ATP6 (19): subunit 6 of ATPase (probably homologous to the oli2 gene in S. cerevisiae).

ATP8 (also called URFx; 19): subunit 8 of ATPase (probably homologous to aapl in S. cerevisiae and URFA6L in human).

MAL (20): This reading frame is similar to the gene encoding the DCCD-binding protein of the ATPase (the olil gene in S. cerevisiae). In Neurospora, the functional gene for this protein is located in the nucleus (21). It is not known if the mitochondrial copy is functional.

S5: This reading frame is located within the intron of the large rRNA gene and encodes the small ribosomal subunit protein S5 (22,23).

ND1 (24), ND2 (25), ND3 (26), ND4 (27) ND4L (26,28), ND5 (28), and ND6 (also called URF6, URFA1 or URFN1; 29): Many of these genes were initially designated URF (unassigned reading frame), but are similar to the corresponding ND genes in mammalian mtDNAs and probably encode mitochondrially translated protein subunits of the NADH dehydrogenase (30).

Unassigned reading frames are present in the introns of ND1 (24), ND4 (27), ND4L (26,28), ATP6 (19,31), both introns of COB (17,18,32) and of ND5 (28). A reading frame with two frame shifts is present in the ND3 intron (26). Methionine-initiated intergenic reading frames longer than 100 amino acids are also included on the map: URFu (14,33), URFj (34), URFk (34), URFL/M (35) and URFN (36).

Transcription of all the presently known genes is clockwise on the above map. Some promoter sites have been identified (37,38). The majority of the genome is transcribed into a small number of long precursor RNAs (35, 39-41).

Introns. All of the introns in 74-OR23-1A mtDNA contain the conserved sequences and potential secondary structures characteristic of Group I introns (42-44; the short sequence thought to be an intron near the start of the ATP6 gene (19) is actually upstream of the coding sequence (31). Group II introns have been found in some wild-type strains (45).

Plasmid DNAs unrelated to the mt chromosome have also been found in several wild-type strains. At least some of these plasmids contain long open reading frames and are transcribed in vivo, suggesting that they are functional genetic elements (46-52).

References

1. Bernard, U. et al. (1976). Nucl. Acids Res. 3:3103-3108.
2. Terpstra, P. et al. (1977). Bioch. Biophys. Acta 475:571-588.
3. Taylor, J.W. and Smolich, B.D. (1985). Curr Genet. 9:597-603.
4. Collins, R.A. and Lambowitz, A.M. (1983). Plasmid 9:53-70.
5. Taylor, J.W. et al. (1986). Evolution 40: 716-739.
6. Terpstra, P. et al. (1977). In: Mitochondria 1977 (Bandlow, W et al., eds) pp.291-302.
7. de Vries, H. et al. (1979). Nucl. Acids Res. 6:1791-1803.
8. Hahn, U. et al. (1979). Cell 17:191-200.
9. Heckman, J.E. and RajBhandary, U.L. (1979). Cell 17: 583-595.
10. Mannella, C.A. et al. (1979). PNAS 76:2635-2639.
11. Heckman, J.E. et al., (1979). J. Biol. Chem. 254:12694-12706.
12. Green, M.R. et al. (1981). J. Biol. Chem. 256:2027-2034.
13. de Jonge, J. and de Vries, H. (1983). Curr. Genet. 7:21-28.
14. Burger, G. et al. (1982). EMBO J. 1:1385-1391.
15. Macino, G. and Morelli, G. (1983). J. Biol. Chem. 258:13230-13235.
16. Browning, K.S. and RajBhandary, U.L. (1982). J. Biol. Chem. 257:5253-5256.
17. Helmer-Citterich, M. et al. (1983). EMBO J. 2:1235-1242.
18. Burke, J.M. et al. (1983). J. Biol. Chem. 259:504-511.
19. Morelli, G. and Macino, G. (1984). J. Mol. Biol. 178:491-507.
20. van den Boogaart, P. et al. (1982). Nature 298:187-189.
21. Jackl, G. and Sebald, W. (1975). Eur. J. Biochem. 54:97-106.
22. Burke, J.M. and RajBhandary, U.L. (1982). Cell 31:509-520.
23. Lambowitz, A.M. et al. (1985). In: Achievements and Perspectives in Mitochondrial Research, vol. II (Quagliariello et al., eds) pp.237-248.
24. Burger, G. and Werner, S. (1985). J. Mol. Biol. 186:231-242.
25. de Vries, H. et al. (1986). EMBO J. 5:779-785.
26. de Vries, H. et al. (1985) In: Achievements and Perspectives in Mitochondrial Research, vol. II (Quagliariello et al., eds) pp. 285-292.
27. Kurdi, M. and Collins, R.A., in preparation.
28. Nelson, M.A. and Macino, G. (1987) Mol. Gen. Genet. 206:307-317.
29. Breitenberger, C.A. and RajBhandary, U.L. (1985). Trends in Bioch. 10:478-483.
30. Ise, W. et al. (1985). EMBO J. 4:2075-2080.
31. Olive, J. and Collins, R.A., in preparation.
32. Collins, R.A. et al. (1988). Nucl. Acids Res. 16:1125-1134.
33. Burger, G. (1985). In: Achievements and Perspectives in Mitochondrial Research, vol. II (Quagliariello et al., eds) pp. 305-316.
34. Collins, R.A., unpublished.
35. Agsterribbe, E. et al. (1989). Curr. Genet. 15:57-62.
36. Burger, G. and Werner, S. (1986). J. Mol. Biol. 191:589-599.
37. Kennell, J.A. and Lambowitz, A.M. (1989). Mol. Cell. Biol 9, 3606-3613.
38. Kubelik, A. and Lambowitz, A.M., personal communication.
39. Collins, R.A. and Lambowitz, A.M. (1985). J. Mol. Biol. 184:413-428.
40. Breitenberger, C.A. et al. (1985). EMBO J. 4:185-195.
41. Burger, G. et al. (1985). EMBO J. 4:197-204.
42. Davies, R.W. et al. (1982). Nature 300:719-724.
43. Michel, F. et al. (1982). Biochimie 64:867-881.
44. Waring, R.B. and Davies, R.W. (1984). Gene 28:277-291.
45. Field, D.J., Sommerfeld, A., Saville, B.J. and Collins, R.A., submitted.
46. Collins, R.A. et al. (1981). Cell 24:443-452.
47. Stohl, L.L. et al. (1982). Nucl. Acids Res. 10:1439-1458.
48. Natvig, D. et al. (1984). J. Bacteriol. 159:288-293.
49. Nargang, F.E. et al. (1984). Cell 38:441-453.
50. Taylor, J.W. et al. (1985). Mol. Gen. Genet. 201:161-167.
51. Nargang, F.E. (1985). Exptl. Mycol. 9:285-294.
52. Nargang, F.E. (1986). Mol. Biol. Evol. 3:19-28.

Neurospora crassa Restriction Polymorphism Map

N = 7 September 1989

Robert L. Metzenberg and Jeff Grotelueschen
Department of Physiological Chemistry
University of Wisconsin
Madison WI 53706, USA

Principle: This is the same as for RFLP maps in other organisms, but the analysis is simple because Neurospora is haploid. Two strains which differ strongly in their genetic background may have DNA sequences differing at many thousands of nucleotide base pairs, and thus at thousands of potential restriction sites. A cross between two such strains can be considered to be heteroallelic, and therefore "marked" at each of these potential restriction site differences. Progeny cultures will inherit the presence or absence of such a restriction site from the parent that contributed it. Progeny are isolated and DNA from them is digested, electrophoresed on gels, and blotted. The blots are probed with the cloned DNA segment of interest to determine which parental allele is present in each progeny.

Sources of data: Most of the data have been obtained from two crosses. Permanent stocks of the progeny of these crosses have been deposited at the Fungal Genetics Stock Center so that all laboratories can contribute to the database. (These stocks are analogous to the lymphoblastoid cell lines used for RFLP mapping in humans (1)). In each of the two crosses, a laboratory strain of N. crassa with a nominally "Oak Ridge" genetic background and carrying several conventional mutant markers was crossed to Mauriceville-1c, a strain with many polymorphic differences from the two laboratory strains. The current results are given in the tables below. Several chromosome assignments might seem questionable on the basis of these data. Such assignments have, however, been confirmed by scoring progeny of other appropriately-marked crosses. Data from confirmatory crosses are available on request. Many of the scorings have been done in laboratories other than that of the authors and generously provided to enrich the map. Limitations of space preclude making separate reference here to the source of mapping data for each locus. Some have been acknowledged in previous versions of these maps (2-5).

The "Big Cross": The larger of the two data sets (henceforth called the "big cross") includes 38 progeny taken from 18 ordered asci. From 17 of these asci, only two non-sister spores from one half of the ascus were taken (the genotypes of the unrepresented half of each ascus can be inferred). The progeny in the table are called A1, A4, B6, etc., where the letter identifies the ascus and the number specifies which of the eight spores of the ascus gave rise to the progeny strain. For Ascus E, all four spore pairs were included, giving rise to E1, E3, E5, and E7. This ascus provides an internal check for polymorphism of the marker and for its Mendelian segregation.

The "Small Cross": This data set includes strains numbered 1 through 20, of which two are the parent strains of the cross. The Oak Ridge-derived parent is numbered 1, and the Mauriceville-1c parent is numbered 6. Strains 2-5 and 7-20 are progeny strains from random ascospores.

Presentation of data: For each linkage group, there is a table for each of the two crosses. The left column is equivalent to a map. It gives the order of DNA fragments and conventional genes; top to bottom on the column equals left to right on the linkage group. The top row gives identification numbers of the individual progeny. For each DNA fragment or conventional gene, "O" means the "allele" that come from the Oak Ridge parent, "M" means the "allele" that came from the Mauriceville parent, and "-" indicates that scoring was not done or was equivocal. "(O)" and "(M)" in strains 1 and 6, respectively, in the "Small Cross" mean that these are the parental strains; the scorings are therefore true by definition (see above). Symbols for conventional genes in the left column are identified in (6) and (7). Other frequently-used abbreviations are as follows. "Fsr" loci are a series of scattered sequences encoding 5S rRNA (8). "Tel" are telomeres (9). Designations of the form number, colon, number, letter (e.g., 22:10G) refer to a particular cosmid in the Vollmer-Yanofsky clonal library (10). Five digit numbers beginning with zeroes (e.g., 00014) are DNA segments the identity of which has not yet been released by the investigator. "con" loci are expressed in connection with conidiation (11). crp-2 codes for a cytoplasmic ribosomal protein homologous to the one in which cryptopleurine resistance mutations occur in yeast (Brett Tyler, unpublished). leu-6 is correctly designated here, but was mistakenly called lys-6 in (5). pma-1 has been removed from the current version of the map because it hybridizes to two different DNA segments, and ambiguity exists as to whether it maps in linkage group I, as previously published, or in linkage group II.

Locating new segments: The unmapped probe of interest is used to score each isolate as "O" or "M", which is printed on a horizontal strip of paper the same width as the body of the tables for the cross used. The tables are then simply scanned for the mapped locus that has scorings most like those of the unmapped segment.

Acknowledgement: This work was supported in part by U.S.P.H.S. grant #GM08995. The authors thank Dorothy Newmeyer and David Perkins for invaluable suggestions.

REFERENCES

1. TANZI, R. E. et al. 1988. *Genomics* 3:129-136.
2. METZENBERG, R. L. et al. 1984. *Neurospora Newsl.* 31:35-39.
3. METZENBERG, R. L. and J. GROTELUESCHEN, 1987. *Fungal Genetics Newsl.* 34:39-44.
4. METZENBERG, R. L. and J. GROTELUESCHEN, 1988. *Fungal Genetics Newsl.* 35:30-35.
5. METZENBERG R. L. and J. GROTELUESCHEN, 1989. *Fungal Genetics Newsl.* 36:51-57.
6. PERKINS, D. D. et al. 1982. *Microbiol. Rev.* 46:426-570.
7. PERKINS, D. D. 1989. *Genetic Maps* 5 (this volume).
8. METZENBERG, R. L. et al. 1985. *Proc. at. Acad. Sci. U. S.* 82:2067-2071.
9. SCHECHTMAN, M. G. 1989. *Fungal Genetics Newsl.* 36:71-73.
10. VOLLMER, S.J. and C. YANOFSKY 1986. *Proc. at. Acad. Sci. U. S.* 83:4869-4873.
11. BERLIN, V. and C. YANOFSKY 1985. *Molec. Cell Biol.* 5:835-848; *ibid.* 849-855.

LG I

	A 1	A 4	B 6	B 7	C 1	C 4	D 5	D 7	E 1	E 3	E 5	E 7	F 1	F 3	G 1	G 4	H 5	H 7	I 6	I 8	J 1	J 4	K 1	K 4	L 1	L 4	M 5	M 8	N 2	N 3	O 2	O 4	P 1	P 4	Q 2	Q 4	R 1	R 4
cyt-21	M	M	M	O	M	O	M	O	M	O	M	O	O	M	M	O	O	M	-	M	M	O	O	-	M	O	-	O	M	M	O	M	O	M	M	O	O	M
Fsr-12	M	O	M	M	M	O	M	O	M	M	O	O	M	M	M	O	O	M	M	M	M	M	O	O	O	O	O	O	O	-	O	M	O	M	M	O	M	M
00014	M	O	M	M	M	O	M	O	M	M	O	O	M	M	M	O	O	M	M	M	M	M	O	O	O	O	O	O	O	M	O	M	O	M	O	O	M	M
sod-1	-	O	M	M	M	M	M	O	M	M	O	O	M	M	M	O	O	M	O	M	M	M	O	O	M	O	O	O	O	M	O	M	O	M	O	M	M	M
A/a, Fsr-30	M	O	M	M	O	O	M	M	M	M	O	O	M	M	M	M	O	M	O	M	M	M	O	O	O	O	O	O	O	O	O	M	O	M	O	O	M	M
22:10G	M	O	M	M	O	O	M	M	M	M	O	-	M	M	M	M	O	M	O	M	M	M	O	O	O	O	O	O	O	O	-	M	O	M	O	O	M	M
un-2, his-2	M	M	M	M	O	O	M	M	M	M	O	O	M	M	M	M	M	M	M	M	M	M	O	O	O	O	O	O	O	O	M	M	O	O	O	O	M	M
lys-4	M	M	M	M	O	O	M	M	M	M	O	O	M	M	M	M	M	M	M	M	M	M	O	O	M	O	O	O	O	O	M	M	O	O	O	O	M	M
Fsr-33	M	M	M	M	O	M	M	M	M	M	O	O	M	M	M	M	M	M	M	M	M	M	O	O	M	O	O	O	O	O	M	M	O	O	O	O	M	M
12:8B	M	M	M	M	M	O	M	M	M	M	O	O	M	M	M	O	M	M	O	M	M	M	O	O	M	-	O	O	O	O	M	M	O	O	M	O	O	M
24:8B	M	M	M	M	M	O	O	M	M	M	O	O	M	O	M	O	M	M	O	M	M	M	O	M	M	O	O	O	O	O	M	M	O	O	O	O	O	M
Fsr-1	M	O	M	M	M	O	O	M	M	M	O	O	-	O	M	O	M	M	M	M	M	M	O	M	M	O	O	O	O	-	M	O	O	M	O	O	O	M
Fsr-18	M	O	M	O	O	O	O	M	M	M	O	O	M	O	O	M	M	M	O	M	M	O	O	O	M	O	M	O	O	M	O	M	O	M	O	O	O	M
21:H5	M	O	M	O	O	O	O	M	O	M	M	O	M	O	O	M	M	M	O	M	M	O	O	O	M	O	M	O	O	M	O	M	O	M	O	O	O	M
arg-13	M	O	M	O	O	O	O	M	O	M	M	O	M	O	O	M	M	M	O	M	O	O	O	O	M	O	M	O	O	M	O	M	O	M	O	O	O	M

	1	2	3	4	5	6	7	8	9	10	11	12	13	14	15	16	17	18	19	20
cyt-21	(O)	O	M	O	O	(M)	M	M	O	M	O	O	O	O	O	O	O	O	-	M
Fsr-12	(O)	O	M	M	M	(M)	M	O	O	M	M	O	O	O	O	O	M	O	-	-
nit-2	(O)	O	M	M	M	(M)	M	O	O	M	M	O	O	O	O	O	O	O	M	O
con-8	(O)	O	M	M	M	(M)	M	O	O	M	O	M	O	O	O	O	O	O	O	O
A/a, arg-3, leu-4	(O)	M	M	M	M	(M)	O	O	O	M	M	M	O	O	O	O	O	O	O	O
Fsr-33	(O)	M	M	M	M	(M)	O	O	O	M	M	M	O	O	O	O	O	O	M	O
Fsr-1	(O)	M	M	M	M	(M)	O	O	O	O	M	M	M	O	O	O	O	O	O	O
al-2, un-7 met-6	(O)	M	M	M	M	(M)	O	O	O	O	M	M	M	O	O	O	O	O	O	O
nic-1	(O)	M	M	M	M	(M)	O	O	O	O	M	M	O	O	O	O	O	O	O	O
Fsr-18	(O)	M	M	M	M	(M)	O	O	O	O	M	M	O	O	O	O	O	O	O	O

LG II

	A1	A4	B6	B7	C1	C4	D5	D7	E1	E3	E5	E7	F1	F3	G1	G4	H5	H7	I6	I8	J1	J4	K1	K4	L1	L4	M5	M8	N2	N3	O2	O4	P1	P4	Q2	Q4	R1	R4
Tel IIL	M	M	O	M	M	O	O	O	M	O	M	–	O	M	M	O	M	O	M	O	O	M	O	M	O	M	O	M	O	M	M	M	M	O	O	M	O	M
00023	M	M	O	M	M	O	O	O	M	O	M	O	O	M	M	O	M	O	M	O	M	M	O	M	O	M	O	M	O	M	O	M	M	O	O	M	O	M
00008	–	–	O	O	–	–	O	O	M	M	O	O	O	O	M	O	M	O	M	O	O	M	O	M	O	M	M	M	O	M	M	O	O	O	M	M	O	O
3:9A	O	M	O	O	M	M	O	O	M	M	O	O	O	O	M	O	M	M	M	O	O	M	O	M	O	–	M	M	O	O	–	M	O	O	M	M	O	O
Fsr-52	O	M	O	O	–	M	O	O	M	M	O	O	O	O	M	M	M	M	M	M	O	M	O	M	O	M	M	M	O	O	–	M	O	O	M	M	O	O
Fsr-32	M	M	O	O	M	M	O	O	M	M	O	O	O	O	M	M	M	M	M	M	M	M	M	M	O	M	M	M	O	O	O	O	O	O	M	M	O	O
arg-5	M	M	O	O	M	M	O	O	M	M	O	O	O	O	M	M	M	M	M	M	M	M	M	M	M	M	M	M	O	O	O	O	O	O	M	M	O	O
cya-4, Fsr-55	M	M	O	O	M	M	O	O	M	M	O	O	O	O	M	M	M	M	M	M	M	M	M	M	M	M	M	M	O	M	O	O	O	O	M	M	O	O
H3H4	O	M	O	O	M	M	O	O	M	M	O	O	O	O	M	M	M	M	O	M	M	M	M	M	M	M	M	M	O	M	–	O	O	O	M	M	O	O
DB0001	O	M	O	O	M	M	O	O	M	O	O	M	O	O	M	M	M	M	O	M	M	M	O	M	M	M	M	M	O	M	O	O	O	O	M	M	O	M
Fsr-3	O	M	O	O	M	M	O	O	M	O	O	M	O	O	M	M	M	M	O	M	M	M	O	M	M	M	M	O	O	M	O	O	O	O	M	M	O	M
ccg-2	O	M	O	O	M	M	O	M	M	O	O	M	O	O	M	M	M	M	O	M	M	M	O	M	M	M	M	O	O	M	O	O	–	O	M	M	O	M
Fsr-34	O	M	O	O	M	M	O	M	M	O	O	M	M	O	M	M	M	O	O	M	M	M	O	M	M	M	M	O	O	M	O	O	O	O	M	M	O	M
Fsr-17	O	M	O	O	M	M	O	M	M	O	O	M	O	M	O	M	M	O	O	O	M	M	O	M	O	M	M	O	–	M	M	O	O	M	M	M	O	M
leu-6	O	M	M	O	M	O	O	M	M	O	M	O	O	M	O	M	M	O	O	O	O	M	O	M	O	M	M	O	–	–	–	O	O	M	O	M	O	M
Tel IIR	O	M	M	O	M	O	O	M	M	O	M	O	O	M	O	M	M	O	O	O	O	M	M	O	M	M	M	O	M	M	O	O	O	M	O	M	O	M

	1	2	3	4	5	6	7	8	9	10	11	12	13	14	15	16	17	18	19	20
cys-3	(O)	O	M	M	O	(M)	O	M	M	O	M	O	O	O	O	M	O	O	M	O
00008	(O)	O	M	M	O	(M)	O	M	O	O	O	O	O	M	O	M	O	M	O	O
Fsr-32, -52	(O)	O	O	M	O	(M)	O	M	O	O	O	O	O	O	O	M	O	O	O	O
Fsr-54	(O)	O	O	M	M	(M)	O	M	O	O	O	O	M	O	O	M	O	O	M	O
hbs	(O)	O	O	M	M	(M)	O	M	O	O	O	O	M	O	O	M	O	M	M	O
con-6	(O)	O	O	M	O	(M)	O	M	O	O	O	O	M	O	O	M	O	O	M	M
Fsr-55	(O)	O	O	M	M	(M)	M	M	O	O	O	O	M	O	O	M	O	O	M	M
arg-12	(O)	O	O	M	M	(M)	M	M	O	O	M	O	M	O	O	M	O	O	M	M
Fsr-21, -34	(O)	O	O	M	M	(M)	M	M	O	O	M	M	M	O	O	M	M	O	M	M
trp-3	(O)	O	O	M	M	(M)	M	O	O	O	M	M	M	O	O	M	O	O	M	M
Fsr-17	(O)	O	O	M	M	(M)	M	O	O	O	M	M	M	M	O	O	M	O	–	–
leu-6	(O)	O	O	M	M	(M)	M	O	O	O	M	M	M	M	O	O	M	O	M	O

LG III

	A1	A4	B6	B7	C1	C4	D5	D7	E1	E3	E5	E7	F1	F3	G1	G4	H5	H7	I6	I8	J1	J4	K1	K4	L1	L4	M5	M8	N2	N3	O2	O4	P1	P4	Q2	Q4	R1	R4
LZC1	0	0	0	0	0	0	M	0	0	0	M	M	M	0	M	M	M	0	0	M	M	M	0	0	0	0	M	M	-	M	0	0	M	0	0	0	M	M
LZE5	-	0	0	0	M	0	M	0	0	0	M	M	M	0	M	M	M	-	-	M	M	M	0	0	M	M	0	0	M	M	M	M	0	0	0	0	M	M
00026	0	0	0	0	-	-	M	-	-	0	M	M	M	M	-	M	M	M	0	M	M	M	0	0	M	M	0	0	M	M	-	M	0	0	-	0	M	M
LZE4	-	0	0	0	-	-	M	0	0	0	M	M	M	0	M	M	-	-	0	M	M	M	0	0	M	M	0	0	M	M	M	M	M	M	0	0	0	0
thi-4	0	0	0	0	M	M	M	0	0	0	M	M	M	M	M	M	M	M	0	0	M	M	0	0	0	0	0	0	M	M	M	M	M	0	M	M	0	0
con-1	0	M	-	0	M	M	M	M	0	0	M	M	M	M	-	M	M	M	0	0	M	0	0	M	0	M	0	0	0	M	M	M	0	0	0	0	0	M
cyt-8	0	0	0	0	M	M	M	M	0	0	M	M	M	M	0	M	M	M	0	0	M	0	0	0	M	M	0	0	M	M	M	M	0	0	0	0	M	M
con-7	0	0	-	0	0	M	M	0	0	0	M	M	M	0	M	M	M	0	0	0	M	M	0	0	0	M	0	0	M	M	M	M	M	0	0	0	M	M
trp-1	0	0	-	0	0	M	M	0	0	0	M	M	M	0	-	M	M	0	0	M	M	M	0	0	0	M	0	0	M	M	M	M	M	0	0	0	M	M
DB002	0	0	0	0	0	M	M	0	0	0	M	M	M	0	M	M	M	0	0	M	M	M	0	0	0	-	0	0	M	M	M	M	-	0	M	0	M	M
32:2G	0	0	0	0	0	M	M	0	-	0	M	M	M	0	M	M	M	0	-	M	M	M	0	0	0	M	0	0	M	0	M	M	M	0	0	0	M	M
Fsr-45	M	0	M	0	0	0	0	M	0	M	0	M	M	0	M	M	0	0	M	0	M	0	0	M	0	0	0	M	0	0	0	M	M	0	0	M	M	0
cat-3	M	0	M	0	0	0	0	M	0	M	0	M	0	M	M	M	0	0	M	0	M	0	M	M	0	0	0	M	0	M	0	M	M	0	0	M	M	0
Tel IIIR	M	0	M	0	0	M	0	M	0	M	0	-	-	M	0	M	0	0	M	0	M	0	0	M	0	0	0	M	0	0	M	M	M	0	0	M	M	0

	1	2	3	4	5	6	7	8	9	10	11	12	13	14	15	16	17	18	19	20
cyt-22	(0)	M	0	M	M	(M)	0	0	M	M	0	M	M	M	0	M	0	M	M	M
LZE4	(0)	M	0	M	M	(M)	0	0	M	M	0	0	M	M	0	M	0	M	0	M
DA25	(0)	M	0	M	M	(M)	0	0	M	M	0	0	M	M	0	M	0	0	0	M
con-1	(0)	0	0	M	M	(M)	0	0	M	M	0	0	M	M	0	M	0	0	0	M
cyt-8	(0)	0	0	M	M	(M)	0	0	M	M	0	0	M	M	0	M	M	0	0	M
DA25	(0)	M	0	M	M	(M)	0	0	M	M	0	0	M	M	0	M	0	0	0	M
con-7	(0)	M	0	M	M	(M)	0	0	M	M	0	0	M	M	0	M	0	M	0	M
trp-1	(0)	M	0	M	M	(M)	0	0	M	M	0	0	M	M	0	M	0	M	0	M
Fsr-45	(0)	0	0	M	0	(M)	0	M	M	0	0	0	M	0	0	M	M	0	0	M

LG IV	A 1	A 4	B 6	B 7	C 1	C 4	D 5	D 7	E 1	E 3	E 5	E 7	F 1	F 3	G 1	G 4	H 5	H 7	I 6	I 8	J 1	J 4	K 1	K 4	L 1	L 4	M 5	M 8	N 2	N 3	O 2	O 4	P 1	P 4	Q 2	Q 4	R 1	R 4
Tel IVL	O	M	O	O	M	O	M	O	O	M	M	O	M	O	M	O	O	O	O	M	M	M	O	M	M	O	M	O	M	O	O	O	O	O	M	M	O	M
SPIAE	O	M	O	O	M	O	M	O	O	M	M	O	M	O	M	M	O	O	O	M	M	M	M	M	M	O	M	O	M	O	M	O	O	O	M	M	O	M
4:3A	M	O	O	O	M	M	O	O	M	M	O	O	O	O	M	M	O	O	M	M	M	M	M	M	M	M	M	M	O	O	O	O	O	O	M	M	M	M
Fsr-63	M	M	M	O	M	M	O	O	M	M	O	O	O	O	M	M	O	O	M	M	M	M	M	M	M	M	M	M	O	-	O	O	M	O	M	M	M	M
pyr-1	M	M	M	O	M	M	O	O	M	M	O	O	O	O	M	M	O	O	M	M	M	M	M	M	M	M	M	M	O	O	O	O	M	O	M	O	M	M
20:11A	M	O	M	O	M	M	O	O	M	M	O	O	O	O	M	M	O	M	M	O	M	O	M	M	M	M	M	M	O	O	O	O	M	O	M	O	M	M
Fsr-62	M	O	M	O	M	O	O	O	M	M	O	O	O	O	M	M	O	M	O	O	M	O	M	M	M	M	M	M	M	O	O	M	M	O	M	O	M	M
Fsr-51	M	O	M	O	M	O	O	O	M	M	O	O	O	O	M	M	O	M	O	O	M	O	M	M	M	M	M	M	O	O	O	M	M	O	M	O	M	M
Fsr-13	M	O	M	O	M	O	O	O	M	M	O	O	O	O	O	M	O	M	O	O	M	O	M	M	M	M	M	M	O	O	O	M	M	O	M	O	M	M
Fsr-4	M	O	M	O	M	O	O	M	M	O	O	M	O	M	O	M	O	O	O	M	O	O	O	M	O	M	O	M	O	-	M	O	O	M	M	M	M	O

	1	2	3	4	5	6	7	8	9	10	11	12	13	14	15	16	17	18	19	20
Tel IVL	(O)	O	M	M	M	(M)	O	M	M	M	O	O	O	O	O	O	O	O	O	M
SPIAE	(O)	O	M	M	M	(M)	O	M	M	M	O	O	O	O	O	O	M	O	O	O
Fsr-63	(O)	M	M	M	O	(M)	M	M	M	M	M	O	O	O	O	O	O	O	O	O
Fsr-62, mtr, 5:3B, 5:4H	(O)	M	M	M	O	(M)	M	M	M	M	M	O	O	O	O	O	O	O	O	O
Fsr-51	(O)	M	M	M	O	(M)	M	M	M	M	O	O	O	O	O	O	O	O	O	O
con-5, con-9 cot-1, con-10b	(O)	M	M	M	M	(M)	M	M	M	M	O	O	O	O	O	O	O	O	O	O
met-5	(O)	M	M	M	M	(M)	M	M	O	M	O	O	O	O	O	O	O	O	O	O
nit-3	(O)	M	M	M	M	(M)	M	M	O	M	O	O	O	M	O	O	O	M	M	O
Fsr-4	(O)	O	M	M	M	(M)	M	M	O	O	-	O	M	M	M	O	M	M	M	M

LG V	A 1	A 4	B 6	B 7	C 1	C 4	D 5	D 7	E 1	E 3	E 5	E 7	F 1	F 3	G 1	G 4	H 5	H 7	I 6	I 8	J 1	J 4	K 1	K 4	L 1	L 4	M 5	M 8	N 2	N 3	O 2	O 4	P 1	P 4	Q 2	Q 4	R 1	R 4
rDNA	O	M	M	O	M	O	M	O	M	O	M	O	M	M	M	O	M	O	O	M	M	O	M	M	M	O	M	O	O	O	M	M	O	M	M	M	O	M
con-2	O	M	O	O	O	O	O	O	M	O	M	O	M	M	O	O	M	O	O	M	M	O	M	M	M	O	M	O	O	O	M	M	O	M	M	M	O	O
Fsr-16	M	M	O	O	O	M	O	O	O	O	M	M	-	M	O	O	M	M	-	-	-	-	M	M	O	O	M	M	O	-	M	M	M	-	-	-	-	-
Fsr-9	M	M	O	O	O	M	O	O	O	O	M	M	M	M	O	O	M	M	O	O	M	M	M	M	O	O	M	M	O	O	M	M	M	M	M	M	O	O
lys-1	M	M	O	O	O	O	O	O	O	O	M	M	M	M	O	O	M	M	O	O	M	M	M	M	O	O	M	M	O	O	M	M	M	M	M	M	O	O
ccg-1	M	O	O	O	M	O	O	O	O	O	M	M	M	O	O	O	M	M	O	O	M	M	M	O	O	O	M	M	O	M	M	M	M	M	O	M	M	O
cyh-2	M	O	O	O	M	O	O	M	O	O	M	M	M	O	O	O	M	M	O	O	O	M	M	O	O	M	M	M	O	M	O	M	M	M	O	M	M	O
al-3	M	M	O	O	M	O	O	M	O	O	M	M	M	O	O	O	M	M	O	O	O	M	M	O	O	M	M	M	O	M	O	M	M	M	O	M	M	O
cya-2	M	M	O	O	M	O	O	M	O	O	M	M	M	O	O	O	M	M	O	M	O	M	O	M	O	M	M	M	O	M	O	M	M	M	O	M	M	O
inl	M	M	O	O	M	O	O	M	O	O	M	M	M	O	O	O	M	M	O	M	O	M	M	O	O	M	M	M	O	M	O	M	M	M	O	M	M	O
Fsr-20	O	M	O	O	M	M	O	M	O	O	M	M	M	M	M	M	M	M	O	M	O	M	M	M	O	M	M	M	M	-	O	M	M	O	O	M	M	O
8:9A	O	M	O	O	M	O	O	M	O	O	M	M	M	M	M	M	-	O	O	M	O	M	M	M	O	M	M	M	M	M	O	M	M	O	O	M	M	O
00015	O	M	O	M	M	O	O	M	O	O	M	M	M	M	M	M	M	O	O	M	O	M	M	M	O	M	M	M	M	M	O	M	M	O	O	M	M	O
Tel VR	O	M	O	M	M	O	O	M	O	M	O	M	M	M	M	M	M	O	O	M	M	M	M	M	O	M	O	M	M	O	M	O	M	O	M	M	M	O

	1	2	3	4	5	6	7	8	9	10	11	12	13	14	15	16	17	18	19	20
rDNA	(O)	M	O	O	O	(M)	O	M	M	O	O	O	M	O	M	O	M	O	M	M
con-2	(O)	M	O	O	O	(M)	O	M	M	M	O	O	M	O	M	O	M	M	M	M
con-4a	(O)	M	O	O	O	(M)	O	M	M	M	M	M	M	O	M	O	M	M	M	M
Fsr-16	(O)	M	O	O	O	(M)	O	M	M	M	M	M	M	O	M	O	O	M	M	M
Fsr-9	(O)	M	O	O	O	(M)	O	M	M	M	M	M	M	O	M	O	O	M	M	M
am	(O)	O	O	M	O	(M)	M	M	M	M	M	M	M	M	M	M	M	M	M	M
inl	(O)	O	O	M	O	(M)	M	M	M	M	M	M	M	M	O	M	O	M	M	M
Fsr-20	(O)	O	O	M	O	(M)	M	O	M	M	M	M	M	M	O	M	M	M	M	M

LG VI

	A1	A4	B6	B7	C1	C4	D5	D7	E1	E3	E5	E7	F1	F3	G1	G4	H5	H7	I6	I8
12:7B	O	M	M	M	M	M	O	M	O	O	M	M	M	M	O	O	O	O	O	O
Fsr-50	M	M	M	M	M	M	M	M	O	O	M	M	M	M	O	O	O	O	O	O
con-11a	M	O	-	M	M	M	M	M	O	O	M	M	M	M	-	O	O	O	M	O

	J1	J4	K1	K4	L1	L4	M5	M8	N2	N3	O2	O4	P1	P4	Q2	Q4	R1	R4
12:7B	O	M	O	M	M	M	O	O	O	M	M	M	M	M	O	O	M	O
Fsr-50	O	O	O	O	M	M	O	O	O	O	M	M	M	M	O	O	M	M
con-11a	O	O	O	O	O	M	O	O	O	M	M	O	M	M	O	O	O	M

	1	2	3	4	5	6	7	8	9	10	11	12	13	14	15	16	17	18	19	20
Bml,con-3	(O)	M	M	M	O	(M)	M	M	O	O	M	M	M	M	M	O	M	M	M	M
Fsr-50	(O)	M	M	M	O	(M)	M	M	O	O	M	M	M	M	M	O	O	M	M	M
00006	(O)	O	M	M	M	(M)	M	M	O	O	M	O	M	M	M	O	M	M	M	M
trp-2	(O)	M	M	M	M	(M)	M	O	O	O	M	O	M	M	M	M	M	O	M	M
con-11a	(O)	M	M	M	M	(M)	M	O	O	O	M	O	M	M	M	M	M	O	M	M

LG VII

	A1	A4	B6	B7	C1	C4	D5	D7	E1	E3	E5	E7	F1	F3	G1	G4	H5	H7	I6	I8
Tel VIIL	M	O	M	M	M	O	M	O	O	M	O	M	O	O	O	M	M	O	M	O
5:5A	M	O	M	M	M	O	M	M	O	M	O	M	O	-	O	M	M	O	M	-
pho-4	M	O	O	O	M	O	M	M	O	O	M	M	O	O	O	O	O	O	M	O
nic-3	M	O	O	O	M	M	M	M	O	O	M	M	O	O	O	M	O	O	M	O
00003	M	-	O	O	M	M	M	M	O	O	M	-	O	O	O	M	O	O	M	O
ars-1	O	O	O	O	M	M	M	M	O	O	M	M	O	O	O	O	O	O	O	O
frq, for	O	O	O	O	M	M	M	M	O	O	M	M	O	O	O	O	O	O	O	O
cat-2	O	O	O	M	M	M	M	M	O	M	M	M	O	O	O	O	O	O	M	O
COXVIII	O	O	O	M	M	M	M	O	O	M	O	M	M	O	M	O	O	O	M	O

	J1	J4	K1	K4	L1	L4	M5	M8	N2	N3	O2	O4	P1	P4	Q2	Q4	R1	R4
Tel VIIL	O	M	O	M	M	M	M	O	M	O	O	M	O	O	M	O	O	O
5:5A	O	M	O	M	M	M	M	O	M	O	O	M	-	-	M	O	O	-
pho-4	O	O	O	O	O	M	M	O	O	O	M	M	O	O	M	M	O	O
nic-3	O	O	O	O	O	M	M	O	O	O	M	M	O	O	M	M	O	O
00003	O	O	O	O	O	M	M	O	O	O	O	M	O	O	M	M	O	O
ars-1	O	O	O	O	M	M	M	M	O	O	M	M	O	O	M	M	O	O
frq, for	O	O	O	O	M	M	O	M	O	O	-	M	O	O	M	M	O	O
cat-2	M	O	M	O	M	M	O	M	M	O	M	O	O	-	M	O	O	M
COXVIII	M	O	M	O	O	-	O	M	O	M	-	O	O	M	M	O	O	M

	1	2	3	4	5	6	7	8	9	10	11	12	13	14	15	16	17	18	19	20
00003	(O)	M	M	O	O	(M)	O	M	-	M	O	M	M	O	O	O	O	-	O	M
un-10	(O)	M	M	M	O	(M)	O	M	O	M	O	O	-	O	O	O	O	M	O	O
COXVIII	(O)	M	M	M	O	(M)	O	O	O	O	M	O	M	O	O	O	O	M	O	O

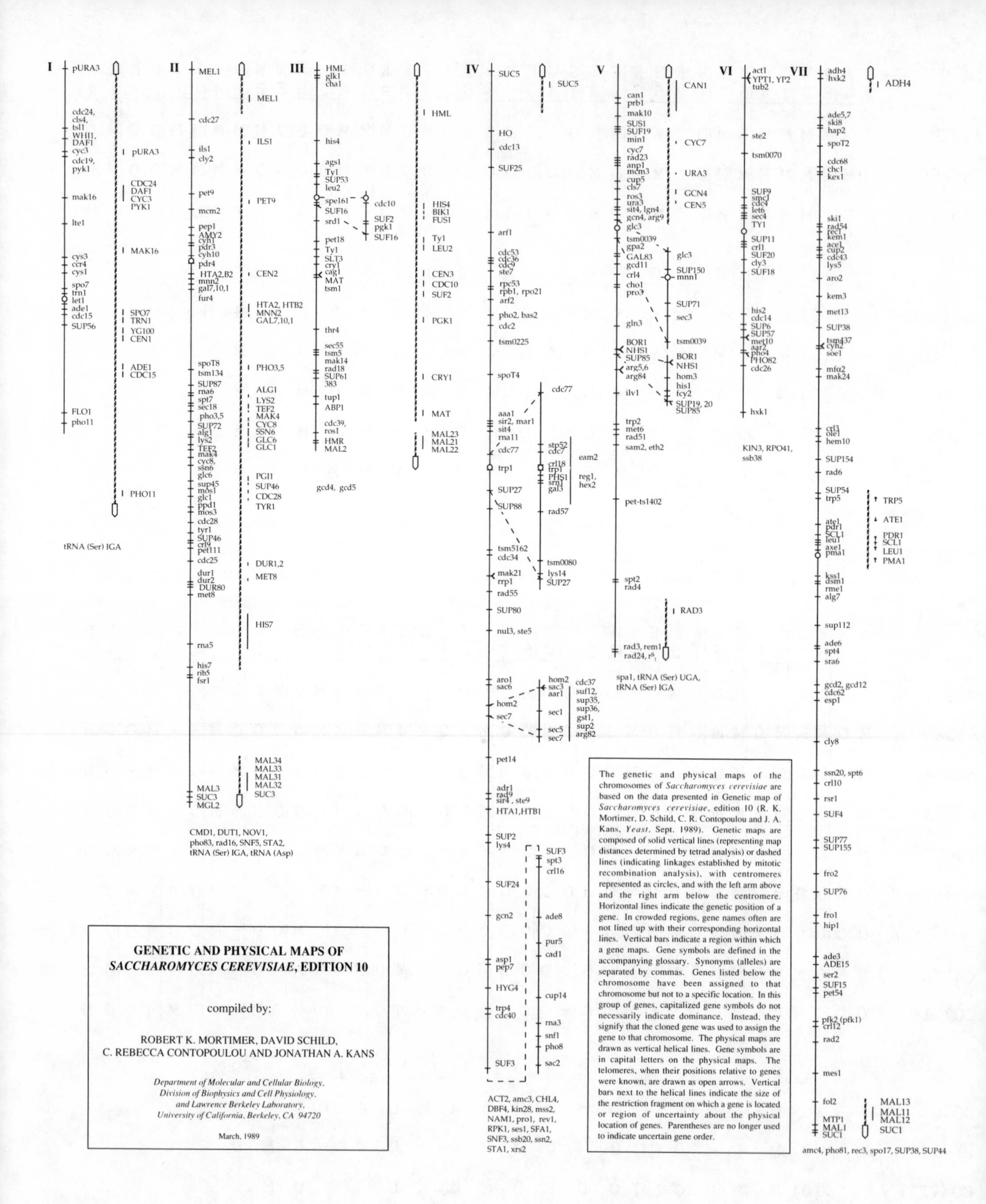

GENETIC AND PHYSICAL MAPS OF *SACCHAROMYCES CEREVISIAE*, EDITION 10

compiled by:

ROBERT K. MORTIMER, DAVID SCHILD,
C. REBECCA CONTOPOULOU AND JONATHAN A. KANS

Department of Molecular and Cellular Biology,
Division of Biophysics and Cell Physiology,
and Lawrence Berkeley Laboratory,
University of California, Berkeley, CA 94720

March, 1989

The genetic and physical maps of the chromosomes of *Saccharomyces cerevisiae* are based on the data presented in Genetic map of *Saccharomyces cerevisiae*, edition 10 (R. K. Mortimer, D. Schild, C. R. Contopoulou and J. A. Kans, *Yeast*, Sept. 1989). Genetic maps are composed of solid vertical lines (representing map distances determined by tetrad analysis) or dashed lines (indicating linkages established by mitotic recombination analysis), with centromeres represented as circles, and with the left arm above and the right arm below the centromere. Horizontal lines indicate the genetic position of a gene. In crowded regions, gene names often are not lined up with their corresponding horizontal lines. Vertical bars indicate a region within which a gene maps. Gene symbols are defined in the accompanying glossary. Synonyms (alleles) are separated by commas. Genes listed below the chromosome have been assigned to that chromosome but not to a specific location. In this group of genes, capitalized gene symbols do not necessarily indicate dominance. Instead, they signify that the cloned gene was used to assign the gene to that chromosome. The physical maps are drawn as vertical helical lines. Gene symbols are in capital letters on the physical maps. The telomeres, when their positions relative to genes were known, are drawn as open arrows. Vertical bars next to the helical lines indicate the size of the restriction fragment on which a gene is located or region of uncertainty about the physical location of genes. Parentheses are no longer used to indicate uncertain gene order.

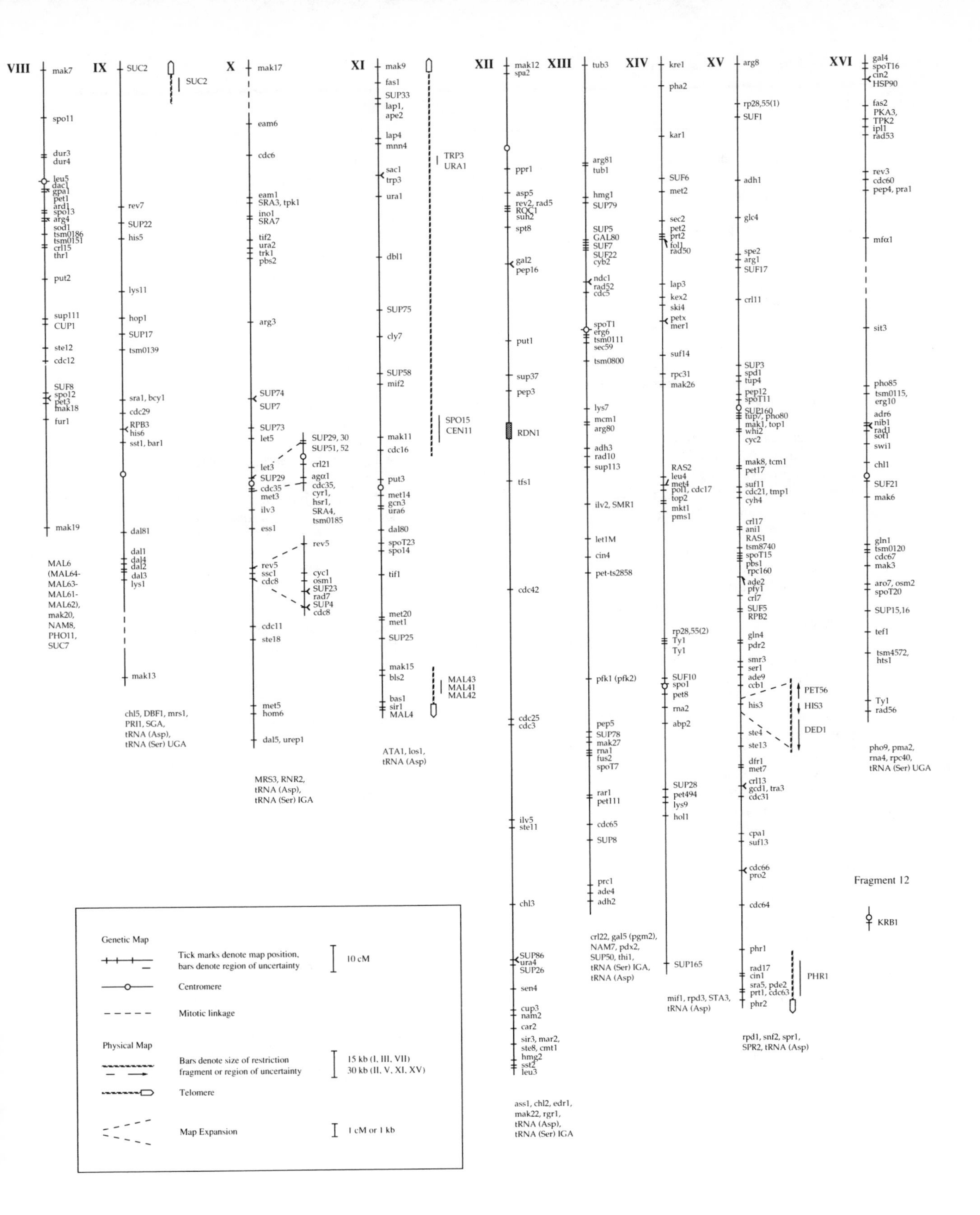

VIII
mak7
spo11
dur3
dur4
leu5
dac1
gpa1
pet1
ard1
spo13
arg4
sod1
tsm0186
tsm0151
crl15
thr1
put2
sup111
CUP1
ste12
cdc12
SUF8
spo12
pet3
mak18
fur1
mak19
MAL6 (MAL64-MAL63-MAL61-MAL62), mak20, NAM8, PHO11, SUC7
IX
SUC2
SUC2
rev7
SUP22
his5
lys11
hop1
SUP17
tsm0139
sra1, bcy1
cdc29
RPB3
his6
sst1, bar1
dal81
dal1
dal4
dal2
dal3
lys1
mak13
chl5, DBF1, mrs1, PRI1, SGA, tRNA (Asp), tRNA (Ser) UGA
X
mak17
eam6
cdc6
eam1
SRA3, tpk1
ino1
SRA7
tif2
ura2
trk1
pbs2
arg3
SUP74
SUP7
SUP73
let5
SUP29, 30
SUP51, 52
let3
SUP29
cdc35
met3
ilv3
crl21
agα1
cdc35, cyr1, hsr1, SRA4, tsm0185
ess1
rev5
rev5
ssc1
cdc8
cyc1
osm1
SUF23
rad7
SUP4
cdc8
cdc11
ste18
met5
hom6
dal5, urep1
MRS3, RNR2, tRNA (Asp), tRNA (Ser) IGA
XI
mak9
fas1
SUP33
lap1, ape2
lap4
mnn4
TRP3
URA1
sac1
trp3
ura1
dbl1
SUP75
cly7
SUP58
mif2
SPO15
CEN11
mak11
cdc16
put3
met14
gcn3
ura6
dal80
spoT23
spo14
tif1
met20
met1
SUP25
mak15
bls2
MAL43
MAL41
MAL42
bas1
sir1
MAL4
ATA1, los1, tRNA (Asp)
XII
mak12
spa2
ppr1
asp5
rev2, rad5
ROC1
suh2
spt8
gal2
pep16
put1
sup37
pep3
RDN1
tfs1
cdc42
cdc25
cdc3
ilv5
ste11
chl3
SUP86
ura4
SUP26
sen4
cup3
nam2
car2
sir3, mar2, ste8, cmt1
hmg2
sst2
leu3
ass1, chl2, edr1, mak22, rgr1, tRNA (Asp), tRNA (Ser) IGA
XIII
tub3
arg81
tub1
hmg1
SUP79
SUP5
GAL80
SUF7
SUF22
cyb2
ndc1
rad52
cdc5
spoT1
erg6
tsm0111
sec59
tsm0800
lys7
mcm1
arg80
adh3
rad10
sup113
ilv2, SMR1
let1M
cin4
pet-ts2858
pfk1 (pfk2)
pep5
SUP78
mak27
rna1
fus2
spoT7
rar1
pet111
cdc65
SUP8
prc1
ade4
adh2
crl22, gal5 (pgm2), NAM7, pdx2, SUP50, thi1, tRNA (Ser) IGA, tRNA (Asp)
XIV
kre1
pha2
kar1
SUF6
met2
sec2
pet2
prt2
fol1
rad50
lap3
kex2
ski4
petx
mer1
suf14
rpc31
mak26
RAS2
leu4
met4
pol1, cdc17
top2
mkt1
pms1
rp28,55(2)
Ty1
Ty1
SUF10
spo1
pet8
rna2
abp2
SUP28
pet494
lys9
hol1
SUP165
mif1, rpd3, STA3, tRNA (Asp)
XV
arg8
rp28,55(1)
SUF1
adh1
glc4
spe2
arg1
SUF17
crl11
SUP3
spd1
tup4
pep12
spoT11
SUP160
tup7, pho80
mak1, top1
whi2
cyc2
mak8, tcm1
pet17
suf11
cdc21, tmp1
cyh4
crl17
ani1
RAS1
tsm8740
spoT15
pbs1
rpc160
ade2
pfy1
crl7
SUF5
RPB2
gln4
pdr2
smr3
ser1
ade9
ccb1
his3
PET56
HIS3
DED1
ste4
ste13
dfr1
met7
crl13
gcd1, tra3
cdc31
cpa1
suf13
cdc66
pro2
cdc64
phr1
rad17
cin1
sra5, pde2
prt1, cdc63
phr2
PHR1
rpd1, snf2, spr1, SPR2, tRNA (Asp)
XVI
gal4
spoT16
cin2
HSP90
fas2
PKA3, TPK2
ipl1
rad53
rev3
cdc60
pep4, pra1
mfα1
sit3
pho85
tsm0115, erg10
adr6
nib1
rad1
sot1
swi1
chl1
SUF21
mak6
gln1
tsm0120
cdc67
mak3
aro7, osm2
spoT20
SUP15,16
tef1
tsm4572, hts1
Ty1
rad56
pho9, pma2, rna4, rpc40, tRNA (Ser) UGA
Fragment 12
KRB1
Genetic Map
Tick marks denote map position, bars denote region of uncertainty
10 cM
Centromere
Mitotic linkage
Physical Map
Bars denote size of restriction fragment or region of uncertainty
15 kb (I, III, VII)
30 kb (II, V, XI, XV)
Telomere
Map Expansion
1 cM or 1 kb

Table 17. Glossary of mapped gene symbols

Symbol	Definition*
aaa	Amino terminal, amino acetyl transferase
aar	Amino acid analog resistance
aas	Amino acid analog sensitive (see also gcn)
abp	Actin binding protein
ace	Activation of CUP expression
act	Actin
ade	Adenine requiring
adh	Alcohol dehydrogenase defective
adr	Dehydrogenase regulation defective
agα	α-cell specific sexual agglutination
ags	Aminoglycoside sensitive
alg	Asparagine-linked glycosylation deficient
amc	Artificial minichromosome maintenance
amy	Antimycin resistance
ani	Anisomycin resistance
anp	ANP and osmotic sensitive
ant	Antibiotic resistance
ape	Aminopeptidase
ard	Arrest at start of cell cycle defective
arf	ADP-ribosylation factor
arg	Arginine requiring
aro	Aromatic amino acid requiring
asp	Aspartic acid requiring
ass	Aspartyl-tRNA synthetase
ata	Sporulation-specific gene characterized by ATA sequences
ate	Arginyl-tRNA-protein transferase deficient
axe	Axenomycin resistance
bar	a cells lack barrier effect on α factor
bas	Basal level control
bcy	Adenylate-cyclase and cAMP-dependent protein kinase deficient
bik	Nuclear fusion (bikaryon)
bls	Blasticidin-S resistance
bor	Borrelidin resistance
cad	Cadmium resistance
cag	Constituvely agglutinable
can	Canavanine resistance
car	Catabolism of arginine defective
ccb	Cross-complementation of budding defect
ccr	Carbon catabolite repression
cdc	Cell division cycle blocked at 36°C
cen	Centromere
cha	Catabolism of hydroxy amino acids
chc	Clathrin heavy chain gene
chl	Chromosome loss
cho	Choline requiring
cin	Chromosome instability
cls	Calcium sensitive
cly	Cell lysis at 36°C
cmd	Calmodulin gene
cmt	Control of mating type
cpa	Arginine requiring in presence of excess uracil
crl	Cycloheximide-resistant temperature-sensitive lethal
cry	Cryptopleurine resistance
cup	Copper resistance
cyb	Cytochrome b_2 deficiency
cyc	Cytochrome c deficiency
cyh	Cycloheximide resistance
cyr	Adenylate cyclase deficient
cys	Cysteine requiring
dac	Division arrest control for mating pheromones
daf	Dominant α factor resistance
dal	Allantoin degradation deficient
dbf	Dumbell formation
dbl	Alcian blue dye binding deficient
ded	Defines essential domain, lethal
dex	Dextran utilization
dfr	Dihydrofolate reductase
dsm	Premeiotic DNA synthesis deficient
dur	Urea degradation deficient
eam	Endogenous ethanolamine biosynthesis
edr	Enhanced delta recombination
erg	Ergosterol biosynthesis defective; many also nystatin resistant
esp	Extra spindle pole bodies
ess	Essential
eth	Ethionine resistance
fas	Fatty acid synthetase deficient
fdp	Unable to grow on glucose, fructose, sucrose or mannose
flk	Flaky
flo	Flocculation
fol	Folinic acid requiring
fro	Frothing
fun	Function unknown
fur	Uracil permease
fsr	Fluphenazine resistance
fus	Fusion defective
gal	Galactose non-utilizer
gcd	General control of amino acid synthesis derepressed
gcn	General control of amino acid synthesis non-derepressible (see also aas)
gcn	Glucosamine accumulation
glc	Glycogen storage
glk	Glucokinase deficient

Table 17. Glossary of mapped gene symbols - continued

Symbol	Definition
gln	Glutamine synthetase non-derepressible
gpα	G protein alpha homologous gene
gst	G -1 to S transition
hap	Global regulator of respiratory genes
hem	Heme synthesis deficient
hex	Hexose metabolism regulation
hip	Histidine specific permease
his	Histidine requiring
hmg	HMG-CoA reductase
hml	Mating type cassette - left
hmr	Mating type cassette - right
ho	Homothallic switching
hol	Histidinol uptake proficient
hom	Homoserine requiring
hop	Homologue pairing
hsp	Heat shock protein
hsr	Heat shock resistance
hta	Histone A genes
htb	Histone B genes
hts	Histidinyl-tRNA synthetase
hxk	Hexokinase deficient
hyg	Hygromycin resitance
ils	Isoleucyl-tRNA synthetase deficient; no growth at 36°C
ilv	Isoleucine-plus-valine requiring
ino	Inositol deficient
ipl	Increase in ploidy
kar	Karyogamy defective
kem	kar enhancing mutation
kex	Killer expression defective
kin	Protein kinase
krb	Suppression of some mak mutations
kre	Killer resistance
kss	Protein kinases
lap	Leucine aminopeptidase deficient
let	Lethal
leu	Leucine requiring
lgn	Sporulation-induced transcripts (see also sit)
los	Loss of suppression and defective in tRNA processing
lte	Low temperature essential
lts	Low temperature sensitive
lys	Lysine requiring
mak	Maintenance of killer deficient
mal	Maltose fermentation
mar	Mating type cassette expression
mat	Mating type locus
mcm	Minichromosome maintenance deficient
mel	Melibiose fermentation
mer	Meiotic recombination
mes	Methionyl-tRNA synthetase deficient; no growth at 36°C
met	Methionine requiring
mfα	α-mating factor
mgl	α-Methylglucoside fermentation
mif	Mitotic frequency of chromosome transmission
min	Methionine inhibited
mkt	Maintenance of K_2 killer factor
mnn	Mannan synthesis defective
mos	Modifier of ochre suppressors
mrs	Mitochondrial RNA splicing
mss	Suppression of a mitochondrial RNA splice defect
mtp	Melezitose fermentation
nam	Nuclear suppressor of mitochondrial mutations
ndc	Nuclear division cycle
nhs	Hydrogen sulfide production inhibitor
nib	Nibbled colony phenotype due to 2μ DNA
nra	Neutral red accumulation
nov	Novobiocin resistance
nul	Non-mater
ole	Oleic acid requiring
oli	Oligomycin resistance
osm	Low osmotic pressure sensitive
pbs	Polymyxin B resistance
pde	Phosphodiesterase (cAMP)
pdr	Pleiotropic drug resistance
pdx	Pyridoxin requiring
pep	Proteinase deficient
pet	Petite; unable to grow on non-fermentable carbon sources
pfk	Phosphofructokinase
pfy	Profilin of yeast
pgi	Phosphoglucose isomerase deficient
pgk	3-Phosphoglycerate kinase deficient
pgm	Phosphoglucomutase deficient
pha	Phenylalanine requiring
pho	Phosphatase deficient
phr	Photoreactivation repair deficient
phs	Hydrogen sulfide production deficient
pka	Protein kinase catalytic subunit
pma	Plasma membrane ATPase mutations
pms	Postmeiotic segregation increased
pol	DNA polymerase 1
ppd	Phosphoprotein phosphatase-deficient
ppr	Defective in pyrimidine biosynthetic pathway regulation
pra	Proteinase A deficient

Table 17. Glossary of mapped gene symbols - continued

Symbol	Definition
prb	Proteinase B deficient
prc	Proteinase C deficient
pri	DNA primase
pro	Proline requiring
prt	Protein synthesis defective at 36°C
pur	Purine excretion
put	Proline nonutilizer
pyk	Pyruvate kinase deficient
rad	Radiation (ultraviolet or ionizing) sensitive
rar	Regulation of autonomous replication
ras	Homologous to RAS proto-oncogene
rdn	Ribosomal RNA structural genes
rec	Recombination deficient
reg	Regulation of galactose pathway enzymes
rev	Revertibility decreased
rgr	Resistant to glucose repression
rib	Riboflavin biosynthesis
rme	Meiosis independent of mating type heterozygosity
rna	RNA synthesis defective; unable to grow at 36°C
rnr	Ribonucleotide reductase
roc	Roccal resistance
ros	Relaxation of sterility
rpo	RNA polymerase II mutants
rpb	RNA polymerase subunit II
rpc	RNA polymerase C
rpd	Reduced potassium dependency
rpk	Regulatory protein kinase
rrp	rRNA processing
$r^{S}1$	Radiation sensitive
rsr	Ras-related
sac	Suppressor of actin mutations
sam	S-adenosyl methionine synthesis
scl	Dominant suppression of t.s. lethality of *crl3*
sec	Secretion deficient
sen	Splicing endonuclease
ser	Serine requiring
ses	Seryl-tRNA synthetase
sfa	Sensitive to formaldehyde
sga	Suppression of growth arrest
sir	Silent mating type information regulation
sit	Suppresion of initiation of transcription
sit	Sporulation-induced transcripts (see also lgn)
ski	Superkiller
slt	Suppression at low temperature
smc	Stability of minichromosomes
smr	Sulfometuron methyl resistance
snf	Deficient in derepression of many glucose-repressible genes
sod	Manganese-superoxide dismutase
soe	Suppresion of *cdc8*
sot	Suppression of deoxythymidine monophosphate uptake
spa	Spindle pole antigen
spd	Sporulation not repressed on rich media
spe	Spermidine resistance
spe	Stationary phase entry
spo	Sporulation deficient
spr	Sporulation regulated genes
spt	Suppressors of Ty transcription
sra	Suppressors of the ras mutation
srd	Suppressor of *rrp1*
srn	Suppressor of yeast *rna1-1*
ssb	Single strand binding protein
ssc	HSP70-related gene
ssn	Suppressor of *snf1*
sst	Supersensitive to α factor
sta	Starch hydrolysis
ste	Sterile
stp	Ste pseudorevertants
suc	Sucrose fermentation
suf	Suppression of frameshift mutation
suh	Suppression of *his2-1*
sup	Suppression of nonsense mutation
sus	Suppression of *ser1*
swi	Homothallic switching deficient
tcm	Tricodermin resistance
tef	Translational elongation factor
tel	Telomere
tfs	Cdc twenty-five suppressor.
thi	Thiamine requiring
thr	Threonine requiring
tif	Translation initiation factor
til	Thioisoleucine resistance
tmp	Thymidine monophosphate requiring
top	Topoisomerase deficient
tpk	Threonine/serine protein kinase
tra	Triazylalanine resistant
trk	Transport of potassium
trn	Proline-tRNA gene
trp	Tryptophan requiring
tsl	Temperature sensitive lethal
tsm	Temperature sensitive lethal mutations
tub	Tubulin; MBC resistance
tup	Deoxythymidine monophosphate uptake positive
tyr	Tyrosine requiring
TY	Transposable element
umr	Ultraviolet mutability reduced
ura	Uracil requiring
urep	Ureidosuccinate permease
whi	Small cell size
YG100	Heat shock gene
ypt	GTP-binding protein
xrs	X-ray sensitive

* Three gene symbols have two definitions: gcn, sit, and spe.

Table 18. List of mapped genes*

Gene / Synonym	Map Position	References**
aaa1	**4L**	F-J. S. Lee, L-W. Lin and J. A. Smith, p.c.
aar1	**4R**	McCusker and Haber (1988a)
aar2	**6R**	McCusker and Haber (1988a)
aas1	4R	
gcn2		
aas2	11R	
gcn3		
aas3	**5L**	Olson *et al.* (1986); Thireos *et al.* (1984); Hinnebusch and Fink (1983)
arg9		
gcn4		
ABP1	**3R**	D. Drubin and D. Botstein, p.c.
abp2	**14R**	K. Wertman and D. Botstein, p.c.
ace1	**7L**	Thiele (1988)
act1	6L	
ACT2	**4**	Schwob *et al.* (1988a and 1988b)
ade1	1R	
ade2	15R	
ade3	7R	
ade4	13R	Schild and Mortimer (1985)
ade5,7	7L	
ade6	7R	
ade8	4R	
ade9	15R	
ADE15	7R	
adh1	15L	
adh2	13R	
adh3	13R	
adh4	**7L**	K. Kaneko and V. Williamson, p.c.
adr1	4R	
adr6	**16L**	Taguchi and Young (1987)
agα1	**10R**	Suzuki and Yanagishima (1986); J. Kurjan, p.c.
ags1	**3L**	Ernst and Chan (1985)
alg1	2R	
alg7	7R	
amc3	**4**	Larionov *et al.* (1988b)
amc4	**7**	Larionov *et al.* (1988b)
AMY1	7L	Saunders and Rank (1981); Balzi *et al.* (1987)
pdr1		
AMY2	2L	Balzi *et al.* (1987)
ani1	**15R**	McCusker and Haber (1988b)
anp1	5L	
ant1	7L	Saunders and Rank (1981); Balzi *et al.* (1987)
pdr1		
ape2	**11L**	Hirsch *et al.* (1988)
lap1		
ard1	8R	
arf1	**4L**	T. Stearns, R. Kahn and D. Botstein, p.c.
arf2	**4L**	T. Stearns, R. Kahn and D. Botstein, p.c.
arg1	15L	
arg3	10L	
arg4	8R	
arg5,6	5R	
arg8	15L	
arg9	**5L**	Olson *et al.* (1986); Hinnebusch and Fink (1983); Thireos *et al.* (1984)
aas3		
gcn4		
arg80	13R	
arg81	13L	
arg82	4R	
arg84	5R	
aro1	4R	
aro2	7L	
aro7	16R	Ball *et al.* (1986)
osm2		
asp1	4R	
asp5	12R	
ass1	**12**	Kolman *et al.* (1988)
ATA1	**11**	M. Breitenbach, p.c.
ate1	7L	Balzi *et al.* (1988)
AXE1	7L	E. Balzi, p.c.
bar1	9L	
sst1		
bas1	**11R**	Arndt *et al.* (1987)
bas2	**4L**	G. R. Fink, p.c.
pho2		
bcy1	**9L**	Matsumoto *et al.* (1982); Toda *et al.* (1987); Cannon and Tatchell (1987)
sra1		
BIK1	**3L**	Trueheart *et al.* (1987)
bls2	**11R**	Ishiguro and Hayashi (1986)
BOR1	5R	
BOR2	7L	Saunders and Rank (1981); Balzi *et al.* (1987)
pdr1		
cad1	**4R**	A. Januska, p.c.
cag1	**3R**	Doy and Yoshimura (1985)
can1	5L	
car2	12R	
ccb1	**15R**	A. Bender, p.c.
ccr4	**1L**	C. Denis, p.c.
cdc2	4L	
cdc3	**12R**	Johnson *et al.* (1987)
cdc4	6L	
cdc5	13L	

Table 18. List of mapped genes - continued

Gene / Synonym	Map Position	References
cdc6	10L	
cdc7	4L	
cdc8	10R	
cdc9	4L	
cdc10	3R	Yeh et al. (1986)
cdc11	10R	
cdc12	8R	
cdc13	4L	B. Garvik and L. Hartwell, p.c.
cdc14	6R	
cdc15	1R	
cdc16	11L	
cdc17	14L	Carson (1987)
pol1		
cdc19	1L	
pyk1		
cdc21	15R	
tmp1		
cdc24	1L	Ohya et al. (1986a)
cls4		
tsl1		
cdc25	12R	Johnson et al. (1987)
cdc25	2R	Portillo and Mazon (1986)
cdc26	6R	
cdc27	2L	J. Truehart, p.c.
cdc28	2R	
cdc29	9L	
cdc31	15R	Schild and Mortimer (1985); Baum et al. (1986)
cdc34	4R	M. G. Goebl, p.c.
cdc35	10R	Boutelet et al. (1985); Casperson et al. (1985)
cyr1		
hsr1		
SRA4		
tsm0185		
cdc36	4L	
cdc37	4R	
cdc39	3R	Jenness et al. (1987)
ros1		
cdc40	4R	Kassir et al. (1985)
cdc42	12R	Johnson et al. (1987)
cdc43	7L	Adams (1984)
cdc53	4L	M.G. Goebl, p.c.
cdc60	16L	Hanic-Joyce (1985)
cdc62	7R	Hanic-Joyce (1985)
cdc63	15R	Hanic-Joyce (1985)
prt1		
cdc64	15R	Hanic-Joyce (1985)
cdc65	13R	G. Johnston, p.c.
cdc66	15R	
cdc67	16R	J. Prendergast, L. Murray, and G. Johnston, p.c.
cdc68	7L	A. Rowley, and G. Johnston, p.c.
cdc77	4L	I. Villadsen, p.c.
ndc2		
CHA1	3L	Bornæs et al. (1988)
CHC1	7L	Payne et al. (1987); S. Lemmon, C. Freund, and E. Jones, p.c.
chl1	16L	
chl2	12R	Kouprina et al. (1988)
chl3	12R	A. Tsouladze and V. Larionov, p.c.; Kouprina et al. (1988)
chl4	4R	Kouprina et al. (1988)
chl5	9L	Kouprina et al. (1988)
cho1	5R	
cin1	15R	T. Stearns and D. Botstein, p.c.
cin2	16L	T. Stearns and D. Botstein, p.c.
cin4	13R	T. Stearns and D. Botstein, p.c.
cls4	1L	Ohya et al. (1986a)
cdc24		
tsl1		
cls7	5L	Ohya et al. (1986b)
cly2	2L	
cly3	6R	
cly7	11L	
cly8	7R	
cmd1	2	Davis et al. (1986)
cmt1	12R	Rine and Herskowitz (1987)
mar2		
sir3		
ste18		
cpa1	15R	
crl1	6R	McCusker and Haber (1988b)
crl3	7L	McCusker and Haber (1988b)
crl4	5R	McCusker and Haber (1988b)
crl7	15R	McCusker and Haber (1988b)
crl9	2R	McCusker and Haber (1988b)
crl10	7R	McCusker and Haber (1988b)
crl11	15L	McCusker and Haber (1988b)
crl12	7R	McCusker and Haber (1988b)
crl13	15R	McCusker and Haber (1988b)
crl15	8R	McCusker and Haber (1988b)
crl16	4R	McCusker and Haber (1988b)
crl17	15R	McCusker and Haber (1988b)
crl18	4L	McCusker and Haber (1988b)
crl21	10R	McCusker and Haber (1988b)
crl22	13R	McCusker and Haber (1988b)

Table 18. List of mapped genes - continued

Gene / Synonym	Map Position	References
cry1	3R	
CUP1	8R	
cup2	7L	Welch et al. (1989)
cup3	12R	Welch et al. (1989)
cup5	5L	Welch et al. (1989)
cup14	4R	Welch et al. (1989)
cyb2	13L	T. Lodi, B. Guiard and I. Ferrero, p. c.
cyc1	10R	
cyc2	15R	
cyc3	1L	
cyc7	5L	Verdiere et al. (1988)
cyp3		
cyc8	2R	
ssn6		
cyc9	3R	
flk1		
tup1		
umr7		
cyh1	2L	
cyh2	7L	
cyh3	7L	Saunders and Rank (1981); Balzi et al. (1987)
pdr1		
cyh4	15R	
cyh10	2L	
cyr1	10R	Casperson et al. (1985)
cdc35		
hsr1		
SRA4		
tsm0185		
cys1	1L	
cys3	1L	
dac1	8R	Fujimura (1989)
gpa1		
DAF1	1L	Cross (1988)
WHI1		
dal1	9R	
dal2	9R	
dal3	9R	
dal4	9R	
dal5	10R	Chisholm et al. (1987)
urep1		
dal80	11R	
dal81	9R	
DBF1	9	J. W. Chapman and L. H. Johnston, p.c.
DBF4	4	J. W. Chapman and L. H. Johnston, p.c.
dbl1	11L	
ded1	15R	Struhl (1985)
dfr1	15R	Barclay et al. (1988)
dsm1	7R	M. Esposito, p.c.
dur1	2R	Genbauffe and Cooper (1986)
dur2	2R	Genbauffe and Cooper (1986)
dur3	8L	
dur4	8L	
DUR80	2R	
DUT1	2	Godsen et al. (1986)
eam1	10L	K. Atkinson, p.c.
eam2	4	K. Atkinson, p.c.
eam6	10L	K. Atkinson, p.c.
edr1	12	W. L. Arthur and R. Rothstein, p.c.
erg6	13R	R. F. Gaber, B. K. Kennedy, D. Copple and M. Bard, p.c.
erg10	16L	Dequin et al. (1988)
tsm0115		
esp1	7R	P. Baum and B. Byers, p.c.
ess1	10R	Hanes (1988)
eth2	5R	Schild and Mortimer (1985)
sam2		
fas1	11L	
fas2	16L	E. Schweizer, p.c.
fcy2	5R	E. Weber, p.c.
fdp1	2R	
flk1	3R	
cyc9		
tup1		
umr7		
FLO1	1R	
fol1	14L	
fol2	7R	
fro1	7R	
fro2	7R	
fsr1	2R	Matsumoto et al. (1986)
fur1	8R	Jenness et al. (1987)
fur4	2R	Weber et al. (1986)
FUS1	3L	Trueheart et al. (1987)
fus2	13R	J. Trueheart, p.c.
gal1	2R	
gal2	12R	
gal3	4R	
gal4	16L	
gal5	13R	G. McKnight, J. Hopper and D. Oh, p.c.
pgm2		Fraenkel (1982)
gal7	2R	

Table 18. List of mapped genes - continued

Gene / Synonym	Map Position	References
gal10	2R	
gal80	**13L**	Schild and Mortimer (1985)
GAL83	5R	
gcd1	15R	
↳ *tra3*		
gcd2	**7R**	Niederberger *et al.* (1986)
↳ ***gcd12***		
gcd4	**3**	Skvirsky *et al.* (1986)
gcd5	**3**	Greenberg *et al.* (1986)
gcd11	**5R**	Harashima and Hinnebusch (1986)
gcd12	**7R**	C. J. Paddon, p.c.
↳ ***gcd2***		
gcn2	4R	
↳ *aas1*		
gcn3	11R	
↳ *aas2*		
gcn4	**5L**	Olson *et al.* (1986); Hinnebusch and Fink (1983); Thireos *et al.* (1984)
↳ ***aas3***		
↳ ***arg9***		
glc1	2R	
glc3	5L	
glc4	15L	
glc6	2R	
glk1	3L	
gln1	16R	
gln3	5R	
gln4	**15R**	Ludmerer and Schimmel (1985)
gpa1	**8R**	Miyajima *et al.* (1987)
↳ ***dac1***		
gpa2	**5R**	Nakafuku *et al.* (1988)
gst1	**4R**	Kikuchi *et al.* (1988); Y. Kikuchi, p.c.
↳ *suf12*		
↳ ***sup2***		
↳ *sup35*		
↳ *sup36*		
hap2	**7L**	Pinkham and Guarente (1985)
hem10	7L	
hex2	**4R**	Niederacher and Entian (1987)
↳ ***reg1***		
hip1	7R	
his1	5R	
his2	6R	
his3	15R	
↳ *his8*		
his4	3L	
his5	9L	
his6	9L	
his7	2R	
hmg1	**13L**	Basson *et al.* (1987)
hmg2	**12R**	Basson *et al.* (1987)
HML	3L	
HMR	3R	
HO	4L	
hol1	14R	
hom2	4R	
hom3	5R	
hom6	10R	Schild and Mortimer (1985)
hop1	**9L**	Hollingsworth and Byers (1989)
HSP90	**16L**	T. Stearns and D. Bostein, p.c.
hsr1	**10R**	H. Iida, p.c.
↳ *cdc35*		
↳ *cyr1*		
↳ ***SRA4***		
↳ *tsm0185*		
HTA2,B2	**2R**	Norris and Osley (1987)
HTA1,B1	**4R**	Norris and Osley (1987)
hts1	**16R**	G. Natsoulis, F. Hilger and G. Fink, p.c.; M. Sandbaken and M. Culbertson, p.c.
↳ *tsm4572*		
hxk1	6R	
hxk2	7L	
HYG4	**4R**	McCusker (1987)
ils1	2L	
ilv1	5R	
ilv2	13R	
↳ *SMR1*		
ilv3	10R	
ilv5	12R	
ino1	10L	
ilp1	**16L**	T. Stearns and D. Botstein, p.c.
kar1	14L	
kem1	**7L**	J. Kim and G. R. Fink, p.c.
kem3	**7L**	J. Kim and G. R. Fink, p.c.
kex1	7L	
kex2	14L	
KIN3	**6**	Jones and Rosamond (1988)
kin28	**4**	Boulet *et al.* (1988)
KRB1	F12	See discussion and Table 16
kre1	**14L**	C. Boone and H. Bussey, p.c.
kss1	**7R**	W. Courchesne and J. Thorner, p.c.
lap1	11L	Hirsch *et al.* (1988)
↳ ***ape2***		
lap3	14L	
lap4	11L	
let1	1R	
let1M	13R	

Table 18. List of mapped genes - continued

Gene / Synonym	Map Position	References
let3	10L	
let5	10L	
let6	6L	
leu1	7L	
leu2	3L	
leu3	**12R**	Brisco *et al.* (1987)
leu4	14L	
leu5	**8C**	Drain and Schimmel (1986)
lgn4	**5L**	Gottlin-Ninfa and Kaback (1986)
sit4		
los1	**11**	Hurt *et al.* (1987)
lte1	**1L**	D. Kaback, p.c.; Wickner *et al.* (1987)
lts1	7L	
lts3	7L	
lts4	4R	
lts10	4R	
lys1	9R	
lys2	2R	
lys4	4R	
lys5	7L	
lys7	13R	
lys9	14R	Borell *et al.* (1984)
lys13		
lys11	9L	
lys13	**14R**	Borell *et al.* (1984)
lys9		
lys14	**4R**	C. R. Contopoulou, p.c.
mak1	15R	Thrash *et al.* (1985)
top1		
mak3	16R	
mak4	2R	
mak5	2R	
mak6	16R	
mak7	8L	
mak8	15R	
tcm1		
mak9	11L	
mak10	5L	
mak11	11L	
mak12	12L	
mak13	9R	
mak14	3R	
mak15	11R	
mak16	1L	
mak17	10L	
mak18	8R	
mak19	8R	
mak20	8	
mak21	4R	
mak22	12	
mak24	7L	
mak26	14L	
mak27	13R	
MAL1	7R	Charron *et al.* (1989)
MAL11		
MAL12		
MAL13		
MAL2	3R	Charron *et al.* (1989)
MAL21		
MAL22		
MAL23		
MAL3	2R	Charron *et al.* (1989)
MAL31		
MAL32		
MAL33		
MAL34		
MAL4	11R	Charron *et al.* (1989)
MAL41		
MAL42		
MAL43		
MAL6	**8**	Needleman *et al.* (1984); Cohen *et al.* (1985); Dubin (1987); Charron *et al.* (1989); Dubin *et al.* (1988)
MAL61		
MAL62		
MAL63		
MAL64		
mar1	4L	
sir2		
mar2	12R	
cmt1		
ste8		
sir3		
MAT	3R	
mcm1	13R	Maine (1984)
mcm2	**2L**	Gibson (1989)
mcm3	**5L**	Gibson (1989)
MEL1	**2L**	Hawthorne (1955) ; Vollrath *et al.* (1988)
mer1	**14L**	Engebrecht and Roeder (1989)
mes1	7R	
met1	11R	
met2	14L	
met3	10R	
met4	14L	
met5	10R	Schild and Mortimer (1985)
met6	5R	Schild and Mortimer (1985)

Table 18. List of mapped genes - continued

Gene / Synonym	Map Position	References
met7	15R	
met8	2R	
met10	6R	
met13	7L	
met14	11R	
met20	11R	
mfα1	**16L**	Flessel *et al.* (1989)
mfα2	7L	
MGL2	2R	
mif1	**14**	M. T. Brown and L. Hartwell, p.c.
MIF2	**11**	M. T. Brown and L. Hartwell, p.c.
min1	5L	
mkt1	**14L**	Wickner (1987)
mnn1	5C	
mnn2	2R	
mnn4	11L	
mos1	**2R**	Gelugne and Bell (1988)
mos3	**2R**	Gelugne and Bell (1988)
mrs1	**9R**	Kreike *et al.* (1986); S. Lotz, p.c.
MRS3	**10**	Schmidt *et al.* (1987)
mss2	**4L**	Boulet *et al.* (1988)
MTP1	**7R**	Perkins and Needleman (1988)
mut1	---	See discussion
mut2	---	See discussion
NAM1	**4**	Altamura *et al.* (1988)
nam2	12R	
NAM7	**13**	Altamura *et al.* (1988)
NAM8	**8**	Altamura *et al.* (1988)
ndc1	**13L**	Thomas and Botstein (1986)
ndc2	**4L**	Villadsen (1988); I. Villadsen, p.c.
cdc77		
NHS1	5R	
nib1	16L	
NOV1	**2**	Pocklington and Orr (1986)
NRA2	**7L**	R. Preston and E. Jones, p.c.
pdr1		
nul3	4R	
ste5		
ole1	7L	
oli1	7L	Saunders and Rank (1981); Balzi *et al.* (1987)
pdr1		
osm1	10R	
osm2	16R	Ball *et al.* (1986)
aro7		
pbs1	**15R**	Boguslawski (1985)
pbs2	**10L**	Boguslawski and Polazzi (1987)
pde2	**15R**	Sass *et al.* (1986); Wilson and Tachell (1988)
sra5		
pdr1	**7L**	Saunders and Rank (1982); Balzi *et al.* (1987)
AMY1		
ant1		
BOR2		
cyh3		
NRA2		
oli1		
smr2		
til1		
pdr2	**15R**	John Golin, p.c.
pdr4	**2C**	R. Preston and E. Jones, p.c.
pdx2	**13R**	G. McKnight, J. Hopper and D. Oh, p.c.
pep1	**2L**	R. Preston, L. Daniels and E. Jones, p.c.
pep3	12R	
pep4	16L	Mechler *et al.* (1987)
pra1		
pep5	**13R**	C. Woolford, C. Dixon, P. Walters and E. Jones, p.c.
pep7	4R	
pep12	15L	
pep16	12R	
pet1	8R	
pet2	14L	
pet3	8R	
pet8	14R	
pet9	2L	
pet11	2R	
pet14	4R	
pet17	15R	
pet18	3R	
pet54	**7R**	M. Contanzo, E. Seaver and T.Fox, p.c.
pet56	**15R**	Struhl (1985)
pet111	**13R**	Poutre and Fox (1987)
pet494	14R	
pet-ts1402	5R	
pet-ts2858	13R	
petx	14L	
pfk1	**13R**	Lobo and Maitra (1983)
pfk2		
pfk2	**7R**	Heinisch and von Borstel (1988)
pfk1		
pfy1	**15R**	A. Adams, V. Oechsner, W. Bandlav and D. Botstein, p.c.; Magdolen *et al.* (1988)
pgi1	2R	

Table 18. List of mapped genes - continued

Gene / Synonym	Map Position	References
pgk1	3R	
pgm2	**13R**	**Fraenkel (1982)**
gal5		
pha2	14L	
pho2	4L	G. R. Fink, p.c.
bas2		
pho3,5	2R	
pho4	6R	
pho8	4R	
pho9	**16**	Yoshida *et al.* (1988) ; K. Yoshida, p.c.
pho11,1	**1R**	D. Kaback, p.c.; de Jonge *et al.* (1988)
pho11,2	**8**	de Jonge *et al.* (1988)
pho80	15R	
tup7		
pho81	**7**	Yoshida *et al.* (1988)
PHO82	6R	
pho83	**2**	Yoshida *et al.* (1988); K. Yoshida, p.c.
pho85	16L	
phr1	15R	
phr2	15R	
PHS1	4R	
PKA3	**16L**	A. Petitjean and K. Tatchell, p.c.
TPK2		
pma1	**7R**	McCusker *et al.* (1987); Ulaszewski *et al.* (1987)
pma2	**16L**	Schlesser and Goffeau (1988)
pms1	14L	Williamson *et al.* (1985)
pol1	**14L**	Budd and Campbell (1987); Lucchini *et al.* (1988)
cdc17		
ppd1	**2R**	Matsumoto *et al.* (1985)
ppr1	12R	
pra1	**16L**	Mechler *et al.* (1987)
pep4		
prb1	5L	Moehle *et al.* (1987)
prc1	13R	
PRI1	**9**	Lucchini *et al.* (1987)
pro1	**4**	M. Brandriss, p.c.
pro2	**15R**	Tomenchok and Brandriss (1987)
pro3	**5R**	Tomenchok and Brandriss (1987)
prt1	15R	
cdc63		
prt2	14L	
prt3	---	See Table 5
pur5	4R	
put1	**12R**	Wang and Brandriss (1986)

Gene / Synonym	Map Position	References
put2	8R	
put3	**11L**	Brandriss (1987)
pyk1	1L	
cdc19		
rad1	16L	
rad2	7R	
rad3	5R	Sitney (1987); Montelone *et al.* (1988)
rem1		
rad4	5R	Sitney (1987)
rad5	12R	
rev2		
rad6	7L	
rad7	10R	
rad9	4R	
rad10	**13R**	Weiss and Friedberg (1985)
rad16	**2R**	J. Game, p.c.
rad17	**15R**	J. Game, p.c.
rad18	3R	
rad23	5L	
rad24	5R	Eckardt-Schupp *et al.* (1987); Sitney (1987)
$r^{s}{}_{1}$		
rad50	14L	
rad51	5R	
rad52	13L	
rad53	**16L**	K. Sitney, p.c.
rad54	7L	
rad55	4R	
rad56	16R	
rad57	4R	
$r^{s}{}_{1}$	5R	
rad24		
xrs2	**4R**	I. A. Zakharov, p.c.
rar1	**13R**	Kearsey and Edwards (1987)
RAS1	15R	
RAS2	14L	
RDN1	12R	
rec1	**7L**	Esposito *et al.* (1988)
rec3	**7L**	Esposito *et al.* (1988)
reg1	**4**	Matsumoto *et al.* (1983); Cannon and Tatchell (1987)
hex2		
rem1	**5R**	Montelone *et al.* (1988); Sitney (1987)
rad3		
rev1	**4**	F. Larimer, p.c.
rev2	12R	
rad5		
rev3	**16L**	C. Lawrence, p. c.
rev5	10R	
rev7	9L	Lawrence *et al.* (1985)

Table 18. List of mapped genes - continued

Gene / Synonym	Map Position	References
rgr1	12	Sakai *et al.* (1988)
rib5	2R	de los Angeles Santos *et al.* (1988)
rme1	7R	
rna1	13R	
rna2	14R	
rna3	4R	
rna4	16	S. Petersen-Bjorn and J. Friesen, p.c.
rna5	2R	
rna6	2R	
tsm7269		
rna11	4L	
rnr2	10C	Elledge and Davis (1987)
ROC1	12R	
ros1	3R	Jenness *et al.* (1987)
cdc39		
ros3	5L	Jenness *et al.* (1987)
rpb1	4L	M. L. Nonet, p.c.; Nonet *et al.* (1987)
rpo21		
RPB2	15R	C. Scafe, M. Nonet and R. Young, p.c.
RPB3	9L	P. Kolodziej and R. Young, p.c.
rpc31	14L	Mosrin *et al.* (1988); P. Thuriaux, p.c.
rpc40	16	C. Mann and I. Treich, p.c.
rpc53	4L	Mann and Treich (1988)
rpc160	15R	Gudenus *et al.* (1988)
rpd1	15	Vidal *et al.* (1988)
rpd3	14	Vidal *et al.* (1988)
RPK1	4	Schwob *et al.* (1988a and 1988b)
rpo21	4L	Himmelfarb *et al.* (1987)
rpb1		
RPO41	6	Greenleaf *et al.* (1986)
rp28,55,1	15L	Papciak and Pearson (1987)
rp28,55,2	14L	Papciak and Pearson (1987)
rrp1	4R	Fabian and Hopper (1987)
rsr1	7R	A. Bender, p.c.
sac1	11L	D. Drubin and D. Bostein, p.c.
sac2	4R	Novick, P., Osmond, B. and D. Botstein, p.c.
sac3	4R	Novick, P., Osmond, B. and D. Botstein, p.c.
sac6	4R	A. Adams, D. Drubin and D. Botstein, p.c.
sam2	5R	
eth2		
SCL1	7L	McCusker *et al.* (1987)
sec1	4R	
sec2	14L	
sec3	5R	
sec4	6L	
sec5	4R	
sec7	4R	
sec18	2R	
sec55	3R	
sec59	13R	
sen1	12R	Winey and Culbertson (1988)
ser1	15R	
ser2	7R	
ses1	4	Kolman *et al.* (1988)
SFA1	4	Mack and M. Brendel, p.c.
SGA	9	Pretorius and Marmur (1988)
sir1	11R	Ivy *et al.* (1985)
sir2	4L	
mar1		
sir3	12R	Ivy *et al.* (1985); Rine and Herskowitz (1987)
mar2		
ste8		
cmt1		
sir4	4R	Ivy *et al.* (1985)
ste9		
sit3	16L	Arndt *et al.* (1989)
sit4	4L	Arndt *et al.* (1989)
sit4	5L	Gottlin-Ninfa *et al.* (1986)
lgn4		
ski1	7L	
ski4	14L	
ski8	7L	Sommer and Wickner (1987)
SLT3	3R	Inge-Vechtomov and Karpova (1984)
smc1	6L	Larionov *et al.* (1985) ; V. Larionov, p.c.
SMR1	13R	
ilv2		
smr2	7L	Saunders and Rank (1981); Balzi *et al.* (1987)
pdr1		
smr3	15R	
snf1	4R	
snf2	15	J. Celenza, L. Neigeborn and M. Carlson, p.c.
SNF3	4	J. Celenza, and M. Carlson, p.c.
SNF5	2	J. Celenza and M. Carlson, p.c.
sod1	8R	van Loon *et al.* (1986)
soe1	7L	J-Y Su and R. A. Sclafani, p.c.
sot1	16L	
spa1	5	Snyder and Davis (1988)
spa2	12L	Snyder (1989)

Table 18. List of mapped genes - continued

Gene / Synonym	Map Position	References
spd1	15L	
spe2	15L	
spe161	**3R**	Crouzet *et al.* (1988)
spo1	14L	
spo7	1L	
spo11	8L	
spo12	8R	
spo13	**8R**	Wang *et al.* (1987)
spo14	11R	
spo15	**11C**	Yeh *et al.* (1986)
spo17	**7L**	Smith *et al.* (1988); Kennedy and Magee (1988)
spoT1	13C	
spoT2	7L	
spoT4	4L	
spoT7	13R	Tanaka and Tsuboi (1985)
spoT8	2R	
spoT11	15L	
spoT15	15R	
spoT16	16L	
spoT20	16R	
spoT23	11R	
spr1	**15**	Primerano *et al.* (1988)
SPR2	**15**	Primerano *et al.* (1988)
spt2	5R	
spt3	4R	
spt4	**7R**	J. Fassler and F. Winston, p.c.
spt6	**7R**	Clark-Adams and F. Winston (1987)
ssn20		
spt7	**2R**	Winston *et al.* (1987)
spt8	**12R**	Winston *et al.* (1987)
sra1	**9L**	Cannon *et al.* (1986); Cannon and Tatchell (1987)
bcy1		
SRA3	**10L**	Cannon *et al.* (1986)
tpk1		
SRA4	**10R**	Cannon *et al.* (1986)
cyr1		
cdc35		
hsr1		
tsm0185		
sra5	**15R**	Wilson and Tatchell (1988)
pde2		
sra6	**7R**	Cannon *et al.* (1986)
SRA7	**10L**	J. F. Cannon, p.c.
srd1	**3R**	G. R. Fabian S. Hess and A. K. Hopper, p.c.
srn1	**4R**	L. S. Nolan, N. S. Atkinson, R. W. Durnst and A. K. Hopper, p.c.
ssb20	**4**	Sugino *et al.* (1986)
ssb38	**6**	Sugino *et al.* (1986)
ssc1	**10R**	Craig *et al.* (1987)
ssn2	4R	
ssn6	2R	
cyc8		
ssn20	**7R**	Neigeborn *et al.* (1987)
spt6		
sst1	9L	
bar1		
sst2	**12R**	Dietzel and Kurjan (1987)
STA1	**4**	Pretorius and Marmur (1988)
DEX2		
MAL5		
STA2	**2**	Pretorius and Marmur (1988)
DEX1		
STA3	**14**	Pretorius and Marmur (1988)
DEX3		
ste2	**6L**	Jenness *et al.* (1987)
ste4	15R	
ste5	4R	
nul3		
ste7	4L	
ste8	12R	
cmt1		
mar2		
sir3		
ste9	4R	
sir4		
ste11	**12R**	Chaleff and Tatchell, (1985) ; Jenness *et al.* (1987); Johnson *et al.* (1987)
ste12	**8R**	Jenness *et al.* (1987)
ste13	15R	
ste18	**10R**	Whiteway *et al.* (1988)
stp52	**4L**	Katz *et al.* (1987)
SUC1	7R	
SUC2	9L	
SUC3	2R	
SUC5	4L	
SUC7	**8**	J. L. Celenza and H. Carlson, p.c.
SUF1	15L	
SUF2	3R	
SUF3	4R	
SUF4	7R	
SUF5	15R	
SUF6	14L	
SUF7	13L	

Table 18. List of mapped genes - continued

Gene / Synonym	Map Position	References
SUF8	8R	
SUF9	**6L**	M. Winey and M. R. Culbertson, p.c.
SUF10	14L	
suf11	15R	
suf12	4R	Kikuchi *et al.* (1988); Y. Kikuchi, p.c.; Ono et al (1984)
gst1		
sup2		
sup35		
sup36		
suf13	15R	
suf14	14L	
SUF15	7R	
SUF16	3R	
SUF17	15L	
SUF18	6R	
SUF19	5L	
SUF20	6R	
SUF21	16R	
SUF22	13L	
SUF23	10R	
SUF24	4R	
SUF25	4L	
suh2	12R	
SUP-1A	---	See Table 11
SUP2	4R	
sup2	4R	Ono *et al.* (1984); Kikuchi *et al.* (1988); Y. Kikuchi, p.c.
gst1		
suf12		
sup35		
sup36		
SUP3	15L	
SUP4	10R	
SUP5	13L	Schild and Mortimer (1985)
SUP6	6R	
SUP7	10L	
SUP8	13R	
SUP11	6R	
SUP15,16	16R	
SUP17	9L	
SUP19	5R	
SUP20		
SUP20	5R	
SUP19		
SUP22	9L	
SUP25	11R	
SUP26	12R	
SUP27	4R	
SUP28	14R	Ono *et al.* (1985)
SUP29	10C	
SUP30		
SUP30	10C	
SUP29		
SUP33	11L	Ono *et al.* (1985)
sup35	4R	Ono *et al.* (1984); Kikuchi *et al.* (1988); Y. Kikuchi, p.c.
gst1		
suf12		
sup2		
sup36		
sup36	4R	Kikuchi *et al.* (1988)
gst1		
suf12		
sup2		
sup35		
SUP37	12R	
SUP38	**7**	All-Robyn *et al.* (1988)
SUP40	**2**	All-Robyn *et al.* (1988)
SUP44	**7**	All-Robyn *et al.* (1988)
sup45	2R	Ono *et al.* (1984)
sup1		
supQ		
sup47		
SUP46	2R	
sup47	2R	Ono *et al.* (1984)
sup1		
sup45		
supQ		
SUP50	**13R**	G. McKnight, J. H. Hopper, and D. Oh, p.c.
SUP51	10C	
SUP52		
SUP52	10C	
SUP51		
SUP53	3L	
SUP54	7L	
SUP56	1R	
SUP57	6R	
SUP58	11L	
SUP61	3R	
SUP71	5R	
SUP72	2R	
SUP73	10L	
SUP74	10L	
SUP75	11L	
SUP76	7R	
SUP77	7R	
SUP78	13R	

Table 18. List of mapped genes - continued

Gene / Synonym	Map Position	References
SUP79	13L	
SUP80	4R	
SUP85	5R	
SUP86	12R	
SUP87	2R	
SUP88	4R	
sup111	8R	Ono *et al.* (1986)
sup112	7R	Ono *et al.* (1986)
sup113	13R	Ono *et al.* (1986)
SUP150	**5L**	Ono *et al.* (1988)
SUP154	**7L**	Ono *et al.* (1988)
SUP155	**7**	Ono *et al.* (1988)
SUP160	**15R**	Ono *et al.* (1988)
SUP165	**14R**	Ono *et al.* (1988)
SUS1	5L	
swi1	16L	
tcm1	15R	
mak8		
tef1	**16R**	Sandbaken and Culbertson (1988)
TEF2	2R	Schirmaier and Philippsen (1984)
tfs1	**12R**	L. C. Robinson and K. Tatchell p.c.
thi1	**13R**	G. McKnight, J. H. Hopper, and D. Oh, p.c.
thr1	8R	
thr4	3R	
tif1	**11R**	P. Müller and P. Linder, p.c.
tif2	**10L**	P. Müller and P. Linder, p.c.
til1	7L	Saunders and Rank (1981); Balzi *et al.* (1987)
pdr1		
tmp1	15R	
cdc21		
top1	**15R**	Thrash *et al.* (1985)
mak1		
top2	14L	Voelkel-Meiman *et al.* (1986)
tpk1	**10L**	Cannon *et al.* (1986)
SRA3		
tra3	15R	
gcd1		
trk1	**10L**	Gaber *et al.* (1988)
TRK2	**16L**	A. Petitjean and K. Tatchell, p.c.
PKA3		
trn1	1R	Cummins *et al.* (1985)
trp1	4R	
trp2	5R	
trp3	11L	
trp4	4R	
trp5	7L	
ura3	5L	
ura4	12R	
ura6	**11R**	Liljelund and Lacroute (1986)
urep1	**10R**	Turoscy and Cooper (1987)
dal5		
WHI1	**1L**	Nash *et al.* (1988)
DAF1		
whi2	**15R**	P. E. Sudbery, p.c.
YG100	**1L**	Ingolia *et al.* (1982); M. Slater, p.c.

Gene / Synonym	Map Position	References
tsl1	1L	Ohya *et al.* (1986a)
cdc24		
cls4		
tsm1	3R	
tsm5	3R	
tsm0039	5R	
tsm0070	6L	
tsm0080	4R	
tsm0111	13R	
tsm0115	16L	Dequin *et al.* (1988)
erg10		
tsm0119	7L	
tsm0120	16R	
tsm134	2R	
tsm0139	9L	
tsm0151	8R	
tsm0185	10R	Boutelet *et al.* (1985)
cdc35		
cyr1		
hrs1		
SRA4		
tsm0186	8R	
tsm0225	4L	
tsm437	7L	
tsm0800	13R	
tsm4572	16R	
hts1		
tsm5162	4R	
tsm7269	2R	
rna6		
tsm8740	15R	
tub1	**13L**	Schatz *et al.* (1986)
tub2	6L	
tub3	**13L**	Schatz *et al.* (1986)
tup1	3R	
cyc9		
flk1		
umr7		
tup4	15L	
tup7	15R	
pho80		
tyr1	2R	
umr7	3R	
cyc9		
flk1		
tup1		
ura1	11L	
ura2	10L	
YP2	**6L**	Segev and Botstein (1987); Galwitz *et al.* (1983); Schmitt *et al.* (1986)
YPT1		
YPT1	**6L**	
YP2		
xrs2	**4R**	I. A. Zakharov, p.c.
383	**3R**	C. Thrash-Bingham and W. L. Fangman, p.c.

* New genes and their positions are indicated in bold. There are three cases of the same gene name being used to describe two gene loci: *cdc25* on chromosomes II and XII, *sit4* on chromosomes IV and V, and *pfk1* or *pfk2* on chromosomes VII and XIII. There are also many cases of gene names with synonyms; for several of these genes, there is lack of agreement about which synonym to use.

** Only new references since the 1985 mapping review (Mortimer and Schild, 1985) have been included.

REFERENCES

Adams, A. E. M. (1984). Cellular morphogenesis in the yeast *Saccharomyces cerevisiae*. Ph. D. thesis, University of Michigan, 1984.

All-Robyn, J. A., Kelley-Geraghty, D. C. and Leibman, S. W. (1988) Cloning of omnipotent suppressors in yeast. *Genome* **30**, Supplement 1, 296

Altamura, N., Ben Asher, E., Dujardin, G, Groudinsky, O., Kermorgant, M. and Slonimski, P. P. (1988). *Yeast* 4, *Special issue*, S209.

Arndt, K. T., Styles, C. and Fink, G. R. (1987). Multiple global regulators control *HIS4* transcription in yeast. *Science* **237**, 874-880.

Arndt, K. T., Styles, C. and Fink, G. R. (1989). A suppressor of a *HIS4* transcriptional defect encodes a protein with homology to the catalytic subunit of protein phosphatases. Cell **56**, 527-537.

Ball, S. G., Wickner, R. B., Cottarel, G., Schaus, M. and Tirtiaux, C. (1986). Molecular cloning and characterization of *ARO7-OSM2*, a single yeast gene necessary for chorismate mutase activity and growth in hypertonic medium. *Mol. Gen. Genet.* **205**, 326-330.

Balzi, E., Chen, W. and Goffeau, A. (1988). The arginyl-tRNA-protein transferase gene *ATE1* of *Saccharomyces cerevisiae*. *Yeast* **4**, *Special issue*, S317.

Balzi, E., Chen, W., Ulaszewski, S., Capjeaux, E. and Goffeau, E. (1987) The multi drug resistance gene *PDR1* from *Saccharomyces cerevisiae*. *J. Biol. Chem.* **262**, 1687-16879.

Barclay, B. J., Huang, T., Nagel, M. G., Misener, V. L., Game, J. C. and Wahl, G. M. (1988). Mapping and sequencing of the dihydrofolate reductase gene (*DFR1*) of *Saccharomyces cerevisiae*. *Gene* **63**, 175-185.

Barratt, R., Newmeyer, D., Perkins, D. and Garnjobst. (1954). Map construction in *Neurospora crassa*. *Adv. Genetics* **6**, 1-91.

Barry, K., Stiles, J. I., Pietras, D. F., Melnick, L. and Sherman, F. (1987). Physical analysis of the COR region: a cluster of six genes in *Saccharomyces cerevisiae*. *Mol. Cell. Biol.* **7**, 632-638.

Basson, M. E., Moore, R. L., O'Rear, J. and Rine, J. (1987). Identifying mutations in duplicated functions in *Saccharomyces cerevisiae*: recessive mutations in *HMG-CoA* reductase genes. *Genetics* **117**, 645-655.

Baum, P., Furlong, C. and Byers, B. (1986). Yeast gene required for spindle pole body duplication: Homology of its product with Ca^{2+} -binding proteins. *Proc. Natl. Acad. Sci. USA* **83**, 5512-5516.

Boguslawski, G. (1985). Effects of polymyxin B sulfate and polymyxin B nonapeptide on growth and permeability of the yeast *Saccharomyces cerevisiae*. *Mol. Gen. Genet.* **199**, 401-405.

Boguslawski, G. and Polazzi, J. O. (1987). Complete nucleotide sequence of a gene conferring polymyxin B resistance on yeast; Similarity of the predicted polypeptide to protein kinases. *Proc. Natl. Acad. Sci. USA* **84**, 5848-5852.

Borell, C. W., Urrestarazu, L. A. and Bhattacharjee, J. K. (1984). Two unlinked lysine genes (*LYS9* and *LYS14*) are required for the synthesis of saccharopine reductase in *Saccharomyces cerevisiae*. *J. Bacteriol.* **159**, 429-432.

Bornæs, C., Holmberg, S. and Petersen, J. G. L. (1988). The yeast *CHA1* gene encodes the catabolic L-serine (L-threonine) dehydratase. *Yeast* **4**, *Special issue*, S321.

Boulet, A., Simon M. and Faye, G. (1988). Isolation and characterization of the cell division cycle *CDC2* gene of *Saccharomyces cerevisiae*. *Yeast* **4**, *Special issue*, S119.

Boutelet, F., Petitjean, A. and Hilger, F. (1985). Yeast *cdc35* mutants are defective in adenylate cyclase and are allelic with *cyr1* mutants while *CAS1*, a new gene, is involved in the regulation of adenylate cyclase. *EMBO J.* **4**, 2635-2641.

Brandriss, M. C. (1987). Evidence for positive regulation of the proline utilization pathway in *Saccharomyces cerevisiae*. *Genetics* **117**, 429-435.

Brisco, P. R. G., Cunningham, R. S. and Kohlaw, G. B. (1987). Cloning, disruption and chromosomal mapping of yeast *LEU3*, a putative regulatory gene. *Genetics* **115**, 91-99.

Budd, M. and Campbell J. L. (1987). Temperature-sensitive mutations in the yeast DNA polymerase I gene. *Proc. Natl. Acad. Sci. USA* **84**, 2838-2842.

Button, L. L. and Astell, C. R. (1986). The *Saccharomyces cerevisiae* chromosome III left telomere has a type X, but not a type Y', *ARS* region. *Mol. Cell. Biol.* **6**, 1352-1356.

Byers, B. and Goetsch, L. (1975). Electron microscopic observations on the meiotic karyotype of diploid and tetraploid *Saccharomyces cerevisiae*. *Proc. Natl. Acad. Sci. USA* **72**, 5056-5060.

Campbell, D., Fogel, S. and Lusnak, K. (1975). Mitotic chromosome loss in a disomic haploid of *Saccharomyces cerevisiae*. *Genetics*. **79**, 383-396.

Cannon, J. F., Gibbs, J. B. and Tatchell, K. (1986). Suppressors of the *ras2* mutation of *Saccharomyces cerevisiae*. *Genetics*, **113**, 247-264.

Cannon, J. F. and Tatchell, K. (1987). Characterization of *Saccharomyces cerevisiae* genes encoding subunits of cyclic AMP-dependent protein kinase. *Mol. Cell. Biol.* **7**, 2653-2663.

Carle, G. F. and Olson, M. (1984). Separation of chromosomal DNA molecules from yeast by orthogonal-field-alteration gel electrophoresis. *Nucl. Acid Res.* **12**, 5647-5664.

Carlson, M., Celenza, J. and Eng, F. (1985). Evolution of the dispersed SUC gene family of *Saccharomyces* by rearrangement of chromosome telomeres. *Mol. Cell. Biol.* **5**. 2894-2902.

Carson, M. J. (1987). CDC17, the structural gene for DNA polymerase I of yeast: mitotic hyperrecombination and effects on telomere metabolism. Ph. D. thesis, University of Washington, Seattle, WA.

Casperson, G. F., Walker, N. and Bourne, H. R. (1985). Isolation of the gene encoding adenylate cyclase in *Saccharomyces cerevisiae*. *Proc. Nat. Acad. Sci. USA* **82**, 5060-5063.

Celenza, J. L. and Carlson, M. (1985). Rearrangement of the genetic map of chromosome VII of *Saccharomyces cerevisiae*. *Genetics* **109**, 661-664.

Chaleff, D. T. and Tatchell K. (1985). Molecular cloning and characterization of the *STE7* and *STE11* genes of *Saccharomyces cerevisiae*. *Mol. Cell. Biol.* **5**, 1878-1886.

Charron, M. J., Read, E., Haut, S. R. and Michels, C. A. (1989). Telomere-associated MAL loci of *Saccharomyces cerevisiae*. *Genetics* in press.

Chen, W., Balzi, E., Capieaux, E. and Goffeau, A. (1988). Complete sequence of a 20 kb DNA segment on chromosome VII from *Saccharomyces cerevisiae*. Yeast **4**, Special Issue, S75.

Chisholm, V. T., Lea, H. Z. Rai, R. and Cooper, T. G. (1987). Regulation of allantoate transport in wild-type and mutant strains of *Saccharomyces cerevisiae*. *J. Bacteriol.* **169**, 1684-1690.

Clark-Adams, C. D. and Winston, F. (1987). The *SPT6* gene is essential for growth and is required for d-mediated transcription in *Saccharomyces cerevisiae*. *Mol. Cell Biol.* **7**, 679-686.

Cohen, J., Goldenthal, M. J., Buchferer, B. and Marmur, J. (1984). Mutational analysis of the *MAL1* locus of *Saccharomyces*: Identification and functional characterization of three genes. *Mol Gen. Genet.* **196**, 208-216.

Cohen, J., Goldenthal, M. J., Chow, T., Buchferer, B. and Marmur, J. (1985). Organization of the MAL loci of *Saccharomyces*: Physical identification and functional characterization of three genes at the *MAL6* locus. *Mol. Gen. Genet.* **200**, 1-8.

Craig, E. A., Kramer, J. and Kosic-Smithers, J. (1987). *SSC1*, a member of the 70-kDa heat shock protein multigene family of *Saccharomyces cerevisiae*, is essential for growth. *Proc. Natl. Acad. Sci. USA* **84**, 4156-4160.

Cross, F. R. (1988). *DAF1*, a mutant gene affecting size control, pheromone arrest, and cell cycle kinetics of *Saccharomyces cerevisiae*. *Mol. Cell. Biol.* **8**, 4675-4684.

Crouzet, M., Bauer, F. and Aigle, M. (1988). Mutants of *Saccharomyces cerevisiae* impaired in stationary phase entry. *Yeast* 4, *Special issue*, S39.

Cummins, C. M., Culbertson, M. R. and Knapp, G. (1985). Frameshift supppressor mutations outside the anticodon in yeast proline tRNAs containing an intervening sequence. *Mol. Cell. Biol.* **5**, 1760-1771.

Davis, T. N., Urdea, M. S., Masiarz, F. R. and Thorner, J. (1986). Isolation of the yeast calmodulin gene: calmodulin is an essential protein. *Cell.* **47**, 423-431.

de Jonge, P., Kaptein, A., Kaback, D. B. and Steensma, H. Y. (1988) Localization of the *Saccharomyces cerevisiae PHO11* gene near the ends of chromosomes I and VIII. *Yeast* **4**, *Special issue*, S79.

de los Angeles Santos, M., Iturriaga, E. and Eslava, A. (1988). Mapping of the *rib5* gene in *Saccharomyces cerevisiae* using UV light as an enhancer of *rad52*-mediated chromosome loss. *Curr. Genet.* **14**, 419-423.

Dequin, S., Gloeckler, R., Herbert, C. J. and Boutelet, F. (1988). Cloning, sequencing and analysis of the yeast *S. uvarum ERG10* gene encoding acetoacetyl CoA thiolase. *Curr. Genet.* **13**, 471-478.

Dietzel, C. and Kurjan, J. (1987). Pheromonal regulation and sequence of the *Saccharomyces cerevisiae SST2* gene: a model for desensitization to pheromone. *Mol. Cell. Biol.* **7**, 4169-4177.

Doi,K. S. and Yoshimura, M. (1985). Alpha mating type-specific expression of mutations leading to constitutive agglutinability in *Saccharomyces cerevisiae*. *J. Bacteriol.* **161**, 596-601.

Drain, P. and Schimmel, P. (1986). Yeast *LEU5* is a *PET*-like gene that is not essential for leucine biosynthesis. *Mol. Gen. Genet.* **204**, 397-403.

Dresser, M. E. and Giroux, C. N. (1988). Meiotic chromosome behavior in spread preparations of yeast. *Cell* **106**, 567-573.

Dubin, R. A. (1987). Molecular organization of the *MAL6* locus of *Saccharomyces carlsbergensis*. *PhD. Thesis*, City University of New York.

Dubin, R. A., Charron, M. J., Haut, S. R., Needleman, R. B. and Michels, C. A. (1988). Constitutive expression of the maltose fermentation enzymes in *Saccharomyces carlsbergensis* is dependent upon the mutational activation of a non-essential homolog of *MAL63*. *Mol. Cell. Biol.* **8**, 1027-1035.

Eckardt-Schupp, F., Siede, W. and Game J. C. (1987). The *RAD24* (=$R_1{}^S$) gene product of *Saccharomyces cerevisiae* participates in two different pathways of DNA repair. *Genetics* **115**, 83-90.

Egerton, M. and Boyd, A. (1988). The *SEC1* and *SEC5* genes of *Saccharomyces cerevisiae*. *Yeast* **4**, *Special issue*, S438.

Elledge, S.J. and Davis, R. W. (1987). Identification and isolation of the gene encoding the small subunit of ribonucleotide reductase from *Saccharomyces cerevisiae*: DNA damage-inducible gene required for mitotic viability. *Mol. Cell. Biol.* **7**, 2783-2793.

Engebrecht, J. and Roeder, S. G. (1989). Yeast *mer1* mutants display reduced levels of meiotic recombination. *Genetics* **121**;, 237-247.

Ernst, J. F. and Chan. R. K. (1985). Characterization of *Saccharomyces cerevisiae* mutants supersensitive to aminoglycoside antibiotics. *J. Bacteriol.* **163**, 8-14.

Esposito, M. S., Brown, J. T. and Rudin, N. (1988). The *REC1* gene of *Saccharomyces cerevisiae* is required for spontaneous mitotic gene conversion, intragenic recombination, intergenic recombination, genomic stability and sporulation. *In vivo and in vitro* properties of the temperature sensitive mutation *rec1-1*. *Yeast* **4**, *Special issue*, S308.

Fabian, G. R. and Hopper, A. K. (1987). *RRP1*, a *Saccharomyces cerevisiae* gene affecting rRNA processing and production of mature ribosomal subunits. *J. Bacteriol.* **169**, 1571-1578.

Fasullo, M. T., Litherland, S. A., Holloman, W. K. and Rothstein, R. J. (1988). Identification and characterization of rec-like proteins in the yeast *Saccharomyces cerevisiae*. *Yeast* **4**, *Special issue*, S295.

Ferguson, B. and Fangman, W. L. (1988) Origin activation in a late replicating region on chromosome V. *Yeast* **4**, *Special issue*, S126.

Flessel, M. C., Brake, A. J. and Thorner, J. (1989). The *MFa1* gene of *Saccharomyces cerevisiae*: genetic mapping and mutational analysis of promoter elements. *Genetics* **121**, 223-236.

Fraenkel, D. (1982). Carbohydrate Metabolism. in The Molecular Biology of the Yeast *Saccharomyces cerevisiae*. Metabolism and Gene Expression. J. N. Strathern, E. W. Jones and J. R. Broach, ed. Cold Spring Harbor.

Fujimura, H-A. (1989). The yeast G-protein homolog is involved in the mating pheromone signal transduction system. *Mol. Cell. Biol.* **9**, 152-158.

Gaber, R F., Styles, C. A. and Fink, G. R. (1988). *TRK1* encodes a plasma membrane protein required for high-affinity potassium transport in *Saccharomyces cerevisiae*. *Mol. Cell. Biol.* **8**, 2848-2859.

Gallwitz, D., Donath, C. and Sander, C. (1983). A yeast gene encoding a protein homologous to the human *c-has/bas* proto-oncogene product. *Nature*, 306, 704-707.

Gelugne, J-P. and Bell, J. B. (1988). Modifiers of ochre suppressors in *Saccharomyces cerevisiae* that exhibit ochre suppressor-dependent amber suppression. *Curr.Genet.* **14**, 345-354.

Genbauffe, F. S. and Cooper, T. G. (1986). Induction and repression of the urea amidolyase gene in *Saccharomyces cerevisiae*. *Mol. Cell. Biol.* **6**, 3954-3964.

Gibson, S. I. (1989). Analysis of *mcm3* in minichromosome maintenance mutant of yeast with a cell division cycle arrest phenotype. *Ph.D.Thesis*, Cornell University.

Godsen, M. H., McIntosh, E. M. and Haynes, R. H. (1986). The isolation of dUTPase gene (DUT1) from *Saccharomyces cerevisiae*. *Yeast* **2**, *Special issue*, S121.

Gottlin-Ninfa, E. and Kaback, K. D. (1986). Isolation and functional analysis of sporulation-induced transcribed sequences from *Saccharomyces cerevisiae*. *Mol. Cell. Biol.* **6**, 2185-2197.

Greenberg, M. L., Myers, P. L., Skvirsky, R. C. and Greer, H. (1986). New positive and negative regulators for general control of amino acid biosynthesis in *Saccharomyces cerevisiae*. *Mol. Cell Biol.* **6**, 1820-1829.

Greenleaf, A. L., Kelly, J. L. and Lehman, I. R. (1986). Yeast *RPO41* gene product is required for transcription and maintenance of the mitochondrial genome. *Proc. Natl. Acad. Sci. USA* **83**, 3391-3394.

Gudenus, R., Mariotte, S., Moenne, A., Ruet, A.,Memet, S., Buhler, J-M., Sentenac, A. and Thuriaux, P. (1988). Conditional mutants of *RPC160*, the encoding the largest subunit of RNA polymerase C in *Saccharomyces cerevisiae*, *Genetics*, **119**, 517-526.

Hanes, S. D., (1988). *Ph.D. Thesis*, Brown University.

Hanic-Joyce, P. J. (1985). Mapping *cdc* mutations in the yeast *S. cerevisiae* by *rad52*-mediated chromosome loss. *Genetics* **110**, 591-607.

Harashima S. and Hinnebusch, A. (1986). Multiple *GCD* genes required for repression of *GCN4*, a transcriptional activator of amino acid biosynthetic genes in *Saccharomyces cerevisiae*. *Mol. Cell. Biol.* **6**, 3990-3998.

Hawthorne, D. C., P (1955). Chromosome mapping in *Saccharomyces*.. *Ph. D. Thesis*, University of Washington.

Hawthorne, D. C. and Leupold, U. (1974). Suppressor mutations in yeast. *Curr. Topics Microbiol. Immunol.* **64**, 1-47.

Heinisch, J. and von Borstel, R. C. (1988). Comparison of the yeast phosphofructokinase sequences and mapping of the genes. *Yeast* **4**, *Special issue*, S335.

Himmelfarb, H. J., Simpson, E.M. and Friesen, J. D.(1987). . Isolation and characterization of temperature sensitive RNA polymerase II mutants of *Saccharomyces cerevisiae*. *Mol. Cell. Biol.* **7**, 2155-2164.

Hinnebusch, A. G. and Fink, G. R. (1983). Positive regulation in the general amino acid control of *Saccharomyces cerevisiae*. *Proc. Natl. Acad. Sci. USA* **80**, 5374-5378.

Hirsch, H. H., Suarez Rendueles, P., Achstetter, T. and Wolf, D. H. (1988). Aminopeptidase yscII of yeast. Isolation of mutants and thier biochemical and genetic analysis. *Eur. J. Biochem.* **173**, 589-598.

Hollingsworth, N. M. and Byers, B. (1989). *HOP1*: a yeast meiotic pairing gene. *Genetics* **121**, 445-462.

Hurt, D. J., Wang, S. S., Lin, Y.-H. and Hopper, A. K. (1987). Cloning and characterization of *LOS1*, a *Saccharomyces cerevisiae* gene that affects tRNA splicing. *Mol. Cell. Biol.* **7**, 1208-1216.

Inge-Vechtomov, S. G. and Karpova, T. S. (1984). Dominant suppressors effective at low temperature (SLT) in *Saccharomyces cerevisiae*. *Genetica* **20**, 1620-1627.

Ingolia, T. D., Slater, M. R. and Craig, E. A. (1982). *Saccharomyces cerevisiae* contains a complete antigene family related to the major heat shock-induced gene of *Drosophila*. *Mol. Cell. Biol.* **2**, 1388-1398.

Ishiguro, J. and Hayashi, M. (1986). Genetic mapping of blasticidin resistant gene, *bls2*, in *Saccharomyces cerevisiae*. *Jpn. J. Genet.* **61**, 529-531.

Ivy, J. M., Hicks, J. B. and Klar, A. J. S. (1985). Map positions of yeast genes *sir1*, *sir3* and *sir4*. *Genetics*, **111**, 735-744.

Jenness, D.D., Goldman, B. S. and Hartwell, L. H. (1987). *Saccharomyces cerevisiae* mutants unresponsive to a-factor pheromone: a-factor binding and extragenic suppression. *Mol. Cell. Biol.* **7**, 1311-1319.

Johnson, D. I. Jacobs, C. W., Pringle, J. R., Robinson, L.C., Carle, G. F. and Olson, M. V. (1987). Mapping of the *Saccharomyces cerevisiae CDC3*, *CDC25*, and *CDC42* genes to chromosome XII by chromosome blotting and tetrad analysis. *Yeast* **3**, 243-253.

Jones, D. G. L., and Rosamond, J. (1988). Identification of a gene encoding a novel protein kinase in yeast. *Yeast* **4**, *Special issue*, S45.

Kaback, D. B., Angerer, L. M. and Davidson, N. (1979). Improved methods for the formation and stabilization of R - loops. *Nucleic Acids Res.* **6**, 2499-2517.

Kaback, D. B., Steensma, H. Y. and de Jonge, P. (1989). Enhanced meiotic recombination on the smallest chromosome of *Saccharomyces cerevisiae*. Proc. Natl. Acad. Sci. U. S. A. **86**, 3694-3698.

Kassir, Y., Kupiec, M., Shalom, A. and Simchen, G. (1985). Cloning and mapping of *CDC40*, a *Saccharomyces cerevisiae* gene with a role in DNA repair. *Genetics* **9**, 253-257.

Katz, M. E., Ferguson, J. and Reed, S. I. (1987). Temperature-sensitive lethal pseudorevertants of ste mutations in *Saccharomyces cerevisiae*. *Genetics* **115**, 627-636.

Kearsey, S. E. and Edwards, J. (1987). Mutations that increase the mitotic stability of minichromosomes in yeast: characterization of *RAR1*. *Mol. Gen. Genet.* **210**, 509-517.

Kikuchi, Y. Shimatake, H. and Kikuchi, A. (1988). A yeast gene required for the G_1-to-S transition encodes a protein containing an A-kinase target site and GTPase domain. *EMBO J.* **7**, 1175-1182.

Klapholz, S. and Easton Esposito, R. (1980). Isolation of *SPO12-1* and *SPO13-1* from a natural variant of yeast that undergoes a single meiotic division. *Genetics*, **96**, 567-588.

Kolman, C. J., Snyder, M. and Soll, D., (1988). Genomic organization of tRNA and aminoacyl-tRNA synthetase genes for two amino acids in *Saccharomyces cerevisiae*. *Genomics* **3**, 201-206.

Kouprina, N. Y., Pashina, O. B., Nikolaishwili, N. T., Tsouladze, A. M. and Larionov., V. L. (1988). Genetic control of chromosome stability in the yeast *Saccharomyces cerevisiae*. *Yeast* **4**, *Special issue*, S87.

Kreike, J., Schulze, M., Pillar, T., Korte, A. and Rodel, G. (1986). Cloning of a nuclear gene *MRS1* involved in the excision of a single group I intron (b13) from the mitochondrial *COB* transcript in *S. cerevisiae*. *Curr. Genet.* **11**, 185-191.

Kunisawa, R., Davis, T. N., Urdea, M. S. and Thorner, J. (1987). Complete nucleotide sequence of the gene encoding the regulatory subunit of 3', 5'-cyclic AMP-dependent protein kinase. *Nucleic Acid Res.* **15**, 368-369.

Kuroiwa, T., Kojima, H., Miyakawa, I. and Sando, N. (1984). Meiotic karyotype of the yeast *Saccharomyces cerevisiae*. *Exp. Cell Res.* **153**, 259-265.

Lam, K. B. and Marmur, J. (1977). Isolation and characterization of *Saccharomyces cerevisiae* glycolytic pathway mutants. *J. Bacteriol.* **130**, 746-749.

Larionov, V. Karpova, T., Kouprina, N. and Gouravleva, G. (1985). A mutant of *S. cerevisiae* with impaired maintenance of centromere plasmids. *Curr. Genet.* **10**, 15-20.

Larionov, V. L., Kouprina, N. Y., Strunnikov, A. V., Vlassov, A. V. and Pirozhkov, V. A. (1988). Direct selection procedure for the isolation of yeast mutants with impaired segregation of artificial minichromosomes. *Yeast* **4**, *Special issue*, S89.

Lawrence, C. W., Das, G. and Christensen, R. B.(1985). *REV7*, a new gene concerned with UV mutagenesis in yeast. *Mol. Gen. Genet.* **200**, 80-85.

Liljelund, P. and Lacroute, F. (1986). Genetic characterization and isolation of the *Saccharomyces cerevisiae* gene coding for uridine monophosphokinase. *Mol. Gen. Genet.* **205**, 74-81.

Lobo, Z. and Maitra, P. K. (1983). Phosphofructokinase mutants of yeast. *J. Biol. Chem.* **258**, 1444-1449.

Lucchini, G., Francesconi, S., Foiani, M., Badaracco, G. and Plevani, P. (1987). Yeast DNA polymerase-DNA primase complex: cloning of *PRI 1*, a single essential gene related to DNA primase activity. *EMBO J.* **6**, 737-742.

Lucchini, G., Mazza, C., Scacheri, E. and Plevani, P. (1988) Genetic mapping of *S cerevisiae* DNA polymerase I gene and characterization of a *pol1* temperature-sensitive mutant altered in the DNA primase-polymerase complex stability. *Mol. Gen. Genet.* **212**, 459-465.

Ludmerer, S. W. and Schimmel, P. (1985). Cloning of *GLN4*: an essential gene that encodes glutaminyl-tRNA synthetase in *Saccharomyces cerevisiae*. *J. Bacteriol.* **163**, 763-768.

Ma, C., and Mortimer, R. (1983). Empirical equation that can be used to determine genetic map distances from tetrad data. *Mol. Cell. Biol.* **3**, 1886-1887.

Magdolen, V., Oechsner, U. Muller, G. and Bandlow, W. (1988). The intron-containing gene for yeast profilin (*PFY*) encodes a vital function. *Mol. Cell. Biol.* **8**, 5108-5115.

Maine, G. T. (1984). *Ph.D. Thesis*. Cornell University.

Matsumoto, K., Uno, I. and Ishikawa, T. (1986). Fluphenazine-resistant *Saccharomyces cerevisiae* mutants defective in the cell division cycle. *J. Bacteriol.* **168**, 1352-1357.

Mastumoto, K., Uno, I., Kato, K. and Ishikawa, T. (1985). Isolation and characterization of a phosphoprotein phosphatase-deficient mutant in yeast. *Yeast* **1**, 25-38.

Matsumoto, K., Uno, I., Oshima, Y. and Ishikawa, T. (1982). Isolation of characterization of yeast mutants deficient in adenylate cylclase and cAMP-dependent protein kinase. *Proc. Natl. Acad. Sci. USA* **79**, 2355-2359.

Matsumoto, K., Yoshimatsu, T. and Oshima, Y. (1983). Recessive mutations conferring resistance to carbon catabolite repression of galactokinase synthesis in *Saccharomyces cerevisiae*. *J. Bacteriol.* **153**. 1405-1414.

McCusker, J. H. and Haber, J. E. (1988a) Mutations in *Saccharomyces cerevisiae* which confer resistance to several amino acid analogs. Submitted to *J. Bacteriology* .

McCusker, J. H. (1987). Pleiotropic drug resistance mutations in *Saccharomyces cerevisiae*. *Ph. D. Thesis*, Brandeis University.

McCusker, J. H. and Haber, J. E. (1988b). Cyclohexamide-resistant temperature-sensitive lethal mutations of *Saccharomyces cerevisiae*. *Genetics* **119**, 303-315.

McCusker, J.H., Perlin, D.S. and Haber, J.E. (1987). Pleiotropic plasma membrane ATPase mutations of *Saccharomyces cerevisiae*. *Mol. Cell. Biol.* **7**, 4082-4088.

Mechler, B., Müller, H. and Wolf, D. H. (1987). Maturation of vacuolar (lysosomal) enzymes in yeast; proteinase yscA and proteinase yscB are catalysts of the processing and activation event of carboxypeptidase yscY. *EMBO J.* **6**, 2157-2163.

Miyajima, I., Nakafuku, M., Nakayama, N., Brenner, C., Miyajima, A., Kaibuchi, K., Arai, K., Kaziro, Y. and Matsumoto, K. (1987). *GPA1*, a haploid-specific essential gene, encodes a yeast homolog of mammalian G protein which may be involved in mating factor signal transduction. *Cell* **50**, 1011-1019

Moehle, C. M, Aynardi, M. W., Kolodny, M. R., Park, F. J. and Jones, E. W. (1987). Protease B. of *Saccharomyces cerevisiae*: isolation & regulation of the *PRB1* structural gene. *Genetics* **115**, 255-263.

Montelone, B. A., Hoekstra, M. F. and Malone, R. E. (1988). Spontaneous mitotic recombination in Yeast; the hyper-recombinational *rem1* mutations are alleles of the *RAD3* gene. *Genetics* **119**, 299-301.

Morrison, A., Lemontt, J. F., Beck, A. K., Bernstine, E. G., Christensen, R. B., Banerjee, S. K. and Lawrence, C. W. (1988). The *REV3* gene of *Saccharomyces cerevisiae* encodes an essential product that appears to be a DNA polymerase. *Yeast* **4**, *Special issue*, S130.

Mortimer, R. K. and Hawthorne, D. C. (1969). "Yeast Genetics." pp. 385-460. In: *The Yeasts*. Edited by A. H. Rose and J. S. Harrison. AcademicPress, London.

Mortimer, R. K. and Hawthorne, D. C. (1975). "Genetic mapping in yeast." pp. 221-233. In: *Methods in Cell Biology. Vol. 11*, Edited by D. M. Prescott, Academic Press, New York.

Mortimer, R. K. and Schild, D. (1980). Genetic map of *Saccharomyces cerevisiae*. *Microbiol. Rev.* **44**, 519-571.

Mortimer, R. K. and Schild, D. (1985). Genetic map of *Saccharomyces cerevisiae*, Edition 9. *Microbiol. Rev.* **49**, 181-212.

Mosrin, C. Moenne, A., Mariotte, S., Sentenac, A. and Thuriaux, P.(1988). Cloning and in vitro mutagenesis of three genes of RNA polymerase c(III) in *S. cerevisiae*. *Yeast* **4**, *Special issue*, S494.

Murray, A. W, Schultes, N. P. and Szostak, J. (1986). Chromosome length controls mitotic chromosome segregation in yeast. *Cell* **45**, 529-536.

Nakafuku, M., Obara, T., Kaibuchi, K., Miyajima, I. Miyajima A., Itoh, H., Nakamura, S. Arai, K., Matsumoto, K. and Kaziro, Y. (1988). Isolation of second G protein homologous gene *(GPA2)* of yeast *Saccharomyces cerevisiae* and studies on its possible functions. *Proc. Natl. Acad. Sci. USA.* **88**, 1374-1378.

Nash, R. , Tokiwa, G., Anand, S., Stojcic, C., Hazlett, M., Erickson, K. and Futcher, B. (1988). Cloning and partial characterization of the *WHI1* gene of *S. cerevisiae*. *Yeast* **4**, *Special issue*, S51.

Needleman, R. B., Kaback, D. B., Dubin, R. A., Perkins, E. L., Rosenberg, N. G., Sutherland, K. A., Forrest, D. B. and Michels, C. A. (1984). *MAL6* of *Saccharomyces*: A complex genetic locus containing three genes required for maltose fermentation. *Proc. Natl. Acad. Sci. USA* **81**, 2811-2815.

Newlon, C. S., Greer, R. P., Hardeman, K. J., Kim, K. E., Lipchitz, L. R., Palzbell, T. G., Synn, S. and Woody, S. T. (1986), Structure and organization of yeast charomosome III. In Yeast Cell Biology, UCLA Symposia on Molecular and Cellular Biology, New Series, Volume 33, J. Hicks ed. Alan R. Liss, N. Y. p. 211-233.

Niederacher, D. and Entian, K-D. (1987). Isolation and characterization of the regulatory *HEX2* gene necessary for glucose repression in yeaast. *Mol. Gen. Genet.* **206**, 505-509.

Niederberger, P., Aebi, M. and Hütter, R. (1986). Identification and characterization of four new *GCD* genes in *Saccharomyces cerevisiae*. *Curr. Genet.* **10**, 657-664.

Neigeborn, L., Celenza, J. L. and Carlson, M. (1987). *SSN20* is an essential gene with mutant alleles that suppress defects in *SUC2* transcription in *Saccharomyces cerevisia*. *Mol. Cell. Biol.* **7**, 672-678.

Nonet, M., Scafe, C., Sexton, J. and Young, R. (1987). Eucaryotic RNA polymerase conditional mutant that rapidly ceases mRNA synthesis. *Mol. Cell. Biol.* **7**, 1602-1611

Norris, D. and Osley, M. A. (1987). The two gene pairs encoding H2A and H2B play different roles in the *Saccharomyces ceresivaie* life cycle. *Mol. Cell. Biol.* **7**, 3473-3481

Ohya, Y., M., Miyamoto, S., Ohsumi, Y. and Anraku, Y. (1986a). Calcium-sensitive *cls4* mutant of *Saccharomyces cerevisiae* with a defect in bud formation. *J. Bacteriol.* **165**, 28-33.

Ohya, Y., Ohsumi, Y. and Anraku Y. (1986b). Isolation and characterization of Ca^{2+}-sensitive mutants of *Saccharomyces cerevisiae*. *J. of Gen. Microbiol.* **132**, 979-988.

Olson, M. V., Dutchik, J. E., Graham, M. Y., Brodeur, G. M., Helms, C., Frank, M., MacCollin, M., Sheinman, R. and Frank, T. (1986b). Random-clone strategy for genomic restriction mapping in yeast. *Proc. Natl. Acad. Sci. USA*, **83**, 7826-7830.

Ono, B-i., Fujimoto, R., Ohno, Y., Maeda, N., Tsuchiya, Y., Usui, T. and Ishino-Arao, Y. (1988). UGA-suppressors in *Saccharomyces cerevisiae;* allelism, action spectra and map positions. *Genetics* **118**, 41-47.

Ono, B-i., Ishino-Arao, Y., Shirai, T., Maeda, N. and Shinoda, S. (1985). Genetic mapping of leucine-inserting UAA suppressors in *Saccharomyces cerevisiae*. *Currr . Genet.* **9**, 197-203.

Ono, B-i., Ishino-Arao, Y., Tanaka, M., Awano, I. and Shinoda, S. (1986). Recessive nonsense suppressors in *Saccharomyces cerevisiae:* Action spectra, complementation groups and map positions. *Genetics* **114**, 363-374.

Ono, B-i., Moriga, N., Ishihara, K., Ishiguro, J., Ishimo, Y. and Shinoda, S. (1984). Omnipotent suppressors effective in y+ strains of *Saccharomyces cerevisiae*: recessiveness and dominance. *Genetics* **107**, 219-230.

Ono, B-i., Stewart, J. W. and Sherman, F. (1979a). Yeast UAA suppressors effective in [y+] strains: serine-inserting suppressors. *J. Mol. Biol.* **128**, 81-100.

Ono, B-i., Stewart, J. W. and Sherman, F. (1979b). Yeast UAA suppressors effective in [y+] strains: leucine-inserting supressors. *J. Mol. Biol.* **132**, 507-520.

Ono, B-i., Wills, N., Stewart, J. W., Gesteland, R. F. and Sherman, F. (1981). Serine-inserting UAA suppression mediated by yeast $tRNA^{Ser}$. *J. Mol. Biol.* **150**, 361-373.

Papciak, S. M. and Pearson, N. J. (1987). Genetic mapping of two pairs of linked ribosomal protein genes in *Saccharomyces cerevisiae*. *Curr. Genet.* **11**, 445-450.

Payne, G. S., Hasson, T. B., Hasson, M. S. and Schekman, R. (1987). Genetic and biochemical characterization of clathrin-deficient *Saccharomyces cerevisiae*. *Mol. Cell. Biol.* **7**, 3888-3898.

Perkins, D. (1949). Detection of linkage in tetrad analysis. *Genetics* **38**, 187-197.

Perkins, E. L. and Needleman, R. B. (1988). $MAL64^{C}$ is a global regulator of a- glucoside fermentation: identification of a new gene involved in melezitose fermentation. *Curr. Genet.* 13, 369-375.

Pinkham, J. L. and Guarente, L. (1985). Cloning and molecular analysis of the *HAP2* locus: a global regulator of respiratory genes in *Saccharomyces cerevisiae*. *Mol. Cell. Biol.* **5**, 3410-3416.

Pocklington, M. and Orr, E. (1986). Novobiocin-resistant mutants in yeast. *Yeast* **2**, *Special issue*, S305.

Portillo, F. and Mazon, M. (1986). The *Saccharomyces cerevisiae* start mutant carrying the *cdc25* mutation is defective in activation of plasma membrane ATPase by glucose. *J. Bacteriol.* **168**, 1254-1257.

Poutre, C. G. and Fox, T. D. (1987). *PET111*, a *Saccharomyces cerevisiae* nuclear gene required for translation of the mitochondrial mRNA encoding cytochrome c oxidase subunit II. *Genetics* **115**, 637-647.

Pretorius, I. S. and Marmur, J. (1988). Localization of yeast glucoamylase genes by PFGE and OFAGE. *Curr. Gen.* **14**, 9 - 13.

Primerano, D., Muthukumar, G., Suhng, S. H. and Magee, P. T. (1988). Molecular characterization of two sporulation regulated (SPR) genes, one of which is involved in spore development. *Yeast* **4**, *Special issue*, S54.

Rine, J. and Herskowitz, I. (1987). Four genes responsible for a position effect on expression from *HML* and *HMR* in *Saccharomyces cerevisiae*. *Genetics* **116**, 9-22.

Sakai, A., Shimizu, Y.and Hishinuma F. (1988). Isolation and characterization of mutants which show an oversecretion phenotype in *Saccharomyces cerevisiae*. *Genetics* **119**, 499-506.

Sandbaken, M. G. and Culbertson, M. R. (1988). Mutations in elongation factor EF-1a affect the frequency of frameshifting and amino acid misincorporation in *Saccharomyces cerevisiae*. *Genetics* **120**, 923-934.

Sass, P., Field, J., Nikawa, J., Toda, T. and Wigler, M. (1986). Cloning and characterization of the high-affinity cAMP phosphodiesterase of *Saccharomyces cerevisiae*. *Proc. Natl. Acad. Sci. USA* **83**, 9303-9307.

Saunders, G. W. and Rank, G. H. (1982). Allelism of pleiotropic drug resistance in *Saccharomyces cerevisiae*. *Can. J. Genet. Cytol.* **24**, 493-503.

Schatz, P.J., Solomon, F. and Botstein, D. (1986). Genetically essential and non-essential a-tubulin genes specify functionally interchangeable proteins. *Mol. Cell. Biol.* **6**, 3722-3733.

Schild, D. and Mortimer, R. K. (1985). A mapping method for *Saccharomyces cerevisiae* using *rad52*--induced chromosome loss. *Genetics*, **110**, 569-589.

Schirmaier, F. and Philippsen, P. (1984). Identification of two genes coding for the translation elongation factor EF-1a of *S. cerevisiae*. *EMBO J.* **3**, 3311-3315.

Schlesser, A. and Goffeau, A. (1988). A second transport-ATPase gene in *Saccharomyces cerevisiae*. *Yeast* **4**, *Special issue*, S359.

Schmidt, C., Söllner, T. and Schweyen, R. J. (1987). Nuclear suppression of a mitochondrial RNA splice defect: nucleotide sequence and disruption of the *MRS3* gene. *Mol. Gen. Genet.* **210**, 145-152.

Schmitt, H. D., Wagner, P., Pfaff, E. and Gallwitz, D. (1986). The *ras*-related *YPT1* gene product in Yeast; A GTP-binding protein that might be involved in microtubule organization. *Cell* **47**, 401-412.

Schwob, E., Alt, G., Andres, S., Dirheimer, G. and Martin, R. P. (1988a). *ACT2*, a novel yeast split gene coding for an actin-like protein. *Yeast* **4**, *Special issue*, S108.

Schwob, E., Andres, S., Alt, G., Dirheimer, G. and Martin, R. P. (1988b). *RPK1*, a new protein kinase gene in *Saccharomyces cerevisiae*. *Yeast* **4**, *Special issue*, S57.

Segev, N. and Botstein, D. (1987). The *ras*-like yeast *YPT1* gene is itself essential for growth, sporulation, and starvation repsonse. *Mol. Cell. Biol.* **7**, 2367-2377.

Sitney, K., (1987). Genetic and molecular studies of the *RAD24* gene of *Saccharomyces*. *Ph. D. Thesis*, University of California, Berkeley.

Sitney, K. C., Budd, M. E. and Campbell, J. L. (1989). DNA polymerase III, a second essential DNA polymerase, is encoded by the *S. cerevisiae CDC2* gene. *Cell* **56**, 599-605.

Skvirsky, R. C., Greenberg, M. L., Myers, P. S. and Greer, H. (1986). A new negative control gene for amino acid biosynthesis in *Saccharomyces cerevisiae*. *Curr. Genet.* **10**, 495-501.

Smith, L. M., Robbins, L. G., Kennedy, A. and Magee, P. T. (1988). Identification and characterization of mutations affecting sporulation in *Saccharomyces cerevisiae*. *Journal Aricle No.* 12122 , *Michigan Agricultural Experiment Station.*.

Sommers, S. and Wickner, R. B. (1987). Gene disruption indicates that the only essential function of the *SKI8* chromosomal gene is to protect *Saccharomyces cerevisiae* from viral cytopathology. *Virology* **157**, 252-256.

Snow, R. (1979). Maximum likelihood estimation of linkage and interference from tetrad data. *Genetics* **92**, 231-245.

Snyder, M. and Davis, R. W. (1988). *SPA1:* A gene important for chromosome segregation and other mitotic functions in *S. serevisiae*. *Cell* **54**, 743-754.

Snyder, M. (1989). The SPA2 protein of yeast localizes to the sites of cell growth. *J. Cell Biol.* **108**, in press.

Struhl, K. (1985). Nucleotide sequence and transcriptional mapping of the yeast *pet56-his3-ded1* gene region. *Nucleic Acids Research* **13**, 8587-6801.

Subik, J., Ulaszewski, S. and Goffeau, A. (1986) Genetic mapping of nuclear mucidin resistance mutations in *Saccharomyces cerevisiae*. *Curr. Genetics* **10**, 665-670.

Sugino, A., Hamatake, R., Eberly, S., Sakai, A., Alexander, P., Desai, R. and Clark, A. (1986). Biochemical and genetical studies of DNA replication porteins in yeast. *Yeast* **2**, *Special issue*, S374.

Suzuki, K. and Yanagishima, N. (1986). Genetic characterization of an a-specific gene responsible for sexual agglutinability in *Saccharomyces cerevisiae:* mapping and gene dose effect. *Curr. Genet.* **10**, 353-357.

Taguchi, A. K. W. and Young, E. T. (1987). Cloning and mapping of *ADR6*, a gene required for sporulation and expression of the alcohol dehydrogenase II isozyme from *Saccharomyces cerevisiae*. *Genetics* **116**, 531-540.

Tanaka, H. and Tsuboi, M. (1985). Cloning and mapping of the sporulation gene, *spoT7*, in *Saccharomyces cerevisiae*. *Mol. Gen. Genet.* **199**, 21-25

Thiele, D. J. (1988). *ACE1* regulates expression of the *Saccharomyces cerevisiae* metallothionein gene. *Mol. Cell. Biol.* **8**, 2745-2752.

Thireos, G., Driscoll, P. M. and Greer, H. (1984). 5' untranslated sequences are required for the translational control of a yeast regulatory gene. *Proc. Natl. Acad. Sci. USA* **81**, 5096-5100.

Thomas, J. H. and Botstein, D. (1986). A gene required for the separation of chromosomes on the spindle apparatus in yeast. *Cell* **44**, 65-76.

Thrash, C., Bankier, A. T., Barrell, B. G. and Sternglanz, R. (1985). Cloning, characterization, and sequenece of the yeast DNA topoisomerase I gene. *Proc. Natl. Acad.Sci. USA* **82**, 4374-4378.

Toda, T., Cameron, S., Sass, P., Zoller, M., Scott, J. D., McMullen, B., Hurwitz, M., Krebs, E. G. and Wigler, M. (1987). Cloning and characterization of *BCY1*, a locus encoding a regulatory subunit of the cyclic AMP-dependent protein kinase in *Saccharomyces cerevisiae*. *Mol. Cell. Biol.* **7**, 1371-1377.

Tomenchok, D. M. and Brandriss, M. C. (1987). Gene-enzyme relationships in the proline biosynthetic pathway of *Saccharomyces cerevisiae*. *J. Bacteriol.* **169**, 5364-5372.

Trueheart, J., Boeke, J. D. and Fink, G.R. (1987). Two genes required for cell fusion during yeast conjugation: evidence for a pheromone-induced surface protein. *Mol. Cell. Biol.* **7**, 2316-2328.

Turoscy, V. and Cooper, T. G. (1987). Ureidosuccinate is transported by the allantoate transport system in *Saccharomyces cerevisiae*. *J. Bacteriol.* **169**, 2598-2600.

Ulaszewski, S., Balzi, E. and Goffeau A. (1987). Genetic and molecular mapping of the *pma1* mutation conferring vanadate resistance to the plasma membrane ATPase from *Saccharomyces cerevisiae*.. *Mol. Gen. Genet.* **207**, 38-46.

van Loon, A. P. G. M., Pesoed-Hurt, B. and Schatz, G. (1986). A yeast mutant lacking mitochondrial manganese-superoxide dismutase is hypersensitive to oxygen. *Proc. Natl. Acad. Sci. USA* **83**, 3820-3824.

Verdiere, J., Gaisne, M., Guiard, B. and Defranous, N. (1988). A single missense mutation in *CYP1 (HAP1)* regulatory gene switches the expression of two structural genes encoding isocytochromes C. *Yeast* **4**, *Special issue*, S425.

Vidal, M., Hilger, F,, Burd, C. G. and Gaber, R. F. (1988). Mutations in *RPD1* and *RPD3* alter potassium transport in yeast. *Yeast* **4**, *Special issue*, S65.

Villladsen, I. S. (1988). *NDC2*, a gene that affects chromosome stability in yeast. *Yeast* **4**, *Special issue*, S98.

Voelkel-Meiman, K., DiNardo, S. and Sternglanz, R. (1986). Molecular cloning and genetic mapping of the DNA topoisomerase II gene of *Saccharomyces cerevisiae*. *Gene* **42**, 193-199.

Vollrath, D., Davis, R. W., Connelly, C. and Hieter, P. (1988). Physical mapping of large DNA by chromosome fragmentation. *Proc. Natl. Acad. Sci. USA*. **85**, 6027-6031.

Wang, H-T., Frackman, S., Kowalisyn, J., Easton Esposito, R. and Elder, R. (1987). Developmental regulation of *SPO13*, a gene required for separation of homologous chromosomes at meiosis I. *Mol. Cell. Biol.* **7**, 1425-1435.

Wang, S.-S. and Brandriss, M.C. (1986). Proline utilization in *Saccharomyces cerevisiae:* analysis of the cloned *PUT1* gene. *Mol. Cell. Biol.* **6**, 2638-2645.

Weber, E., Jund, R. and Chevallier, M-R. (1986). Chromosomal mapping of the uracil permease gene of *Saccharomyces cerevisiae*. *Curr. Genet.* **11**, 93-96.

Weiss, W. A. and Friedberg, E.C. (1985). Molecular cloning and characterization of the yeast *RAD10* gene and expression of *RAD10* protein in *E. coli*. *EMBO J.* **4**, 1575-1582.

Welch, J. W., Fogel, S., Buchman, C. and Karin, M. (1989). The *CUP2* gene product regulates the expression of *CUP1* gene coding for yeast metalothionine. *EMBO J.* **8**, in press.

Whiteway, M., Hougan, L. and Thomas, D. Y. (1988). Expression of *MFα1* in *MATα* cells supersensitivie to a-factor leads to self-arrest. Mol. Gen. Genet. **214**, 85-88.

Whiteway, M. and Szostak, J. W. (1985). The *ARD1* gene of yeast functions in the switch between the mitotic and alternative developmental pathways. Cell **43**, 483-492.

Wickner, R. (1979). Mapping chromosomal genes of *Saccharomyces cerevisiae* using an improved genetic mapping method. *Genetics* **92**, 803-821.

Wickner, R. B., Koh, T. J., Crowley, J. C , O'Neil, J. and Kaback. D. (1987). Molecular cloning of chromosome I DNA from *Saccharomyces cerevisiae*: isolation of the *MAK16* gene and analysis of an adjacent gene essential for growth at low temperatures. *Yeast* **3**, 51-57.

Wickner, R.B. (1987). *MKT1*, a non-essential *Saccharomyces cerevisiae* gene with a temperature-dependent effect on replication of M_2 double-stranded RNA. *J. Bact.*.**169**, 4941-4945.

Williamson, M. S., Game, J. C. and Fogel, S. (1985). Meiotic gene conversion mutants in *Saccharomyces cerevisiae*. I. Isolation and characterization of *psm1-1* and *pms1-2*. *Genetics* **110**, 609-646.

Wilson, R. B. and Tatchell, K. (1988). *SRA5* encodes the low-K_m cyclic AMP phophodiesterase of *Saccharomyces cerevisiae*. *Mol. Cell. Biol.* **8**, 505-510.

Winey, M. and Culbertson, M. R. (1988). Mutations affecting the t-RNA-splicing endonuclease activity of *Saccharomyces cerevisiae*. *Genetics* **118**, 609-617.

Winston, F., Dollard, C., Malone, E. A., Clare, J., Kapakos, J. G., Farabaugh, P. and Minehart, P.L. (1987). Three genes are required for trans-activation of TY transcription in yeast. *Genetics* **115**, 649-656.

Whiteway, M., Hougan, L. and Thomas, D. Y. (1988). Expression of *MFa1* in *MATa* cells supersensitive to a-factor leads to self-arrest. *Mol. Gen. Gent.* **214**, 85-88.

Yeh, E., Carbon, J. and Bloom K. (1986). Tightly centromere-linked gene (*SPO15*) essential for mieosis in the yeast *Saccharomyces cerevisiae*. *Mol. Cell. Biol.*, **6**, 158-167.

Yoshida, K., Ogawa, N., Okada, S., Hiraoka, E. and Oshima. (1988). Regularoty circuit for phosphatase synthesis in *Saccharomyces cerevisiae*. *Yeast* **4**, *Special issue*, S374.

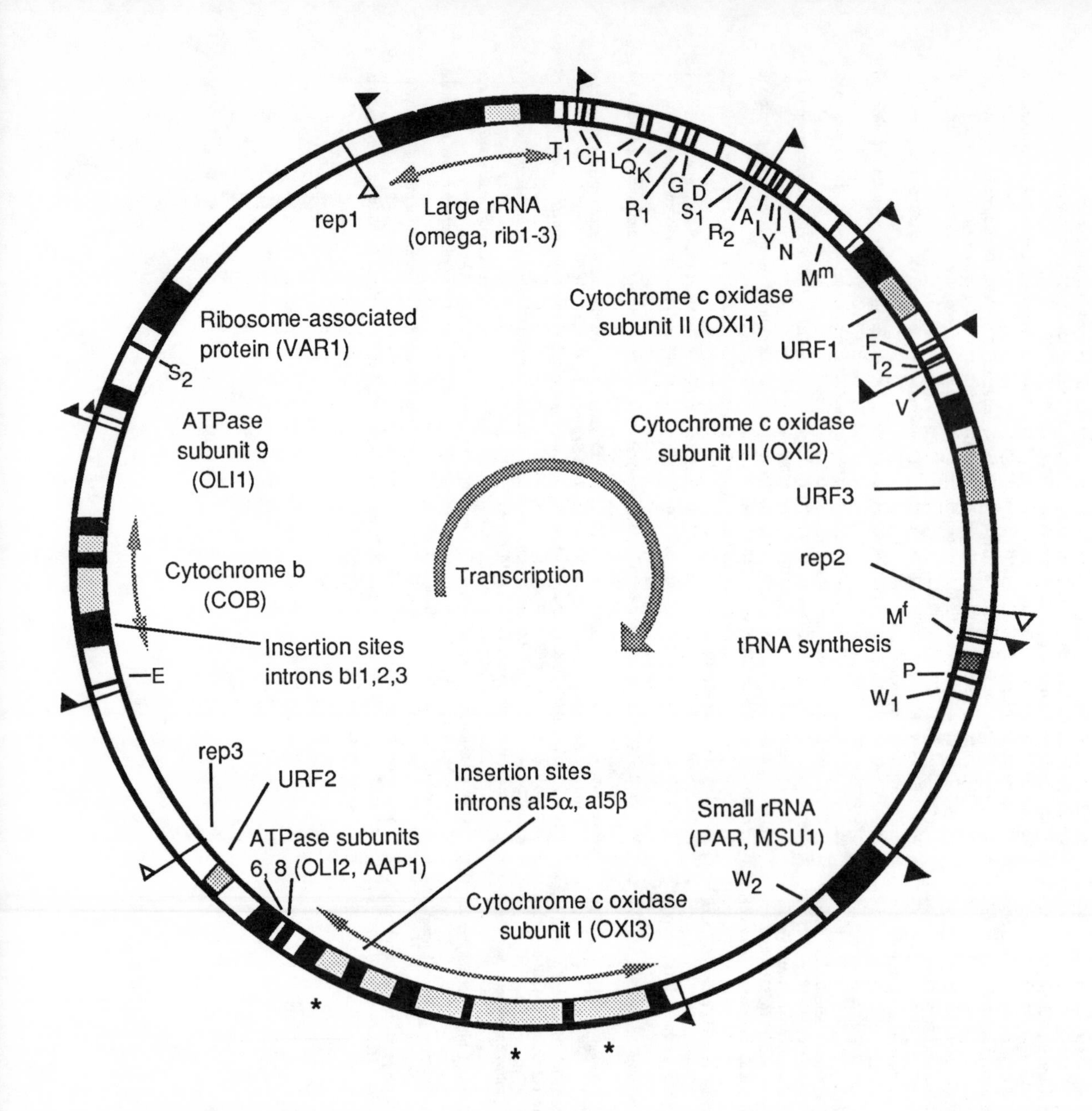

Legend to Figure 1

Genetic and physical maps of mtDNA from *S. cerevisiae* D273-10B
The positions of genetic loci and genes are based on data presented in Tables I and III and taken from refs. 88-95. Genes for rRNAs and known proteins are represented by black bars, those for URFs by dark-grey shaded areas. Note that URFs 2 and 3 consist of a series of overlapping reading frames linked by +1 and -1 frameshifts. For ease of presentation, these sequences are shown as continuous areas on the map. tRNA genes are identified by means of the cognate amino acid, using the one-letter code. Subscripts indicate iso-acceptors. Introns where present are indicated by a lighter shade of grey and those of the group II type are additionally indicated by an asterisk. D273-10B contains a medium-sized mtDNA, which lacks a number of introns present in the genes for apocytochrome *b* and COI in so-called long strains (see Table III). Insertion sites of the missing introns are indicated. Sites of transcription-initiation are indicated by (▶), with (▷) indicating sites associated with replication origins (rep,ori). For all genes but one ($tRNA_2^{Thr}$), transcription is in a clockwise direction.

MITOCHONDRIAL DNA IN THE YEAST *Saccharomyces cerevisiae*

August 1989

L.A. GRIVELL
Section for Molecular Biology, Department of Molecular Cell Biology, University of Amsterdam, Kruislaan 318, 1098SM Amsterdam, The Netherlands.

MtDNA in *S. cerevisiae* is a circular molecule, which, dependent on strain, varies in size from about 74 to 85 kb [reviewed in refs 1-5]. In common with the much smaller mtDNAs from metazoa, it contains genes coding for the two ribosomal RNAs, a set of tRNAs and some components of the respiratory enzymes of the inner membrane (Table I). In addition, there are genes without counterparts in metazoan mtDNAs. These code for at least two endonucleases involved in intron transposition; proteins required for RNA splicing (RNA maturases); a protein associated with the small subunit of the mitochondrial ribosome and an RNA that forms part of an RNAse P-like tRNA processing enzyme. Three reading frames identified by DNA sequence analysis (URFs 1-3) have not yet been assigned to known protein products, but bear some resemblance to the RNA maturase family [46]. Genes notably absent, but present in mtDNAs from sources as diverse as mammals, plants, fungi and trypanosomes, are those for subunits of the complex I type of mitochondrial NADH:Q reductase. This absence correlates with the fact that S. cerevisiae appears to lack this type of NADH dehydrogenase [83,84].
A compilation of all published DNA sequence data covers an estimated 92% of the genome, with few large gaps remaining [5]. It therefore seems unlikely that other genes remain to be discovered. Since mtDNA from S. cerevisiae strain D273-10B is the single most extensively characterized genome, this DNA is used as the reference for the gene map shown in Fig. 1 and the landmark information presented in Table II.
Yeast mtDNA contains 19-20 transcriptional initiation sites scattered around the genome. [85,89]. Transcription starts within the sequence (A/T)TATAAGTA, with the last A being the site of initiation. Sequences outside this motif probably also contribute to promoter activity, as there are several sites in the genome, for which no corresponding capped transcripts are found [85,95].
Little is known of the sequences that signal termination of transcription. Although the 3'-ends of many mRNAs coincide with a dodecamer motif AAUAAUAUUCUU (Table II), this sequence is, however, primarily a site for endonucleolytic cleavage.
Three genes are split, the number of introns being strain-dependent (Table III). The introns can be divided into two groups (I and II) on the basis of differences in their predicted secondary structure and in the mechanism by which self-splicing occurs in vitro (see ref 108 for review). Self-splicing of group I introns occurs via a transesterification mechanism that is initiated by binding of a guanosine nucleotide to a site within the intron. The reaction can result in circularisation of the excised intron. Transesterification also occurs in the case of group II introns, but in this case, the reaction is initiated by nucleophilic attack of an intron-internal nucleotide on the 5'-exon/intron junction. The excised intron therefore accumulates in the form of a lariat containing a 5'-3', 2' branched nucleotide.
Without exception, expression of the remaining genes requires some form of RNA processing, which is dependent on the action of nuclear-encoded enzymes. Modifications include cleavage of individual transcripts from multigenic precursors, trimming at the 5'- or 3'-terminus and in the case of tRNAs, -CCA addition and base modification.
The mRNAs in yeast mitochondria are unlikely to be capped. They also lack post-transcriptionally added poly(A) sequences [109]. Most contain long, AU-rich untranslated 5'-leader and 3'-trailer sequences. Virtually nothing is known of the mechanisms whereby mRNAs are selected for translation, nor how the initiator AUG (which is seldom the first in the RNA) is distinguished from others in its immediate surroundings.

Acknowledgements
Work carried out in the author's laboratory was supported in part by grants from the Netherlands Foundation for Chemical Research (SON) with financial aid from the Netherlands Organization for the Advancement of Research (NWO).

TABLE I - INVENTORY OF GENE PRODUCTS OF YEAST MITOCHONDRIAL DNA

Gene/product	Associated genetic locus	Size	Footnote	Reference(s)
Cytochrome *c* oxidase:				
Subunit I	OXI3	510 aa (56,000)	a	6-8
Subunit II	OXI1	251 aa (28,480)		6, 9-12
Subunit III	OXI2	269 aa (30,340)		6,13,14
ATPase complex:				
Subunit 6	OLI2, OLI4, PHO1	259 aa (28,257)		15-17
Subunit 8	AAP1	48 aa (5870)	a	18
Subunit 9	OLI1, PHO2	76 aa (8970)		19-21
Ubiquinol-cytochrome *c* reductase:				
Cytochrome *b*	COB (box4,5,8,1,2,6, ant, diu,fun,muc)	385 aa (44,000)	a	6,22-29
Ribosome-assoc. protein	VAR1	40,000-44,000	b	30-33
Intron-encoded proteins with RNA maturase or DNA endonuclease activity				
Cytochrome b:		423 aa		
Intron 2 (bI2)	box3		c	34
Intron 3 (bI3)	box10	384 aa		35-37
Intron 4 (bI4)	box7			38-41
Subunit I cytochrome *c* oxidase:				
Intron 1 (aI1)		777 aa		7,42
Intron 2 (aI2)		785 aa		7
Intron 3 (aI3)		334 aa		7
Intron 4 (aI4)	mim2-1	316 aa		7,43-45
Intron 5α (aI5α)		310 aa	d	46
Intron 5β (aI5β)		320 aa	d	46
21S rRNA:				
Intron 1	fit1	235 aa		47,48
Unassigned reading frames				
URF1		386 aa		12,49,50
URF2			e	15,51
URF3			e	49
Other		80 aa; 46 aa	f	5
RNAs				
25 tRNAs	Various syn$^-$;	69-76 nt		12, 52-66
Large rRNA	cap, ery, spi rib1-3; omega various syn$^-$;	3273 nt		67-74
Small rRNA	MSU1, par	1686 nt		75-80
tRNA syn. locus		±450 nt		81,82

a) These products display anomalous migration in SDS-polyacrylamide gels. True molecular weights (given in brackets) are much higher than predicted from mobility.

b) The VAR1 product displays strain-dependent length variation (33). The molecular weight range given is based on the electrophoretic mobilities of different variants in the strains so far examined.

c) The bI2 (box3) and bI3 (box10) maturases may both be active as fusion proteins coded in part by the first exons of the cytochrome *b* gene and in part by the intron. For the bI4 (box7) and aI1 maturases, evidence has been presented for the liberation of an active, intron-encoded maturase by proteolytic cleavage of the fusion protein (7,42).

d) Intron absent from D273-10B, but present in strains with longer forms of the gene (see Table III).

e) Families of overlapping reading frames related by +1 or -1 frameshifts.

f) The shorter of these two reading frames is initiated by ATA (5). It is not known whether either is expressed.

ABLE II - LANDMARKS ON YEAST MITOCHONDRIAL DNA (S. CEREVISIAE D273-10B)

ene/Sequence	Coordinates (bp) from	to	Reference(s)
RNA^{Thr}_{UGU} (T)	1379	1451	58
ranscription initiation site (tRNA^{Cys}, tRNA^{His})	1915		96
RNA^{Cys}_{GCA} (C)	1930	2002	53,58
RNA^{His}_{GUG} (H)	2071	2142	53,58
RNA^{Leu}_{UAA} (L)	3507	3588	57
RNA^{Gln}_{UUG} (Q)	3623	3693	57
RNA^{Lys}_{UUU} (K)	4476	4547	59,65
RNA^{Arg}_{UCU} (R)	4719	4791	59,65
RNA^{Gly}_{UCC} (G)	4876	4947	55,56,59,65
RNA^{Asp}_{GUC} (D)	5710	5781	59,65
RNA^{Ser}_{GCU} (S)	6653	6735	55,56,59,65
RNA^{Arg}_{ACG} (R)	6739	6809	55,59,65
RNA^{Ala}_{UCG} (A)	7246	7318	59
RNA^{Ile}_{GAU} (I)	$\pm$7526	$\pm$7596	57,66,95
RNA^{Tyr}_{GUA} (Y)	$\pm$7876	$\pm$7946	57,95,97
RNA^{Asn}_{GUU} (N)	8359	8428	12
$RNA^{Met.m}_{CAU}$ (M)	9464	9537	12,66
ubunit 2 cyt. c oxidase (COII)	10,539	11,294	9,11,12
ATAATATTCTT; end of COII mRNA	11,357	11,368	12
JRF1	11,536[a]	12,696	12
ATAATATTCTT	12,910	12921	
ranscription initiation site (tRNA^{Phe}, tRNA^{Val}, COIII, URF3)	13,690		86,90
RNA^{Phe}_{GAA} (F)	13,715	13,786	55,56,61
RNA^{Thr}_{UAG} (T)[b]	14,046	14,116	54

Table II - continued

Gene/Sequence	Coordinates (bp)		Reference(s)
	from	to	
$tRNA^{Val}_{UAC}$ (V)	14,485	14,555	54,55,61
Subunit 3 cyt. c oxidase (COIII)	15,168	15,977	13
URFs 3a-d	±16,047	17,509	49
AATAATATTCTT	16,962	16,973	
AATAATATTCTT	17,632	17,643	
rep2, ori5	±19,600	±20,000	98-100
Transcription initiation site ($tRNA^{Met.F}$,tRNA syn. locus , $tRNA^{Pro}$)	21,251		81,90
$tRNA^{Met.F}_{CAU}$ (M)	21,279	21,353	62,90
tRNA synthesis locus	±21,610	±21,953	81,82
$tRNA^{Pro}_{UGG}$ (P)	22,686	22,757	62,90
$tRNA^{Trp}_{UCA}$ (W)	±25,200		62
15S rRNA: start of transcription	27,407		77,78
15S rRNA	27,484	29,169	76,80
$tRNA^{Trp}_{UCA}$ (W)	30,484	30,551	60
Transcription initiation site (COI, ATPase subunits 6,8)	34,139		93
Subunit I cyt. c oxidase (COI); exon 1	34,681	34,849	7
Intron 1	34,850	37,297	7
Exon 2	37,298	37,333	7
Intron 2	37,334	39,843	7
Exon 3	39,844	39,884	7
Intron 3	39,885	41,398	7
Exon 4	41,399	41,878	7
Intron 4	41,879	42,888	7
Exon 5[c]	42,889	43,299	7
Insertion site intron 5α (1365 bp)	43,138	43,139	46
Insertion site intron 5β (1545 bp)	43,275	43,276	46
Intron 5(γ)[d]	43,300	44,186	7
Exon 6[e]	44,187	44,659	7
AATAATATTCTT; 3'-end COI mRNA	44,732	44,746	93
ATPase subunit 8	45,616	45,762	18
ATPase subunit 6	46,432	47,211	15
URF2	47,288	49,829	15,51
AATAATATTCTT	49,921	49,932	
Transcription initiation site ($tRNA^{Glu}$, apocytochrome b)	53,204		87
$tRNA^{Glu}_{UUC}$ (E)	53,599	53,667	63
Apocytochrome b, exon 1[f]	54,755	55,511	26
Insertion site intron 1 (bI1; 764 bp)	55,169	55,170	34
Insertion site intron 2 (bI2; 1394 bp)	55,183	55,184	34
Insertion site intron 3 (bI3; 1691 bp)	55,260	55,261	101
Intron 1 (bI4)[g]	55,512	56,928	26
Exon 2 (5)[g]	56,929	56,979	26
Intron 2 (bI5)[g]	56,980	57,712	26
Exon 3 (6)[g]	57.713	58,062	26

Table II - continued

Gene/Sequence	Coordinates (bp) from	to	Reference(s)
AATAATATTCTT; 3'-end cyt. b mRNA	58,160	58,171	29,93
Transcription initiation site (ATPase subunit 9, tRNA^{Ser}, VAR1)			
Major start	60,647		88
Minor start	60,725		88
ATPase subunit 9	61,288	61,518	19,20
AATAATATTCTT; 3'-end ATPase 9 mRNA	61,618		21
$tRNA^{Ser}_{UGA}$ (S)	62,773	62,859	55
Ribosome-associated protein (VAR1)[h]	63,473	64,663	30
AATAATATTCTT; 3'-end VAR1 mRNA	65,200	65,211	31
rep1, ori3	±68,800	±69,600	102
21S rRNA, 5'-exon	71,721	74,408	74
Intron 1	74,409	75,550	47,74
21S rRNA 3'-exon	75,551	530	74

Listing of landmarks follows the direction of transcription (clockwise in Fig. 1). Coordinates are given relative to the position of the single recognition site for SalI at 0/75,606. Genes for tRNAs are identified by their cognate amino acid and anticodon (5'-3')

a) The start position of this URF is that given in ref. 12. More recent analysis (50) has led to the suggestion that initation may in fact occur within the C-terminal coding region of the gene for COII (position 11,276).
b) Gene located on opposite strand.
c) Equivalent to exons 5-7 in long versions of the gene.
d) Intron 7 in long versions of the gene
e) Originally thought to be 3 separate exons. Comparison with other fungi (103) suggests, however, that the low degree of amino acid similarity found with mammalian versions of the proteins is due to a low degree of sequence of conservation rather than the presence of introns.
f) Equivalent to exons 1-4 in long versions of the gene.
g) Numbers in brackets refer to the designation of the intron in the long version of the gene.
h) Sequences given are for the S. cerevisiae petite strain a17-10, which transmits the VAR1 (40.0) allele. D273-10B transmits the (42.0) allele (ref.30 and refs. therein)

TABLE III-THE DISTRIBUTION OF OPTIONAL INTRONS IN SOME COMMONLY USED LABORATORY YEAST STRAINS

Class	Strain	Introns: 21S rRNA	Cytochrome *c* oxidase subunit 1: 1	2	3	4	5α	5β	5γ	Cytochrome *b*: 1	2	3	4	5	Ref.
I	*S. carlsbergensis*	-	-	+	+	-	-	-	+	-	-	-	+	+	a
II	*S. cerevisiae* D273-10B, JS1-3D, DP1-1B	+	+	+	+	+	-	-	+	-	-	-	+	+	b
III	*S. cerevisiae* 55RC-3C, D85-2D, 41-6/161, M12/154	-	+	+	+	+	+	+	+	+	+	+	+	+	c
IV	*S. cerevisiae* KL14-4A, MH41-7B, 777-3A	+	+	+	+	+	+	+	+	+	+	+	+	+	d

a: refs 8, 104,105; b: refs 7, 26, 68,104,106; c: refs 104, 107; d: 8, 46, 104, 106.

REFERENCES

1. Grivell, L.A. (1989) Eur. J. Biochem., 182, 477-493.
2. Attardi, G. and Schatz, G. (1988) Ann. Rev. Cell Biol., 4, 289-333.
3. Wolf, K. and Del Giudice, L. (1988) Adv.in Genetics, 25, 185-308.
4. Tzagoloff, A. and Myers, A.M. (1986) Ann. Rev. Biochem., 55, 249-285.
5. De Zamaroczy, M. and Bernardi, G. (1986) Gene 47, 155-177.
6. Slonimski, P.P. and Tzagoloff, A. (1976) Eur. J. Biochem., 61, 27-41.
7. Bonitz, S.G., Coruzzi, G., Thalenfeld, B.E., Tzagoloff, A. and Macino, G. (1980) J. Biol. Chem., 255, 11927-11941.
8. Hensgens, L.A.M., Arnberg, A.C., Roosendaal, E., Van der Horst, G., Van der Veen, R., Van Ommen, G.J.B. and Grivell, L.A. (1983) J. Mol. Biol., 164, 35-58.
9. Fox, T.D. (1979) Proc. Natl. Acad. Sci. USA, 76, 6534-6538.
10. Weiss-Brummer, B., Guba, R., Haid, A. and Schweyen, R.J. (1979) Curr. Genet., 1, 75-83.
11. Coruzzi, G. and Tzagoloff, A. (1979) J. Biol. Chem., 254, 9324-9330.
12. Coruzzi, G., Bonitz, S.G., Thalenfeld, B.E. and Tzagoloff, A. (1981) J. Biol. Chem., 256, 12780-12787.
13. Thalenfeld, B.E. and Tzagoloff, A. (1980) J. Biol. Chem., 255, 6173-6180.
14. Baranowska, H., Szczesniak, B., Ejchart, A., Kruzewska, A. and Claisse, M. (1983) Curr. Genet., 7, 225-233.
15. Macino, G. and Tzagoloff, A. (1980) Cell, 20, 507-517.
16. Cobon, G.S., Beilharz, M.W., Linnane, A.W. and Nagley, P. (1982) Curr. Genet., 5, 97-107.
17. Simon, M. and Faye, G. (1984) Mol. Gen. Genet., 196, 266-274.
18. Macreadie, I.G., Novitski, C.E., Maxwell, R.J., Ulrik, J., Ooi, B-G., McMullen, G.L., Lukins, H.B., Linnane, A.W. and Nagley, P. (1983) Nucl. Acids Res., 11, 4435-4451.
19. Hensgens, L.A.M., Grivell, L.A., Borst, P. and Bos, J.L. (1979) Proc. Natl. Acad. Sci. USA, 76, 1663-1667.
20. Tzagoloff, A., Nobrega, M., Akai, A. and Macino, G. (1980) Curr. Genet., 2, 149-157.
21. Thalenfeld, B.E., Bonitz, S.G., Nobrega, F.G., Macino, G. and Tzagoloff, A. (1983) J. Biol. Chem., 258, 14065-14068.
22. Church, G.M., Slonimski, P.P. and Gilbert, W. (1979) Cell, 18, 1209-1215.
23. Alexander, N.J., Perlman, P.S., Hanson, D.K. and Mahler, H.R. (1980) Cell, 20, 199-206.
24. Haid, A., Schweyen, R.J., Bechmann, H., Kaudewitz, F., Solioz, M. and Schatz, G. (1979) Eur. J. Biochem., 94, 451-464.
25. Nobrega, F.G. and Tzagoloff, A. (1980) J. Biol. Chem., 255, 9828-9837.
26. Slonimski, P.P., Pajot, P., Jacq, C., Foucher, M., Perrodin, G., Kochko, A. and Lamouroux, A. (1978) In: Biochemistry and Genetics of Yeasts (M. Bacila, B.L. Horecker and A.O.M. Stoppani, eds), Academic Press, 1978, pp. 339-368.
27. Van Ommen, G.J.B., Boer, P.H., Groot, G.S.P., De Haan, M., Roosendaal, E., Grivell, L.A., Haid, A. and Schweyen, R.J. (1980) Cell, 20, 173-183.
28. Bonitz, S.G., Homison, G., Thalenfeld, B.E., Tzagoloff, A. and Nobrega, F.G. (1982) J. Biol. Chem., 257, 6268-6274.
29. Di Rago, J.P. and Colson, A.-M (1988) J. Biol. Chem., 263, 12564-12570.
30. Hudspeth, M.E.S., Ainley, W.M., Shumard, D.S., Butow, R.A. and Grossman, L.I. (1982) Cell, 30, 617-626.
31. Farrelly, F., Zassenhaus, H.P. and Butow, R.A. (1982) J. Biol. Chem., 257, 6581-6587.
32. Zassenhaus, H.P. and Perlman, P.S. (1982) Curr. Genet., 6, 179-188.
33. Hudspeth, M.E.S., Vincent, R.D., Perlman, P.S., Shumard, D.S., Treisman, L.O. and Grossman, L.I. (1984) Proc. Natl. Acad. Sci. USA, 81, 3148-3152.
34. Lazowska, J., Jacq, C. and Slonimski, P.P. (1980) Cell, 22, 333-348.
35. Lazowska, J., Claisse, M., Gargouri, A., Kotylak, Z., Spyridakis, A. and Slonimski, P.P. (1989) J. Mol. Biol., 205, 275-289.
36. Delahodde, A., Goguel, V., Becam, A.M., Creusot, F., Perea, J., Banroques, J. and Jacq, C. (1989) Cell, 56, 431-441.
37. Goguel, V., Bailone, A., Devoret, R. and Jacq, C. (1989) Mol. Gen. Genet., 216, 70-74.
38. Anziano, P.Q., Hanson, D.K., Mahler, H.R. and Perlman, P.S. (1982) Cell, 30, 925-932.
39. Hanson, D.K., Lamb, M.R., Mahler, H.R. and Perlman, P.S. (1982) J. Biol. Chem., 257, 3218-3224.
40. De la Salle, H., Jacq, C. and Slonimski, P.P. (1982) Cell, 28, 721-732.
41. Weiss-Brummer, B., Rödel, G., Schweyen, R.J. and Kaudewitz, F. (1982) Cell, 29, 527-536.
42. Carignani, G., Groudinsky, O., Frezza, D., Schiavon, E., Bergantino, E. and Slonimski, P.P. (1983) Cell, 35, 733-742.
43. Dujardin, G., Jacq, C. and Slonimski, P.P. (1982) Nature, 298, 628-632.
44. Netter, P., Carignani, G., Jacq, C., Groudinsky, O., Clavilier, L. and Slonimski, P.P. (1982) Mol. Gen. Genet., 188, 51-59.
45. Wenzlau, J.M., Saldanha, R.J., Butow, R.A. and Perlman, P.S. (1989) Cell, 56, 421-430.
46. Hensgens, L.A.M., Bonen, L., De Haan, M., Van der Horst, G. and Grivell, L.A. (1983) Cell, 32, 379-389.
47. Dujon, B. (1980) Cell, 20, 185-197.
48. Zinn, A.R. and Butow, R.A. (1985) Cell, 40, 887-895.
49. Michel, F. (1984) Curr. Genet., 8, 307-317.
50. Bordonné, R., Dirheimer, G. and Martin, R.P. (1988) Curr.Genet., 13, 227-233.
51. Séraphin, B., Simon, M. and Faye, G. (1985) Nucl. Acids Res., 13, 3005-3014.
52. Bolotin-Fukuhara, M., Faye, G. and Fukuhara, H. (1977) Mol. Gen. Genet., 152, 295-305.
53. Bos, J.L., Osinga, K.A., Van der Horst, G. and Borst, P. (1979) Nucl. Acids Res., 6, 3255-3266.
54. Li, M. and Tzagoloff, A. (1979) Cell, 18, 47-53.
55. Martin, N.C., Miller, D.L., Donelson, J.E., Sigurdson, C., Hartley, J.L., Moynihan, P.S. and Pham, H-D. (1979) In: Extrachromosomal DNA. ICN-UCLA Symposia on Molecular and Cellular Biology Vol. XV (D.J. Cummings, P. Borst, I.B. Dawid, S.M. Weissman and C.F. Fox, eds), Academic Press, pp.357-375.

56. Miller, D.L., Martin, N.C., Pham, H-D. and Donelson, J.E. (1979) J. Biol. Chem., 254, 11735-11740.
57. Berlani, R.E., Bonitz, S.G., Coruzzi, G., Nobrega, M. and Tzagoloff, A. (1980) Nucl. Acids Res., 8, 5017-5030.
58. Berlani, R.E., Pentella, C., Macino, G. and Tzagoloff, A. (1980) J. Bact., 141, 1086-1097.
59. Bonitz, S.G. and Tzagoloff, A. (1980) J. Biol. Chem., 255, 9075-9081.
60. Martin, R.P., Sibler, A-P., Bordonné, R., Canaday, J. and Dirheimer, G. (1980) In: The Organization and Expression of the Mitochondrial Genome (A.M. Kroon and C. Saccone, eds) Elsevier/North-Holland, pp. 311-314.
61. Miller, D.L., Sigurdson, C., Martin, N.C. and Donelson, J.E. (1980) Nucl. Acids Res., 8, 1435-1442.
62. Newman, D., Pham, H-D., Underbrink-Lyon, K. and Martin, N.C. (1980) Nucl. Acids Res., 8, 5007-5016.
63. Nobrega, F.G. and Tzagoloff, A. (1980) FEBS Lett., 113, 52-54.
64. Wesolowski, M., Monnerot, M. and Fukuhara, H. (1980) Curr. Genet., 2, 121-129.
65. Miller, D.L. and Martin, N.C. (1981) Curr. Genet., 4, 135-143.
66. Sibler, A.P. Dirheimer, G. and Martin, R.P. (1985) Nucl. Acids Res., 13, 1341-1345.
67. Bolotin-Fukuhara, M. (1979) Mol. Gen. Genet., 177, 39-46.
68. Heyting, C. and Menke, H.H. (1979) Mol. Gen. Genet., 168, 279-291.
69. Julou, C. and Bolotin-Fukuhara, M. (1982) Mol. Gen. Genet., 188, 256-260.
70. Knight, J.A., Courey, A.J. and Stebbins, B. (1982) Curr. Genet., 5, 21-27.
71. Merten, S., Synenki, R.M., Locker, J., Christianson, T. and Rabinowitz, M. (1980) Proc. Natl. Acad. Sci. USA, 77, 1407-1421.
72. Locker, J. and Rabinowitz, M. (1981) Plasmid, 66, 302-314.
73. Sor, F. and Fukuhara, H. (1982) Nucl. Acids Res., 10, 6571-6577.
74. Sor, F. and Fukuhara, H. (1982) Nucl. Acids Res., 11, 339-348.
75. Tabak, H.F., Hecht, N.B., Menke, H.H. and Hollenberg, C.P. (1979) Curr. Genet., 1, 33-43.
76. Sor, F. and Fukuhara, H. (1980) C.R. Acad. Sci. Paris, 291, serie D, 933-936.
77. Osinga, K.A., Evers, R.F., Van der Laan, J.C. and Tabak, H.F. (1981) Nucl. Acids Res., 9, 1351-1364.
78. Christianson, T., Edwards, J., Levens, D., Locker, J. and Rabinowitz, M. (1982) J. Biol. Chem., 257, 6494-6500.
79. Shen, Z. and Fox, T.D. (1989) Nucl. Acids Res., 17, 4535-4539.
80. Li, M., Tzagoloff, A., Underbrink-Lyon, K. and Martin, N.C. (1982) J. Biol. Chem., 257, 5921-5928.
81. Miller, D.L. and Martin, N.C. (1983) Cell, 34, 911-917.
82. Hollingsworth, M.J. and Martin, N.C. (1986) Mol. Cell. Biol., 6, 1058-1064.
83. De Vries, S. and Grivell, L.A. (1988) Eur. J. Biochem., 176, 377-384.
84. De Vries, S. and Marres, C.A.M. (1988) Biochim. Biophys. Acta, 895, 205-239.
85. Edwards, J.C., Levens, D. and Rabinowitz, M. (1983) Cell, 31, 337-346.
86. Osinga, K.A., De Haan, M., Christianson, T. and Tabak, H.F. (1982) Nucl. Acids Res., 10, 7993-8006.
87. Christianson, T., Edwards, J.C., Mueller, D.M. and Rabinowitz, M. (1983) Proc. Natl. Acad. Sci. USA, 80, 5564-5568.
88. Edwards, J.C., Osinga, K.A., Christianson, T., Hensgens, L.A.M., Janssens, P.M., Rabinowitz, M. and Tabak, H.F. (1983) Nucl. Acids Res., 11, 8269-8282.
89. Christianson, T. and Rabinowitz, M. (1983) J. Biol. Chem., 258, 14025-14033.
90. Miller, D.L., Underbrink-Lyon, K., Najarian, D.R., Krupp, J. and Martin, N.C. (1983) In: Mitochondria 1983: Nucleo-Mitochondrial Interactions (R.J. Schweyen, K. Wolf and F. Kaudewitz, eds), De Gruyter, Berlin, pp. 151-164.
91. Osinga, K.A., De Vries, E., Van der Horst, G.T.J. and Tabak, H.F. (1984) Nucl. Acids Res., 12, 1889-1900.
92. Tabak, H.F., Osinga, K.A., De Vries, E., Van der Bliek, A.M., Van der Horst, G.T.J., Groot-Koerkamp, M.J.A., Van der Horst, G., Zwarthoff, E.C. and MacDonald, M.E. (1983) In: Mitochondria 1983: Nucleo-Mitochondrial Interactions (R.J. Schweyen, K. Wolf and F. Kaudewitz, eds), De Gruyter, Berlin, pp. 79-93.
93. Osinga, K.A., De Vries, E., Van der Horst, G. and Tabak, H.F. (1984) EMBO J., 3, 829-834.
94. Palleschi, C., Franscisci, S., Zennaro, E. and Frontali, L. (1984) EMBO J., 3, 1389-1395.
95. Bordonné, R., Dirheimer, G. and Martin, R.P. (1987) Nucl.Acids Res., 15, 7381-7394.
96. Palleschi, C., Francisci, S., Bianchi, M.M. and Frontali, L. (1984) Nucl. Acids Res., 12, 7317-7326.
97. Sibler, A-P., Dirheimer, G. and Martin, R.P. (1983) FEBS Lett., 152, 153-156.
98. De Zamaroczy, M., Marotta, R, Faugeron-Fonty, G., Coursot, R., Mangin, M., Baldacci, G. and Bernardi, G. (1981) Nature, 292, 75-78.
99. Goursot, R., Mangin, M. and Bernardi, G. (1982) EMBO J., 1, 705-711.
100. Baldacci, G. and Bernardi, G. (1982) EMBO J., 1, 987-994.
101. Lazowska, J., Jacq, C. and Slonimski, P.P. (1981) Cell 27, 12-14.
102. Blanc, H. and Dujon, B. (1980) Proc. Natl. Acad. Sci. USA, 77, 3942-3946.
103. De Jonge, J.C. and De Vries, H. (1982) Curr. Genet., 7, 21-28.
104. Sanders, J.P.M., Heyting, C., Verbeet, M.P., Meijlink, F.C.P.W. and Borst, P. (1977) Mol. Gen. Genet., 157, 239-261.
105. Bos, J.L., Osinga, K.A., Van der Horst, G., Hecht, N.B., Tabak, H.F., Van Ommen, G.J.B. and Borst, P. (1980) Cell, 20, 207-214.
106. Morimoto, R. and Rabinowitz, M. (1979) Mol. Gen. Genet., 170, 25-48.
107. Dujon, B., Colson, A.M. and Slonimksi, P.P. (1977) In: Mitochondria 1977. Genetics and Biogenesis of Mitochondria (W. Bandlow, R.J. Schweyen, K. Wolf and F. Kaudewitz, eds), De Gruyter/Berlin-New York, pp.579-669.
108. Tabak, H.F., Van der Horst, G., Winter, A.J., Smit, J., Van der Veen, R., Kwakman, J.H.J.M., Grivell, L.A. and Arnberg, A.C. (1988) Cold Spring Harbor Symposia on Quantitative Biology, 52, 213-221.
109. Moorman, A.F.M., Van Ommen, G.J.B. and Grivell, L.A. (1978) Mol. Gen. Genet., 160, 13-24.

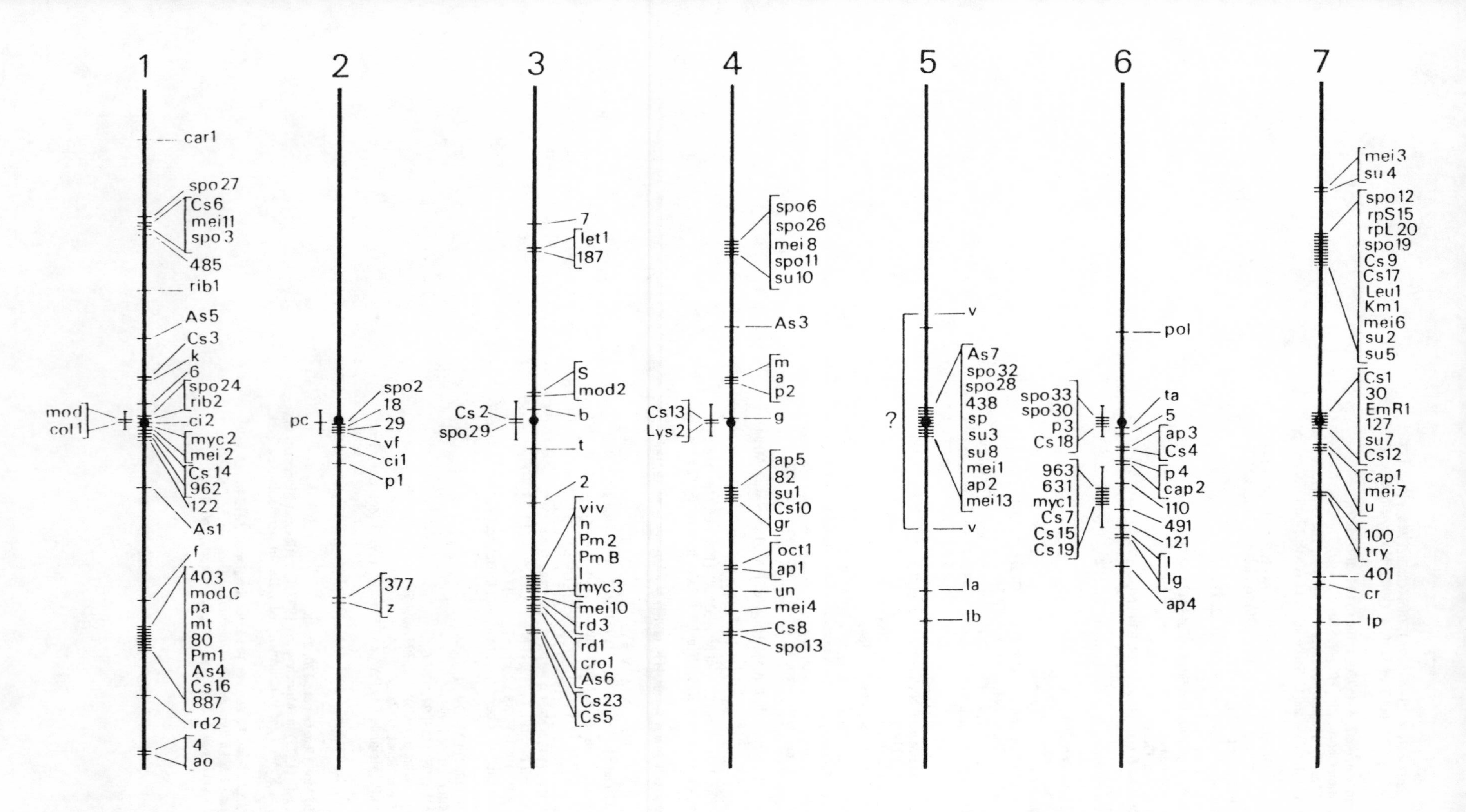

GENETIC MAP OF PODOSPORA ANSERINA

GENETIC MAP OF PODOSPORA ANSERINA

Denise MARCOU, Marguerite PICARD-BENNOUN and Jean-Marc SIMONET
Laboratoire de Génétique, Bât. 400, Université de Paris-Sud
F-91405 ORSAY, France

February 1982

Podospora anserina (=Pleurage anserina (Ces.) Kuntze) is a coprophilous fungus belonging to the Ascomycetes-Pyrenomycetes group (Family: Sordariaceae, order: Sphaeriales). The media and methods for Podospora culture and manipulation have been summarized by ESSER. In: Handbook of Genetics (R.C. King, ed.). New-York : Plenum Press 1974 .

After a postmeiotic mitosis in which the four spindles are nearly perpendicular to the long axis of the ascus, the 8 haploid meiotic products are distributed into 4 ascospores. Each spore contains 2 different meiotic products, arising from the same half tetrad. It is thus easy to detect genetic exchanges and to carry out complementation tests.

The haploid number of chromosome is 7 and accordingly, there are 7 well defined linkage groups. Owing to the particular modalities of meiotic recombination (e.g. very strong chiasma interference in some chromosome segments), the second division segregation frequencies on several chromosome arms reach a maximum (often higher than 0.66, and even up to 0.98) and then decrease (35,63). For the same reason, some genes still remain difficult to localize and will not be mentionned in this paper.

All the mutants have been isolated from the same geographical race (46), though they now bear either the S or s allele of the allelic system of cytoplasmic incompatibility. Most of the genes have been named according to the phenotypic characteristics of the mutant alleles, with the exception of most of the genes involved in spore colour, which are designated by numbers. Genes concerning incompatibility systems have a double symbol : a small letter, according to ESSER, and a capital letter given in brackets (BERNET's nomenclature). Where the order of a group of genes, or of genes with a centromere, is unknown, they are bracketed together. Prof. K. ESSER kindly provided us his unpublished data concerning the localization of the loci marked with ⋆ in the Table. The loci marked with ⋆⋆ have been mapped by L. BELCOUR and with ⋆⋆⋆ by M. PERROT (personal communications).

The strains are kept in the author's collection or with Prof. K. ESSER at : Lehrstuhl für allgemeine Botanik, Ruhr-Universität , 4630 Bochum-Querenburg, Postfach 102148, R.F.A.
or Prof. J. BEGUERET and Prof. J. BERNET at : Laboratoire de Génétique, Faculté des Sciences (3ème Tranche), F-33405 Talence, France.

GENE SYMBOL	N° OF ALLELES	% OF SDS	NAME AND CHARACTERISTICS	REFERENCES
			LINKAGE GROUP 1	
carl	4	59.3±3	caryogamy deficient	35,58,61,63,68
spo27	2	87±7	sporulation deficient	63,70
Cs6	2	85±3.6	cold-sensitive growth	35,44,63
meill	2	85	meiotic block at metaphase I	61,63,68
spo3	2	86±8	sporulation deficient	63,70
485	2	88±3	green spores	63
rib1	2	66±16	riboflavin requiring, yellow spores pale and sterile mycelium	63
AS5	2	40	informational antisuppressor	63 , 77
Cs3	2	16±3.4	cold-sensitive growth	44,69
k(Z)	2	12	incompatibility factor	10
6	2	6	brown spores	63
rib2	2	1	riboflavin requiring, dark green spores, mycelium ♀ sterile	63
spo24	2	0.3-3	sporulation deficient	70
col1	2	0-12	colonial mycelium	63
mod*	2	0	modifies mutant Incoloris (LG3) to pink sterile mycelium	
ci2	2	≤0.1	circulosa : clock mutant	35,63
mei2	2	≤0.2	meiotic block before pachytene, US sensitive, mutator	35,42,58,60, 61,63,68
myc2	2	≤0.2	mycelium not dense, poorly fertile, slow growth	35,63
Cs14	2	3	cold-sensitive growth and methionine requiring	44
962	2	≤3	auxotrophic	63
122	28	3.3±0.3	ascospores, perithecia and mycelium colorless	2,8,35,59,63
AS1	6	35±7	reduces or nullifies the effects of ribosomal suppressors, antisuppressor	35,42,43,63 ,72,77
f	5	83	sinuous hyphae, mycelium flat flexuosa	23,26,29,35 50,63
403	2	90	female sterility	
mod c	4	93 ± 2	specific suppressor of incompatibility factors R and V, female sterility	75

80	2	98	female sterility	
Pm1	7	98	paramomycin resistant	16,72,78
AS4	46	98	informational antisuppressor	42,43,72,77
Ĉs16	2	95	cold-sensitive growth	44
887 = PmA	2	98	paromomycin hypersensitive	63,78
rd2	2	83±	round ascospores	35,63
4	2	58.8±1.4	green ascospores	34,35,63
ao	2	59	albo-lana : mycelium velvety and white	35,63

LINKAGE GROUP 2

spo2	2	0.5	sporulation deficient	58,70
18 ("locus 14")	>100	0.3-0.6	spores brown and small	32
29 =as ("locus 14")	> 80	0.15±0.09 1.3 ±0.2	spores white or green according to alleles = albospora the segment 29 is itself constitued of three cistrons	2,23,32,34, 39,40,41,59, 64,65,66
vf ("locus 14")	> 40	0.50±0.04	"vert-foncé":dark green spores	32,33,67
pc	4	0-1.8	perithecia and mycelium pale	
ci1	2	10	circulosa : clock mutant	
pl	2	18	pumila, small ascospores	29,33,34,50
z	2	83	zonata : brown mycelium ♀ sterile, clock mutant	13,20,22,23,25,26 28,29,30,38,63,67
385 =377	4	84±4	green ascospores	63

LINKAGE GROUP 3

7	2	82.5±2	brown and irregular ascospores, brown mycelium, ♀ sterile	63
187	2	76.3±2.7	green ascospores	63
let1	2	75	lethal : spores homocaryotic never formed	63
mod2	2	11	modifier of incompatibility in non allelic systems	5
s/S	4	11	heterogenic incompatibility, allelic mechanism	4,8,10,18,19 21,24,29,36, 51,52,53,54,55 55,56,57
su-m	2	4	suppressor for m mutant (LG4)	24
b(C)	16	4	heterogenic incompatibility, non-allelic mechanism	3,6,7,8,9,10, 11,12,19,21,27 29
spo29	2	0.2-5.8	sporulation deficient	
Cs2	2	3±1.6	cold-sensitive growth	44

GENE SYMBOL	N° OF ALLELES	% OF SDS	NAME AND CHARACTERISTICS	REFERENCES
su-m	2	4	suppressor for m mutant (LG4)	24
b(C)	16	4	heterogenic incompatibility, non-allelic mechanism	3,6,7,8,9,10 11,12,19,21,27,29
spo29	2	0.2-5.8	sporulation deficient	
Cs2	2	3±1.6	cold-sensitive growth	44
t(B)	2	13	heterogenic incompatibility allelic mechanism	9,10,19,29, 52,53
2	2	35±4	brown spores, no germination	34
myc3	2	50	colorless mycelium, slow growth, ♀ sterile	61,63
Pm2	7	70	paromomycin resistant	16 ,72,78
PmB	48	70	paromomycin hypersensitive	78
n	2	69	nivea : mycelium white, serial hyphae reduced, many aborted protoperithecia	67
viv	2	74	vivax : rythmic growth, increases longevity asssociated with I, reduced growth on minimal medium	67
I	36	80	Incoloris : mycelium colorless, nearly ♀ sterile, mutant dominant	29,31,36,45, 46,47,48,50,67
rd3	3	72±3	round ascospores	61,63
mei10	2	82.5	meiotic block at Metaphase I	61,63
AS6	2	70	ribosomal antisuppressor, cold-sensitive growth	44,63,72,77
rd1=96	2	84±1	round ascospores	34,59,62,63
cro1	2	82±10	drastic modification of crosier's nuclei	61,63,68
Cs23	2	80±7.6	cold-sensitive growth	44,63
Cs5	2	78±7	cold-sensitive growth	44,62,63

LINKAGE GROUP 4

GENE SYMBOL	N° OF ALLELES	% OF SDS	NAME AND CHARACTERISTICS	REFERENCES
spo6	2	84±6	sporulation deficient	63,70
spo26	2	62±7	sporulation deficient	63,70
mei8	2	68±4.5	meiotic block at diplotene	61,62,63,68,69
spo11	2	63±12	sporulation deficient	63,70
su10	15	58±1.4	nonsense suppressor	63,72
AS3	6	40	ribosomal antisuppressor	14,42,43, 72,77
a(E)	16	18	heterogenic incompatibility, non-allelic mechanism	3,6,8,9,10,11, 12,19,21,27,29
m	2	17	minor : small perithecia	5,24,29
p2=49	2	17	pumila : small spores	2

g	2	0.2	glaber : smooth mycelium	29
Cs13	2	0-3.8	cold-sensitive growth	44
lys2	2	unknow	lysine requiring	72
su1	>70	35±1.5	ribosomal suppressor	14,41,42,43,63,72,73
gr	2	35	grisea : grey-green spores, aerial hyphae reduced, block in perithecia development	45,67
ap5	2	42±3.5	spores with persistent appendix	
82	4	43±2.5	green spores	2,34,59,63
Cs10	2	43±8.3	cold-sensitive growth	44
ap1=64	3	60	ascospores with persistent appendix	2,34,59
oct1	2	68	octospora : when homozygous, sporulation deficient. When heterozygous, many asci with 8 monokaryotic spores	13,63
un	2	75	undulata : clock mutant	25,29,37,63
mei4	2	82	meiotic block at metaphase I, recombination modificator	34,58,61,63,68,69
Cs8	2	89±3.4	cold-sensitive growth	44,63
spo13	2	93±4	sporulation deficient	63,70

LINKAGE GROUP 5

v(R)	2	40	heterogenic incompatibility, allelic and nonallelic mechanism	8,10,19,21,27,29,52,53
AS7	3	0	informational antisuppressor paromomycin resistant sporulation deficient	72,73
spo32	2	0	sporulation deficient colonial mycelium	
spo28	2	0	sporulation deficient	
su3	5	≤0.1	informational suppressor	14,41,43,63
su8	4	≤0.1	non-sense suppressor	63,72
438	2	0.04-0.6	spores white or absent	63
ap2=154	3	0.006-1.3	ascospores with persistent appendix	2,34,63
sp	2	0.2	splendida : mycelium glossy, sterile	29
mei1	5	≤1	meiotic block before pachytene, UV sensitive, mutator	58,60,61,63,68,69
mei13	2	0.8-10.9	abnormal meiosis sporulation deficient	
la = mi	4	77	lanosa : mycelium velvety	1,29,34,63,67
lb	2	87	lano-alba: mycelium white,velvety	

GENE SYMBOL	N° OF ALLELES	% OF SDS	NAME AND CHARACTERISTICS	REFERENCES
			LINKAGE GROUP 6	
pol*	2	38	politor : reduced aerial hyphae, ♀ sterile, slow growth on minimal medium	
spo30	2	0	sporulation deficient	
spo33	2	0	sporulation deficient reduced germination	
p3	2	0.4-3.9	pumila : very small spores	63
Cs18	2	0.02-6.7	cold-sensitive growth	44
ta	2	0-2.2	tarda : clock mutant	29,35,63,67
5	5	0.2	brown spores, perithecia and mycelium	34,35,63
ap3=111	3	4-10	ascospores with persistent appendix	2,34,63
Cs4	2	8±3.6	cold-sensitive growth	44
p4=68	2	11-21	pumila : small ascospores, no germination	2,34,35,63
myc1	2	15	slow growth, mycelium glossy and sterile	63
cap2	2	22	chloramphenicol resistant	63
110	3	13-25	yellow ascospores, perithecia nearly empty, slow growth	2,34,63
631	2	15	auxotrophic, perhaps adenine requiring	63
491	2	36±1.9	green spores	34,63
963	2	35±10	reduces the pigmentation of slightly coloured spore mutants	42
121	3	40±2	yellow spores	34,35,63
Cs7	2	35±6	cold-sensitive growth	44
Cs15	2	34±8	cold-sensitive growth	44
Cs19	2	33±9	cold-sensitive growth	44
ap4=63	2	30-53	spores with persistent appendix no germination	34,35,63
1	3	32-47	lenta : mycelium velvety, sterile, very slow growth	29
1g	3	45-63	lanuginosa : mycelium velvety, sterile, slow growth	29

LINKAGE GROUP 7

mei3 = Pm3	7	65	meiotic block at pachytene	35,58,61,63,68,72
su4	7	76±1.9	nonsense suppressor	35,41,42,58,63,72
leu-1	2	69±8	leucine requiring	15,72
Cs9	2	74±7	cold-sensitive growth	44,63
Cs17	2	71±17	cold-sensitive growth	44,63
su2=AS2	> 20	70	informational suppressor	14,34,41,42,43 63,72,77
su5	2	70	informational suppressor	43,63
Km1	2	70	kanamycine hypersensitive	63
mei6	3	65±6.5	meiotic block before pachytene, antimutator	61,63,68,69
spo19	2	69±7	sporulation deficient	70
rpS15	4	58±8	structural gene of ribosomal protein S15	74
rpL20***	2	very close to rp S15	structural gene of ribosomal protein L20	
spo12	2	55±18	sporulation deficient	70
Cs1	2	0-3.6	cold-sensitive growth	44,63
Cs12	2	0-3.4	cold-sensitive growth	44,63,70
su7	3	0	informational suppressor	14,43,63,71
127	3	0.0015	green ascospores	63
30	11	0.0376±0.003	green and small ascospores	2
u(Q)	2	5	heterogenic incompatibility, allelic mechanism	8,10,19,29, 52,53,63
EmR1	75	3±3	Emetin resistant Anisomycin hypersensitive	74
cap1	2	8	chloramphenicol resistant	63
mei7	2	11±4	meiotic block ar pachytene	61,68
100	2	34±6	yellow ascospores	2,34,63
try	2	36	tryptophane requiring, slow growth	
401	2	66	green spores, white mycelium nearly sterile	63
cr	2	70	crispa : aerial hyphae curved	45,67
lp	3	86	lano-pallida : mycelium velvely and white	

REFERENCES

1. AUVITY, M. 1970. C.R. Acad. Sc. Paris. 270:935-937.
2. BEGUERET, J. 1967. C.R. Acad. Sc. Paris. 264:462-465.
3. BEGUERET, J. 1969. C.R. Acad. Sc. Paris. 269:458-461.
4. BEISSON-SCHECROUN, J. 1962. Ann. Genet. 4:3-50.
5. BELCOUR, L., BERNET, J. 1969. C.R. Acad. Sc. Paris. 269:712-714.
6. BELCOUR, L. 1971. Molec. gen. Genet. 112:263-274.
7. BERNET, J. 1963. C.R. Acad. Sc. Paris. 256:771-773.
8. BERNET, J. 1963. Ann. Sc. Nat. Bot. Biol. Veg. 4:205-223.
9. BERNET, J. 1965. Ann. Sc. Nat. Bot. Biol. Veg. 6:611-768.
10. BERNET, J. 1967. C.R. Acad. Sc. Paris. 265:1330-1333.
11. BERNET, J., BELCOUR, L. 1967. C.R. Acad. Sc. Paris. 265:1536-1539.
12. BERNET, J., BEGUERET, J. 1968. C.R. Acad. Sc. Paris. 266:716-719.
13. BLAICH, R., ESSER, K. 1969. Arch. Mikrobiol. 68:201-209.
14. COPPIN-RAYNAL, E. 1977. J. of Bact. 131:876-883.
15. CROUZET, M., PERROT, M., NOGUEIRA, M., BEGUERET, J. 1978. Biochem. Gen. 16:271-286.
16. DEQUARD, M., COUDERC, J-L., LEGRAIN, P., BELCOUR, L., PICARD-BENNOUN, M. 1980. Biochem. Gen. 18:263-280.
17. DEQUARD, M., BELCOUR, L. 1981. Exp. Mycology. 5:363-368.
18. ESSER, K. 1954. C.R. Acad. Sc. Paris. 238:1731-1733.
19. ESSER, K. 1956. Z. Abstammungs-und Vererbungsl. 87:595-624.
20. ESSER, K. 1956. Naturwissenschaften. 43:284-285.
21. ESSER, K. 1959. Z. Vererbungsl. 90:29-52.
22. ESSER, K., DEMOSS, J.A., BONNER, D.M. 1960. Z. Vererbungsl. 91:291-299.
23. ESSER, K. 1966. Z. Vererbungsl. 97:327-344.
24. ESSER, K. 1968. Genetics. 60:281-288.
25. ESSER, K. 1969. Mycologia. 51:1008-1011.
26. ESSER, K., MINUTH, W. 1970. Genetics. 64:441-458.
27. ESSER, K., BLAICH, R. 1973. Genetics. 17:107-152.
28. HERZFELD, F., ESSER, K. 1969. Arch. Mikrobiol. 65:146-162.
29. KUENEN, R. 1962. Z. Vererbungsl. 93:66-108.
30. LYSEK, G., ESSER, K. 1970. Arch. Mikrobiol. 73:224-230.
31. MARCOU, D. 1961. Ann. Sc. Nat. Bot. 2:653-764.
32. MARCOU, D., PICARD, M. 1967. C.R. Acad. Sc. Paris. 265:1962-1965.
33. MARCOU, D. 1969. C.R. Acad. Sc. Paris. 269:2362-2365.
34. MARCOU, D. 1979. Molec. gen. Genet. 173:299-305.
35. MARCOU, D., MASSON, A., SIMONET, J.M., PIQUEPAILLE, G. 1979. Molec. gen. Genet. 176:67-79.
36. MONNOT, F. 1953. C.R. Acad. Sc. Paris. 236:2330-2332.
37. NGUYEN VAN HUONG. 1962. C.R. Acad. Sc. Paris. 254:2646-2648.
38. NGUYEN VAN HUONG. 1967. C.R. Acad. Sc. Paris. 264:280-283.
39. PICARD, M. 1970. C.R. Acad. Sc. Paris. 270:498-501.
40. PICARD, M. 1971. Molec. gen. Genet. 111:35-50.
41. PICARD, M. 1973. Genet. Res. 21:1-15.
42. PICARD-BENNOUN, M. 1976. Molec. gen. Genet. 147:299-306.
43. PICARD-BENNOUN, M., COPPIN-RAYNAL, E. 1977. Physiol. Veg. 15:481-489.
44. PICARD-BENNOUN, M., LE COZE, D. 1980. Genet. Res. 36:289-297.
45. PRILLINGER, H., ESSER, K. 1977. Molec. gen. Genet. 156:333-345.
46. RIZET, G. 1939. C.R. Acad. Sc. Paris. 209:771-773.
47. RIZET, G. 1941. Bull. Soc. Bota. France. 88:517-520.
48. RIZET, G. 1941. C.R. Acad. Sc. Paris. 213:42-45.
49. RIZET, G. 1943. Bull. Soc. Linnéenne de Normandie. France 3:14-15.
50. RIZET, G., ENGELMANN, C. 1949. Rev. Cytol. Biol. Veg. 11:201-304.
51. RIZET, G. 1952. Rev. Cytol. Biol. Veg. 13:51-92.
52. RIZET, G. 1953. C.R. Acad. Sc. Paris. 237:666-668.
53. RIZET, G., ESSER, K. 1953. C.R. Acad. Sc. Paris. 237:760-761.
54. RIZET, G., SCHECROUN, J. 1959. C.R. Acad. Sc. Paris. 249:2392-2394.

55. RIZET, G., SICHLER, G. 1950. C.R. Acad. Sc. Paris. 231:719-721.
56. SCHECROUN, J. 1958. C.R. Acad. Sc. Paris. 246:1268-1270.
57. SCHECROUN, J. 1959. C.R. Acad. Sc. Paris. 248:1394-1397.
58. SIMONET, J-M., ZICKLER, D. 1972. Chromosoma. 37:327-351.
59. SIMONET, J-M. 1973. Molec. gen. Genet. 123:263-281.
60. SIMONET, J-M. 1976. Mut. Res. 41:225-232.
61. SIMONET, J-M., ZICKLER, D. 1978. Molec. gen. Genet. 162:237-242.
62. SIMONET, J-M., ZICKLER, D. 1979. Molec. gen. Genet. 175:359-367.
63. SIMONET, J-M., MARCOU, D., PICARD-BENNOUN, M., LE COZE, D., PIQUEPAILLE, G. 1982. submitted to Genetics.
64. TOURE, B. 1972. Molec. gen. Genet. 117:267-280.
65. TOURE, B., MARCOU, D. 1970. C.R. Acad. Sc. Paris. 270:619-621.
66. TOURE, B., PICARD, M. 1972. Genet. Res. Camb. 19:313-319.
67. TUDZYNSKI, P., ESSER, K. 1979. Molec. gen. Genet. 173:71-84.
68. ZICKLER, D., SIMONET, J-M. 1979. Soc. Bot. Fr. Actual. Bot. 2:222.
69. ZICKLER, D., SIMONET, J-M. 1980. Genetics. In press.
70. ZICKLER, D., SIMONET, J-M. 1980. Exp. Mycology.4:191-206.

ADDENDUM

71. COPPIN-RAYNAL, E. 1980. Anal. Biochem. 109:395-398.
72. COPPIN-RAYNAL, E. 1981. Bioch. Genet. 19:729-740.
73. COPPIN-RAYNAL, E. 1982. Current Genetics. In Press.
74. CROUZET, M., BEGUERET, J. 1982. Submitted to Current Genetics.
75. LABARERE, J., BERNET, J. 1977. Genetics. 87:249-257.
76. LABARERE, J., BERNET, J. 1979. J. of Gen. Microbiol. 113:19-27.
77. PICARD-BENNOUN, M. 1981. Molec. gen. Genet. 183:175-180.
78. RANDSHOLT, N., IBARRONDO, F., DEQUARD, M., PICARD-BENNOUN, M. 1982. Bioch. Genet. In Press.

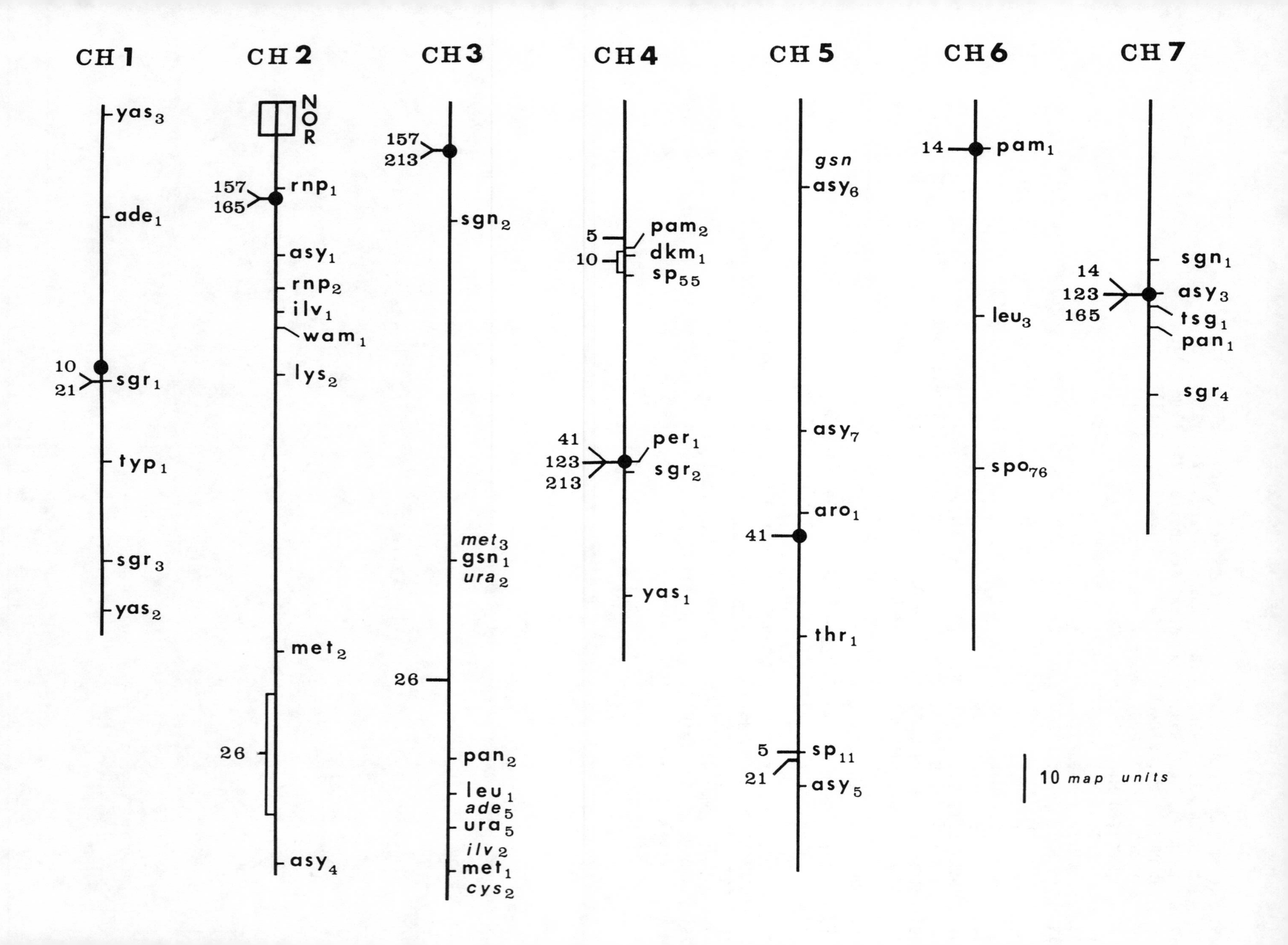
CH 1
yas_3
ade_1
10
21
sgr_1
typ_1
sgr_3
yas_2
CH 2
NOR
157
165
rnp_1
asy_1
rnp_2
ilv_1
wam_1
lys_2
met_2
26
asy_4
CH 3
157
213
sgn_2
met_3
gsn_1
ura_2
26
pan_2
leu_1
ade_5
ura_5
ilv_2
met_1
cys_2
CH 4
5
10
pam_2
dkm_1
sp_{55}
41
123
213
per_1
sgr_2
yas_1
CH 5
gsn
asy_6
asy_7
aro_1
41
thr_1
5
21
sp_{11}
asy_5
CH 6
14
pam_1
leu_3
spo_{76}
CH 7
sgn_1
14
123
165
asy_3
tsg_1
pan_1
sgr_4
10 map units

3.68

SORDARIA MACROSPORA GENETIC MAPS

G. LEBLON, V. HAEDENS, A. D. HUYNH, L. LE CHEVANTON, P. J. F. MOREAU and D. ZICKLER Laboratoire IMG, Batiment 400, Universite Paris-Sud, Centre d'Orsay, 91405 Orsay, Cedex, France

The map figures show the seven linkage groups of Sordaria macrospora. Mapping is based on tetrad analysis (meiotic crossing over). Since each linkage group has been assigned to its respective chromosome, it is named by its corresponding chromosome number. The arbitrarily designated left arm of each group is drawn above the centromere and the right arm below; the centromere is represented by a circle. The reciprocal translocation breakpoints which have been used for establishing linkage group chromosome correlations are shown at the left of each linkage group. Genes whose map positions with respect to neighboring markers are not precisely established, are shown by italicized symbols in their approximate position. Several additional auxotrophic loci are not displayed, because their relative order to all the neighboring markers that are shown is not known: lys1, cys1, cys3, ade3, leu2 and nic1 (chromosome 1), ilv3 (ch5), ade4 and his1 (ch6), lys3asco and cys4 (ch7).

The comparison of 15 homologous loci or probably homologous loci between Sordaria macrospora and Neurospora crassa suggests that many linkages could have been conservéd: ade1,typ1 (ch1 of Sordaria macrospora) and ade3,phe1 (linkage group LGI of Neurospora crassa), NOR,ilv1,lys2,met2 (ch2) and NOR,ilv1or2,lys2,met3 (LGV), met3,ura2,pan2,leu1,ura5,met1 (ch3) and met1,pyr1,pan1,leu2,pyr2,met5 (LGIV) aro1,thr1 (ch5) and arom,thr2or3, (LGII), pan1,lys3asco (ch7) and pan2,lys5asco (LGVI).

The meanings of symbols are as follows: ade--adenine requirement, aro--requirement for aromatic amino acids plus p-aminobenzoic acid, asy--asynaptic meiosis, CH--chromosome, cys--cysteine requirement, dkm--dark mycelium, ilv--isoleucine plus valine requirement, gsn--gray ascospore, slow growing and no spore, leu--leucine requirement, lys--lysine requirement, lysasco--lysine requirement and autonomous ascospore semilethal, met--methionine requirement, NOR--nucleolus organizer region, pam--pink ascospore and mycelium, pan--panthothenic acid requirement, per--colonial on sorbose medium, ura--uridine or uracile requirement, rnp--reduced number of perithecia, sgn--slow growing and no spores, sgr--slow growing, sp--self sterile, spo--rare sporulation, thr--threonine requirement, tsg--temperature sensitive growth, typ--tyrosine or phenylalanine requirement, wam--white ascospore and mycelium, yas--yellow ascospore.

REFERENCES

HAEDENS, V. 1985 These doct. Univers. Paris-sud n 295.
HUYNH, A.D., G. LEBLON and D. ZICKLER 1986 Current Genetics 10: 545-555.
LEBLON, G., D. ZICKLER and S. LEBILCOT 1986 Genetics 112: 183-204.
PERKINS, D.D., A. RADFORD, D. NEWMEYER and M. BJORKMAN 1982 Microbiological reviews 426-570.
ZICKLER, D., G. LEBLON, V. HAEDENS, A. COLLARD and P. THURIAUX 1984 Current Genetics 8: 57-67.

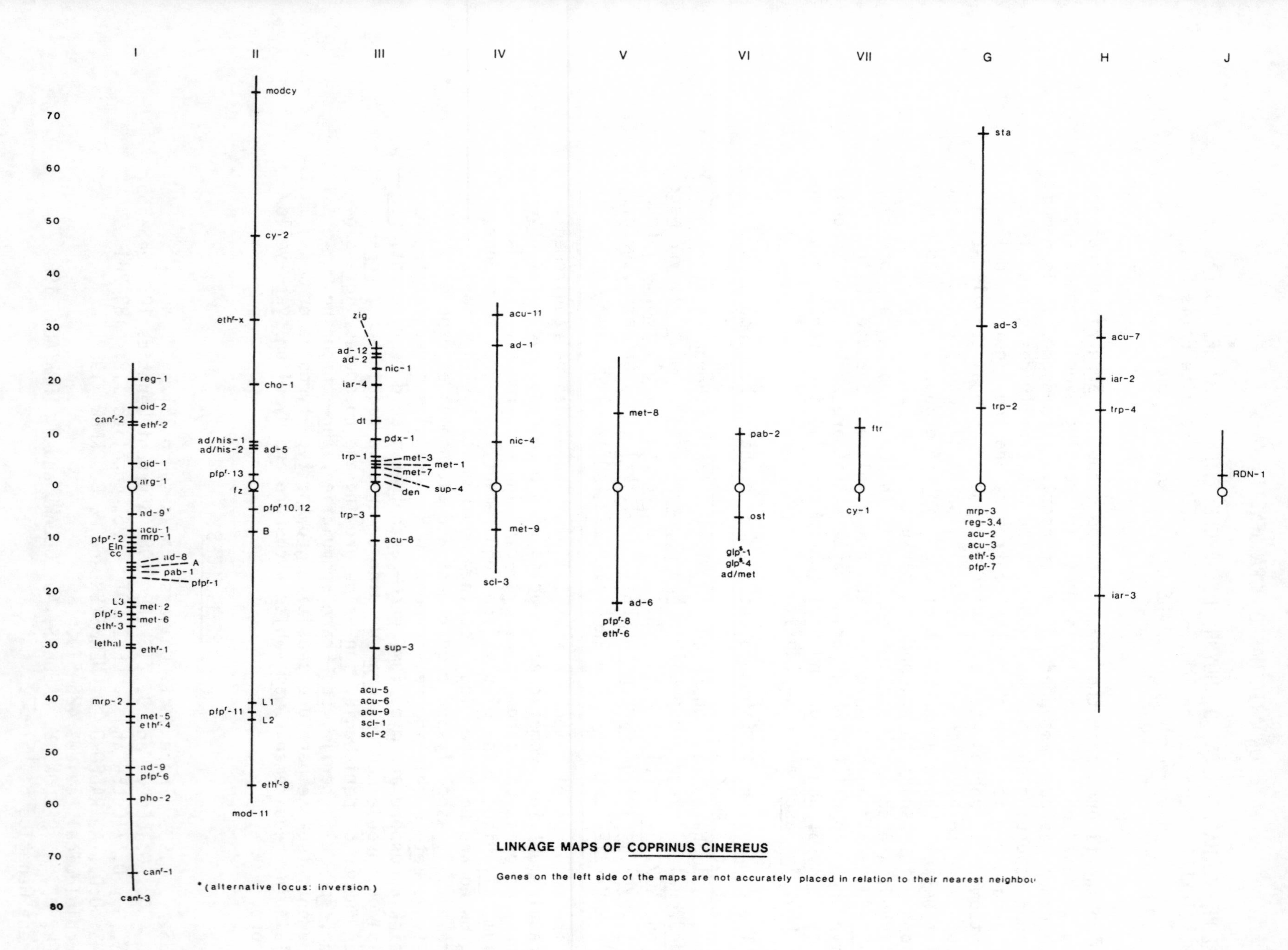

LINKAGE MAPS OF COPRINUS CINEREUS

Genes on the left side of the maps are not accurately placed in relation to their nearest neighbou

Linkage map of *Coprinus cinereus* (Schaeff. ex Fr.) S.F. Gray,
Ink cap. N = 12 (ref 6)

Date compiled: October 1986

Compiled by: Dr. Jane North,
Department of Biological Sciences, City of London Polytechnic,
Calcutta House, Old Castle Street, London E1 7NT, UK.

A NOTE ON THE NOMENCLATURE AND LIFE CYCLE OF THE SPECIES

Coprinus cinereus is the name now used for cultures isolated from leaf litter, manure heaps, mushroom compost and enriched straw throughout the world, and identified variously as *C. lagopus*, *C. cinereus*, *C. fimentarius and C. macrorhizus f. microsporus* Hongo(1-3, 21-3, 43). These strains, including a number held by the ATCC, are interfertile(25,26,50) and are typical of the species *C. cinereus* which forms a unique type of microsclerotium in culture(25). A genetic map of *C. macrorhizus* has been published in Japan(24); where possible, markers from this have been incorporated into figure.

C. cinereus shows considerable polymorphism with respect to morphogenesis(27,28) and biochemistry(4,15,29,30). Variation in restriction enzyme cleavage sites in mitochondrial DNA has been reported(8); the nuclear DNA has extensive insertion/deletion polymorphism within single-copy sequences(31).

The life cycle is typical of the agaricales(5,38). Forced heterokaryons(10) and stable diploids(11) can be made, allowing studies of gene expression in different cell-types(12,29,30). Dikaryons can be resolved directly from chlamydospores(5) or veil cells(7). Mitotic recombination can be induced in diploids by ultraviolet light (9); haploidisation of diploids can be achieved using griseofulvin(35).

Mutants are normally selected from the uninucleate oidia. A replica-plating technique using Velcro as a transfer material is available(14).

ACKNOWLEDGEMENTS

The author wishes to thank all who have so generously provided unpublished data and expresses her gratitude to Dr. D. Moore and Dr. P.R. Day for the use of their linkage maps, on which the mapping in *Coprinus cinereus* is largely based.

TABLE 1 MORPHOLOGICAL MUTANTS, LINKAGE GROUPS, AND RELEVANT MAP DISTANCES

Mutant Linkage Group		Phenotypic Expression Relevant Map Distances	Reference
den	spon	Thick, undulating hyphae, excessively branched; stunted growth	7
Group III		2.4 units *met-1*, 14 units *ad-12*	40,47,48
mrp-1	UV	Dense, compact colony	15
Group I		4.5 units mating type A, 27.5 units *reg-1*	15
mrp-2	spon	Thin colony with "wet" center of collapsed aerial hyphae	54
Group I		23 units mating type A, 15 units *eth*^{r}*-3*	54
mrp-3	spon	Normal, but half growth rate	54
Group G		5 units *eth*^{r}*-5*	53,54
oid-1	UV	Excess production of oidia	44
Group I		27 units mating type A, 17.3 units *ad-8*, 15 units *acu-1*	44,51
oid-2	UV	Excessive production of oidia	51
Group I		27.4 units *ad-8*, 21 units *acu-1*	51
ost	NTG	Ostrich-like, feathery mycelium	19
Group VI		15.5 units *pab-2*	19
scl-1	wild	Production of sclerotia	27
Group III		28.6 units *scl-2*	27
scl-2	wild	Production of sclerotia	28
Group III		26.2 units *pdx*	28
scl-3	wild	Production of sclerotia	28
Group IV		28.1 units *nic-4*	28
cc	spon	Concentric	36
Group I		3.2 units proximal* to mating type A	24,36
fz	UV	Frizzy	36
Group II		8.9 units from mating type B, probably on centromere*	24
dt	UV	Dichotomous growth	36
Group III		3.5 units *pdx*	24
Eln	UV,NTG	Elongation of basidiocarp stipe	37
Group I		3.8 units proximal* to mating type A	24
zig	TEPA	Zig-zag branched mycelium	9
Group III		1.8 units distal to ad-2	9
sta	TEPA	Star-shaped colonies	9
Group G		37.7 units distal to ad-3	9

*inferred from relationship between maps in refs.24 and 38.

Abbreviations: spon = spontaneous; UV = ultraviolet light; NTG = nitrosoguanidine; TEPA = tris, 1-azindinyl phosphine oxide

TABLE 2
AUXOTROPHIC AND BIOCHEMICAL MUTANTS, LINKAGE GROUPS, AND RELEVANT MAP DISTANCES

Mutant Linkage Group		Phenotypic Expression Relevant Map Distances	Reference
Amino Acids			
arg-1	UV	Arginine requirement	16
Group I		0.1 units centromere	17
arg-2	UV	Arginine requirement	22
Group unknown			
met-1	UV	Methionine requirement	16
Group III		4.2 units *pdx-1*, 2.0 units *sup-4*, 32 units *sup-3*	5,17
met-2	UV	Methionine requirement	44
Group I		6 units *pab-1*, 2 units *met-6*	3
met-3	UV	Methionine requirement	54
Group III		0.14 units *met-1*	17
met-4	UV	Methionine requirement	21
Group unknown			
met-5	UV	Methionine requirment	16
Group I		27 units *met-6*	3
met-6	UV	Methionine requirement	16
Group I		2 units *met-2*	3
met-7	UV	Methionine requirement	45
Group III		0.27 units *met-3*, 0.14 units *met-1*	17
met-8	UV	Methionine requirement	16
Group V		34 units *ad-6*, 14.5 units centromere	17
met-9	UV	Methionine requirement	45
Group IV		7.7 units centromere	17
trp-1	UV	Tryptophan requirement	41
Group III		0.5 units *met-1*	41
trp-2	UV	Tryptophan requirement	41
Group G		16 units proximal to *ad-3*	41,9
trp-3	UV	Tryptophan requirement	41
Group III		9.5 units *me-1* on *acu-8* arm	41
trp-4	UV	Tryptophan requirement	41
Group H		5.8 units *iar-2*; 34.9 *iar-3*	51,57
Nucleic Acid Bases			
ad-1	UV	Adenine requirement	45,54
Group IV		18.5 units *nic-4*	17
ad-2	UV	Adenine requirement	16,44,45
Group III		2.3 units *nic-1*	17
ad-3	UV	Adenine requirement	44,45,54
Group G		29 units centromere, 16 units distal to trp-2	17,9
ad-4	UV	Adenine requirement	44
Group unknown			
ad-5	UV	Adenine requirement	44,45
Group II		11.7 units *cho-1*, 18.2 units mating type B	44
ad-6	UV	Adenine requirement	16
Group V		22 units centromere	17

Table 2 continued

Mutant Linkage Group		Phenotypic Expression Relevant Map Distances	Reference
ad-8	UV	Adenine requirement	16,45,54
Group I		1.3 units mating type A,	
		12.7 units centromere	3
ad-9	spon	Adenine requirement	54
Group I		9 units mating type A, 25 units *eth*^{r}*-1*	54,56
ad-12	spon	Adenine requirement	47
Group III		12-15 units *den*, 0.2 units *ad-2*	47,48,49
ad/his-1	spon	Double requirement, adenine and histidine	47
Group II		Very close to *ad-5* and *ad/his-2*	47
ad/his-2	UV	Double requirement, adenine and histidine	44
Group II		Very close to *ad-5* and *ad/his-1*	47
ad/met	spon	Alternative requirement for adenine or methionine	46
Group VI		4.7 units *pab-2*	46
Vitamins			
cho-1	UV	Choline requirement	3
Group II		30.2 units mating type B	3
cho-2	UV	Choline requirement	3
Group II		Variable linkage (inversion)	17
nic-1		Nicotinic acid requirement	44
Group III		2.3 units *ad-2*, 12.8 units *pdx-1*	17
nic-4	UV	Nicotinic acid requirement	45
Group IV		8.5 units centromere	17
pab-1	UV	*para*-Aminobenzoic acid requirement	16
Group I		6.0 units *met-2*, 0.5 units mating type A	3
pab-2	UV	*para*-Amimobenzoic acid requirement	45
Group VI		9.5 units centromere	17
pdx-1	UV	Pyridoxine requirement	45
Group III		4.3 units *met-1*, 12.8 units *nic-1*	17
Carbon Source			
acu-1	NTG,UV	Nonutilization of acetate	42
Group I	spon	5 units *ad-8*, 13-17 units *oid-1*	42
acu-2	NTG	Nonutilization of acetate	42
Group G		29 units *ad-3*, 2 units *acu-3*	42
acu-3	NTG,UV	Nonutilization of acetate	42
Group G		25-28 units *ad-3*	42
acu-5	NTG	Nonutilization of acetate	42
Group III		9 units *met-1*	42
acu-6	NTG	Nonutilization of acetate	42
Group III		5.5 units *met-1*	42
acu-7	UV	Nonutilization of acetate	42
Group H		7.8 units *iar-2*	51,57
acu-8	UV	Nonutilization of acetate	42
Group III		17.7 units *met-1*	42
acu-11	spon	Resistance to fluoroacetate	42
Group IV		22.2 units *nic-4*, 38 units *met-9*	42
ftr-1	spon,NTG	Fructose transport	19
Group VII		11.4 units centromere	19

Mutant Linkage Group		Phenotypic Expression Relevant Map Distances	Reference
Phosphatase			
pho-1	UV	Lack of alkaline phosphatase	15
Group unknown		No linkage	
pho-2	UV	Lack of alkaline phosphatase	15
Group I		40 units mating type A, free	
		recombination *acu-1*	15
reg-1	UV	Regulator of alkaline phosphatase	15
		27.5 units *mrp-1*	15
reg-2	UV	Regulator of alkaline phosphatase	15
Group unknown		5 units *reg-5*	15
reg-3,4	UV	Regulator of alkaline phosphatase	15
Group G		23 units *ad-3*, 5 units *acu-3*	15
reg-5	UV	Regulator of alkaline phosphatase	15
Group unknown		5 units *reg-2*	15
gp^s*-1*	wild	Inhibition by ß-glycerophosphate	15
Group VI		15 units *gp*^{s}*-4*, 18.0 units *pab-2*	15
gp^s*-4*	spon	Inhibition by ß-glycerophosphate	15
Group VI		3 units *pab-2*	15
Suppressors			
sup-3	spon,UV	Suppression of *met-1*	5
Group III		28 units *met-1*	5,17
sup-4	spon,UV	Suppression of *met-1*	5
Group III		1.5 units *met-1*	5,17
Color			
pur-3	ß-PL	Secretion of purple quinone-type pigment	20,44
Group I		Close to mating type A	20
Resistance to Amino Acid and other Analogues			
can^r*-1*	wild	Resistance to 10^{-4}M L-canavanine	29
Group I		39 units *eth*^{r}*-1*, 50 units *ad-8*	29
can^r*-2*	wild	Resistance to 10^{-4}M L-canavanine	29
Group I		26.5 units from *ad-8*; unlinked *me-5*	29
can^r*-3*	wild	Resistance to 10^{-4}M L-canavanine	29
Group I		14.0 units *can*^{r}*-1*	29
eth^r*-1*	UV,spon	Resistance to 10^{-3}M L-ethionine	
	wild	(methionyl tRNA synthetase)	4,54
Group I		10 units *met-2*, 9-18 units mating type A	4
eth^r*-1*	UV	Resistance to 10^{-3}M L-ethionine	54
Group I		24 units *ad-8*, 50 units *eth*^{r}*-1*	54
eth^r*-3*	spon	Resistance to 10^{-3}M L-ethionine	54
Group I		6 units mating type A, 4 units *pfp*^{r}*-5*,	
		7 units *eth*^{r}*-1*	53,54

Mutant Linkage Group		Phenotypic Expression Relevant Map Distances	Reference
eth^r-4	UV	Resistance to 10^{-3}M L-ethionine	54
Group I		32 units mating type A, 0.5 units *met-5*, 8 units *pfp^r-6*	53,54
eth^r-5	UV	Resistance to 10^{-3}M L-ethionine	54
Group G		22 units *ad-3*, 13.5 units *pfp^r-7*	53,54
eth^r-6	spon	Resistance to 10^{-3}M L-ethionine	54
Group V		13 units *pfp^r-8*	53,54
eth^r-8	UV	Resistance to 10^{-3}M L-ethionine	58
Group unknown			
eth^r-9	UV	Resistance to 10^{-3}M L-ethionine	58
Group II		12.5 units distal to *L-2*	58
eth-x	UV	Complementary gene to *eth^r-8*	58
Group II		26.5 units distal to *adhi-1*	58
pfp^r-1	UV	Resistance to 10^{-4}M DL-*para*-fluorophenylalanine (permease)	54
Group I		2 units mating type A, 16 units *eth^r-1*	54
pfp^r-1	UV,wild	Resistance to 10^{-4}M PFP (permease)	52,55
Group I		4 units mating type A, 6 units *pfp^r-1*	52,54
pfp^r-3		Resistance to 10^{-4}M PFP, semi-dominant,	51,52
Group possibly VI		overproduction of aromatic compounds	
pfp^r-4	spon	Resistance to 10^{-4}M PFP (phenylalanine	
Group unknown		tRNA synthetase)	52,55
pfp^r-5	spon	Resistance to 10^{-4}M PFP	54
Group I		9 units mating type A, 3.4 units *eth^r-3*	53,54
pfp^r-6	UV	Resistance to 10^{-4}M PFP	54
Group I		8-16 units *eth^r-4*	53,54
pfp^r-7	UV	Resistance to 10^{-4}M PFP	54
Group G		12 units *ad-3*	53,54
pfp^r-8	spon	Resistance to 10^{-4}M PFP	54
Group V		13 units *eth^r-6*, 18 units *ad-6*	53,54
pfp^r-10	UV	Resistance to 10^{-4}M PFP	29
Group II		4.1 units proximal to mating type B	58
pfp^r-11	UV	Resistance to 10^{-4}M PFP	29
Group II		33.7 units distal to mating type B	29
pfp^r-12	UV	Resistance to 10^{-4}M PFP	58
Group II		Very close to *pfp^r-10*	58
pfp^r-13	UV	Resistance to 10^{-4}M PFP	58
Group II		11.0 units mating type B on *cho-1* arm	58
mod-10	wild	Modifies dominance of *pfp^r-10*	29
Group unknown			
mod-11	wild	Modifies dominance of *pfp^r-11*	29
Group II		20 units *pfp^r11*	29,58
L-1	UV	Lethal gene associated with dominant PFP resistance (*pfp^r-12)*	58
Group II		32.0 units distal to mating type B	58
L-2	UV	Lethal gene associated with dominant PFP resistance (*pfp^r-13)*	58
Group II		35.0 units distal to mating type B	58
L-3	UV	Lethal gene associated with dominant PFP resistance (*pfp^r-10*)	29
Group I		6.6 units distal to mating type A	56
iar-2	UV,spon	Indole analogue resistance	51,57
Group H		5.8 units *trp-4*; 7.8 units *acu-7*	51,57
iar-3	UV,spon	Indole analogue resistance	51,57
Group H		34.9 *trp-4* away from *acu-7*	51,57
iar-4	UV,spon	Indole analogue resistance	51,57
Group III		12.6 units *me-1*; 4.2 units *ad-2*	51,57
Resistance to Antibiotics			
cy-1	wild	Resistance to 3.6 μM cycloheximide	30
Group VII		36.5 units *ftr*	30
cy-2	UV	Resistance to 3.6 μM cycloheximide	30
Group II		27.9 units distal to *cho-1*	30
modcy	UV	Modifies dominance of *cy-2*	30
Group II		27.0 units distal to *cy-2*	30
Mating type			
A	wild	Mating type A	
Group I		15.0 units centromere	17
B	wild	Mating type B	
Group II		8.5 units centromere	17
Ribosomal RNA			
RDN-1	spon	Ribosomal RNA gene cluster	39
Group unknown (J) not I,II,III,IV or G		3.0 units centromere	39

Abbreviations: spon = spontaneous; UV = ultraviolet light; NTG = nitrosoguanidine; PL = propiolactate.

EFERENCES

ublished Data

1. Lange, J.E., *Flora Agaricina Danica, Copenhagen, 1-5* (1935-1940).
2. Orton, P.D., *Trans. Br. Mycol. Soc., 40,* 263 (1957).
3. Day, P.R., and Anderson, G.E., *Genet. Res., 2,* 414 (1961).
4. Lewis, D., *Nature 200,* 151 (1963).
5. Lewis, D., *Genet. Res., 2,* 141 (1961).
6. Lu, B.C., and Raju, N.B., *Chromosoma, 29,* 305-316 (1970).
7. Cowan, J.W., *Nature, 204,* 1113 (1964).
8. Casselton, L.A. and Economou, A., p.213-219 *in* Developmental Biology of Higher Fungi, eds. Moore, D. *et al.* Cambridge University Press, Cambridge, UK (1985).
9. Amirkhanian, J.D., and Cowan, J.W., *Journal of Heredity 76,* 348-354 (1985).
0. Swiezynski, K.M., and Day, P.R., *Genet. Res., 1,* 114 (1960).
1. Casselton, L.A., *Genet. Res., 6,* 190 (1965).
2. Casselton, L.A., and Lewis, D., *Genet. Res., 9,* 63 (1967).
3. Casselton, L.A., and Casselton, P.J., *Trans. Br. Mycol. Soc., 49,* 569 (1966).
4. Kaplan, P.J., and Walls, D., *Genet. Res., 17,* 279 (1971).
5. North, J., and Lewis, D., *Genet. Res., 18,* 153 (1971).
6. Anderson, G.E., *Heredity, 13,* 411 (1959).
7. Moore, D., *Genet. Res., 9,* 331 (1967).
8. Moore, D., *J. Gen. Microbiol., 74,* 163 (1967).
9. Moore, D., and Stewart, G.R., *Genet. Res., 18,* 341 (1971).
0. Morgan, D.H., *Genet. Res., 7,* 195 (1966).
1. Chang-Ho, Y., and Yee, N.T., *Trans. Br. Mycol. Soc., 68,* 167-172 (1977).
2. North, J., *Trans. Br. Mycol. Soc., 75,* 161-162 (1980).
3. Morimoto, N., Suda, S., and Sagara, N., *Plant and Cell Physiology, 22,* 247-252 (1981).
4. Takemaru, T., p.356, *in* Experimental Methods in Microbial Genetics (ed. Ishikawa, T.); Kyoritsu Publishing Co. Ltd., Tokyo, Japan (1982).
5. Kemp, R.F.O., *Trans. Br. Mycol. Soc., 65,* 375-388 (1975).
6. Moore, D., Elhiti, M.M.Y., and Butler, R.D., *New Phytologist, 83,* 695-722 (1979).
7. Waters, H., Moore, D., and Butler, R.D., *New Phytologist, 74,* 207-213 (1979).
8. Hereward, F.V., and Moore, D., *J. Gen. Microbiol. 113,* 13-18 (1979).
9. SenathiRajah, S., and Lewis, D., *Genet. Res., 25,* 95-107 (1975).
0. North, J., *J. Gen. Microbiol., 128,* 2747-2753 (1982).
1. Wu, M.M.J., Cassidy, J.R., and Pukkila, P.J., *Current Genetics, 7,* 385-392 (1983).
32. Rao, P.S., and Niederpruem, D.J., *J. of Bacteriology, 100,* 1222-1228 (1969).
33. Raju, N.B., and Lu, B.C., *Canadian Journal of Botany, 48,* 2183-2186 (1970).
34. Pukkila, P.J., Binninger, D.M., Cassidy, J.R., Yashar, B.M. and Zolan, M.E., p.499-512 *in* Developmental Biology of Higher Fungi, Eds. Moore, D. *et al.* Cambridge University Press, Cambridge, UK. (1985).
35. North, J., *J. Gen. Microbiol., 98,* 529-534 (1977).
36. Takemaru, T., Kamada, T., and Murakami, S., *Rept. Tottori. Mycol. Inst. (Japan), 10,* 377-382 (1973).
37. Takemaru, T., and Kamada, T., *Rept. Tottori. Mycol. Inst. (Japan), 9,* 21-35 (1971).
38. Lewis, D., and North, J., p.691-699, *in* Handbook of Microbiology *4* (ed. Laskin, A.I., and Lechevalier, H.A.) CRC Press, Cleveland, Ohio (1974).
39. Cassidy, J.R., Moore, D., Lu, B.C., and Pukkila, P.J., *Current Genetics, 8,* 607-613 (1984).
40. Lu, B.C., *Canadian Journal of Genetics and Cytology, 11,* 834-847 (1969).
41. Tilby, M.J., *J. Gen. Microbiol., 93,* 126-132 (1976).
42. King, H.R., and Casselton, L.A., *Mol. and Gen. Genetics, 157,* 319-325 (1977).
43. Kimura, K., *Botanical Magazine, 65,* 232-235 (1952).

Unpublished Data

44. Day, P.R., Plant Breeding Institute, Cambridge, UK.
45. Morgan, D.H., John Innes Institute, Norwich, Norfolk, UK.
46. North, J., Department of Biological Sciences, City of London Polytechnic, London, UK.
47. Cowan, J.W., Department of Botany, King's College, London, UK.
48. Greenaway, W., Department of Botany, King's College, London, UK.
49. Rahman, M.A., Department of Botany, King's College, London, UK.
50. Moore, D., Department of Botany, University of Manchester, Manchester, UK.
51. Casselton, L.A., School of Biological Sciences, Queen Mary College, London, UK.
52. Barker, C., University College, London, UK.
53. Talmud, P.J., Department of Biochemistry, Queen Mary's Hospital Medical School, London, UK.
54. Lewis, D., School of Biological Sciences, Queen Mary College, London, UK.
55. Maggs, G., University College, London, UK.
56. SenathiRajah, S., Thesis, University of London, UK (1973).
57. Veal, D., School of Biological Sciences, Queen Mary College, London, UK.
58. Jehan, M., Thesis, University of London, UK (1979).

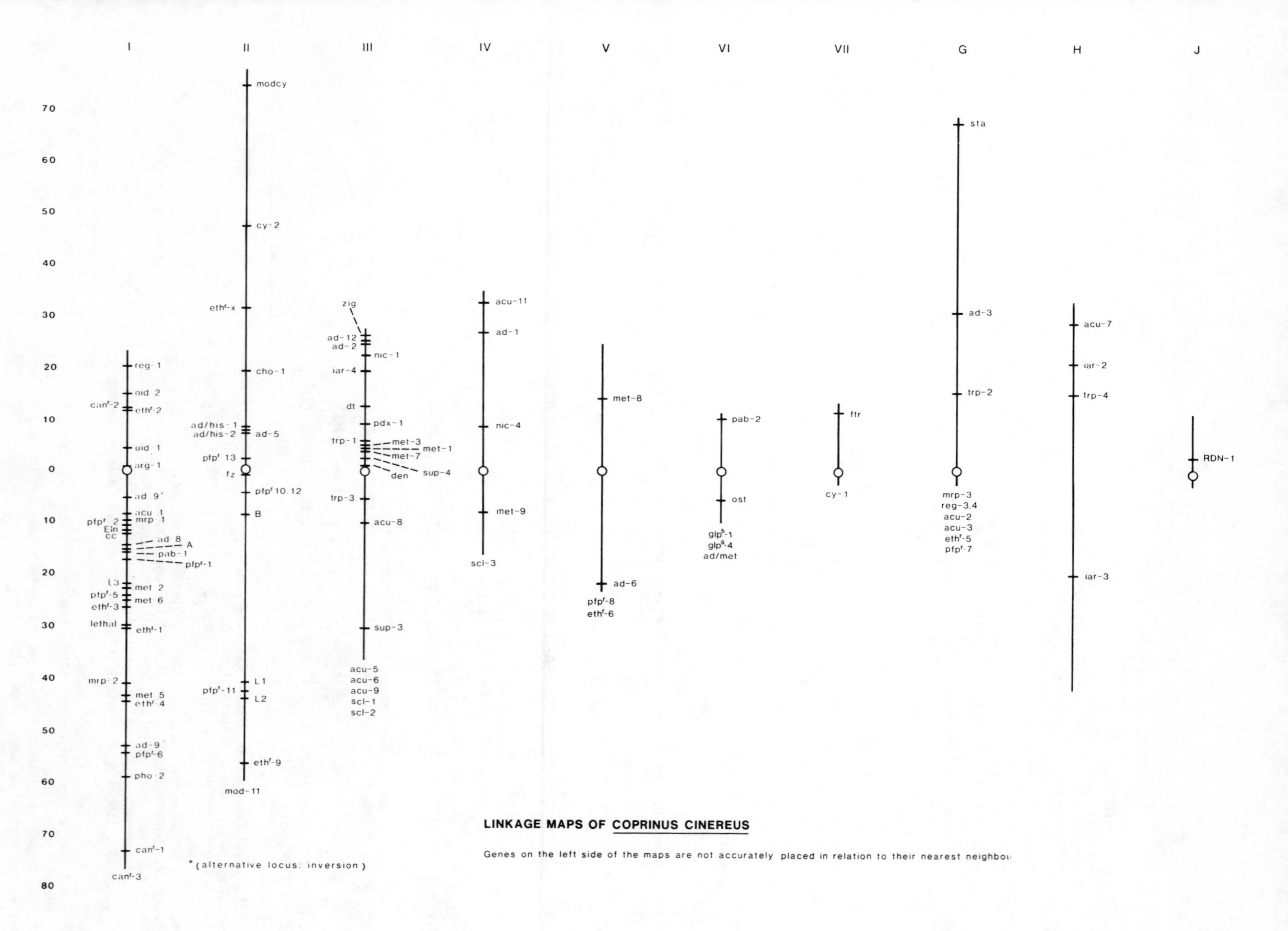

LINKAGE MAPS OF COPRINUS CINEREUS

Genes on the left side of the maps are not accurately placed in relation to their nearest neighbou

Linkage map of *Coprinus cinereus* (Schaeff. ex Fr.) S.F. Gray,
Ink cap. N = 12 (ref 6)

Date compiled: October 1986

Compiled by: Dr. Jane North,
Department of Biological Sciences, City of London Polytechnic,
Calcutta House, Old Castle Street, London E1 7NT, UK.

A NOTE ON THE NOMENCLATURE AND LIFE CYCLE OF THE SPECIES

Coprinus cinereus is the name now used for cultures isolated from leaf litter, manure heaps, mushroom compost and enriched straw throughout the world, and identified variously as *C. lagopus*, *C. cinereus*, *C. fimentarius and C. macrorhizus f. microsporus* Hongo(1-3, 21-3, 43). These strains, including a number held by the ATCC, are interfertile(25,26,50) and are typical of the species *C. cinereus* which forms a unique type of microsclerotium in culture(25). A genetic map of *C. macrorhizus* has been published in Japan(24); where possible, markers from this have been incorporated into figure.

C. cinereus shows considerable polymorphism with respect to morphogenesis(27,28) and biochemistry(4,15,29,30). Variation in restriction enzyme cleavage sites in mitochondrial DNA has been reported(8); the nuclear DNA has extensive insertion/deletion polymorphism within single-copy sequences(31).

The life cycle is typical of the agaricales(5,38). Forced heterokaryons(10) and stable diploids(11) can be made, allowing studies of gene expression in different cell-types(12,29,30). Dikaryons can be resolved directly from chlamydospores(5) or veil cells(7). Mitotic recombination can be induced in diploids by ultraviolet light (9); haploidisation of diploids can be achieved using griseofulvin(35).

Mutants are normally selected from the uninucleate oidia. A replica-plating technique using Velcro as a transfer material is available(14).

ACKNOWLEDGEMENTS

The author wishes to thank all who have so generously provided unpublished data and expresses her gratitude to Dr. D. Moore and Dr. P.R. Day for the use of their linkage maps, on which the mapping in *Coprinus cinereus* is largely based.

TABLE 1 MORPHOLOGICAL MUTANTS, LINKAGE GROUPS, AND RELEVANT MAP DISTANCES

Mutant Linkage Group		Phenotypic Expression Relevant Map Distances	Reference
den	spon	Thick, undulating hyphae, excessively branched; stunted growth	7
Group III		2.4 units *met-1*, 14 units *ad-12*	40,47,48
mrp-1	UV	Dense, compact colony	15
Group I		4.5 units mating type A, 27.5 units *reg-1*	15
mrp-2	spon	Thin colony with "wet" center of collapsed aerial hyphae	54
Group I		23 units mating type A, 15 units *eth*^{r}*-3*	54
mrp-3	spon	Normal, but half growth rate	54
Group G		5 units *eth*^{r}*-5*	53,54
oid-1	UV	Excess production of oidia	44
Group I		27 units mating type A, 17.3 units *ad-8*, 15 units *acu-1*	44,51
oid-2	UV	Excessive production of oidia	51
Group I		27.4 units *ad-8*, 21 units *acu-1*	51
ost	NTG	Ostrich-like, feathery mycelium	19
Group VI		15.5 units *pab-2*	19
scl-1	wild	Production of sclerotia	27
Group III		28.6 units *scl-2*	27
scl-2	wild	Production of sclerotia	28
Group III		26.2 units *pdx*	28
scl-3	wild	Production of sclerotia	28
Group IV		28.1 units *nic-4*	28
cc	spon	Concentric	36
Group I		3.2 units proximal* to mating type A	24,36
fz	UV	Frizzy	36
Group II		8.9 units from mating type B, probably on centromere*	24
dt	UV	Dichotomous growth	36
Group III		3.5 units *pdx*	24
Eln	UV,NTG	Elongation of basidiocarp stipe	37
Group I		3.8 units proximal* to mating type A	24
zig	TEPA	Zig-zag branched mycelium	9
Group III		1.8 units distal to ad-2	9
sta	TEPA	Star-shaped colonies	9
Group G		37.7 units distal to ad-3	9

*inferred from relationship between maps in refs.24 and 38.

Abbreviations: spon = spontaneous; UV = ultraviolet light; NTG = nitrosoguanidine; TEPA = tris, 1-azindinyl phosphine oxide

TABLE 2
AUXOTROPHIC AND BIOCHEMICAL MUTANTS, LINKAGE GROUPS, AND RELEVANT MAP DISTANCES

Mutant Linkage Group		Phenotypic Expression Relevant Map Distances	Reference
Amino Acids			
arg-1	UV	Arginine requirement	16
Group I		0.1 units centromere	17
arg-2	UV	Arginine requirement	22
Group unknown			
met-1	UV	Methionine requirement	16
Group III		4.2 units *pdx-1*, 2.0 units *sup-4*, 32 units *sup-3*	5,17
met-2	UV	Methionine requirement	44
Group I		6 units *pab-1*, 2 units *met-6*	3
met-3	UV	Methionine requirement	54
Group III		0.14 units *met-1*	17
met-4	UV	Methionine requirement	21
Group unknown			
met-5	UV	Methionine requirment	16
Group I		27 units *met-6*	3
met-6	UV	Methionine requirement	16
Group I		2 units *met-2*	3
met-7	UV	Methionine requirement	45
Group III		0.27 units *met-3*, 0.14 units *met-1*	17
met-8	UV	Methionine requirement	16
Group V		34 units *ad-6*, 14.5 units centromere	17
met-9	UV	Methionine requirement	45
Group IV		7.7 units centromere	17
trp-1	UV	Tryptophan requirement	41
Group III		0.5 units *met-1*	41
trp-2	UV	Tryptophan requirement	41
Group G		16 units proximal to *ad-3*	41,9
trp-3	UV	Tryptophan requirement	41
Group III		9.5 units *me-1* on *acu-8* arm	41
trp-4	UV	Tryptophan requirement	41
Group H		5.8 units *iar-2*; 34.9 *iar-3*	51,57
Nucleic Acid Bases			
ad-1	UV	Adenine requirement	45,54
Group IV		18.5 units *nic-4*	17
ad-2	UV	Adenine requirement	16,44,45
Group III		2.3 units *nic-1*	17
ad-3	UV	Adenine requirement	44,45,54
Group G		29 units centromere, 16 units distal to trp-2	17,9
ad-4	UV	Adenine requirement	44
Group unknown			
ad-5	UV	Adenine requirement	44,45
Group II		11.7 units *cho-1*, 18.2 units mating type B	44
ad-6	UV	Adenine requirement	16
Group V		22 units centromere	17

Table 2 continued

Mutant Linkage Group		Phenotypic Expression Relevant Map Distances	Reference
ad-8	UV	Adenine requirement	16,45,54
Group I		1.3 units mating type A,	
		12.7 units centromere	3
ad-9	spon	Adenine requirement	54
Group I		9 units mating type A, 25 units *eth*^{r}*-1*	54,56
ad-12	spon	Adenine requirement	47
Group III		12-15 units *den*, 0.2 units *ad-2*	47,48,49
ad/his-1	spon	Double requirement, adenine and histidine	47
Group II		Very close to *ad-5* and *ad/his-2*	47
ad/his-2	UV	Double requirement, adenine and histidine	44
Group II		Very close to *ad-5* and *ad/his-1*	47
ad/met	spon	Alternative requirement for adenine or methionine	46
Group VI		4.7 units *pab-2*	46
Vitamins			
cho-1	UV	Choline requirement	3
Group II		30.2 units mating type B	3
cho-2	UV	Choline requirement	3
Group II		Variable linkage (inversion)	17
nic-1		Nicotinic acid requirement	44
Group III		2.3 units *ad-2*, 12.8 units *pdx-1*	17
nic-4	UV	Nicotinic acid requirement	45
Group IV		8.5 units centromere	17
pab-1	UV	*para*-Aminobenzoic acid requirement	16
Group I		6.0 units *met-2*, 0.5 units mating type A	3
pab-2	UV	*para*-Amimobenzoic acid requirement	45
Group VI		9.5 units centromere	17
pdx-1	UV	Pyridoxine requirement	45
Group III		4.3 units *met-1*, 12.8 units *nic-1*	17
Carbon Source			
acu-1	NTG,UV	Nonutilization of acetate	42
Group I	spon	5 units *ad-8*, 13-17 units *oid-1*	42
acu-2	NTG	Nonutilization of acetate	42
Group G		29 units *ad-3*, 2 units *acu-3*	42
acu-3	NTG,UV	Nonutilization of acetate	42
Group G		25-28 units *ad-3*	42
acu-5	NTG	Nonutilization of acetate	42
Group III		9 units *met-1*	42
acu-6	NTG	Nonutilization of acetate	42
Group III		5.5 units *met-1*	42
acu-7	UV	Nonutilization of acetate	42
Group H		7.8 units *iar-2*	51,57
acu-8	UV	Nonutilization of acetate	42
Group III		17.7 units *met-1*	42
acu-11	spon	Resistance to fluoroacetate	42
Group IV		22.2 units *nic-4*, 38 units *met-9*	42
ftr-1	spon,NTG	Fructose transport	19
Group VII		11.4 units centromere	19

Mutant Linkage Group		Phenotypic Expression Relevant Map Distances	Reference
Phosphatase			
pho-1	UV	Lack of alkaline phosphatase	15
Group unknown		No linkage	
pho-2	UV	Lack of alkaline phosphatase	15
Group I		40 units mating type A, free	
		recombination *acu-1*	15
reg-1	UV	Regulator of alkaline phosphatase	15
		27.5 units *mrp-1*	15
reg-2	UV	Regulator of alkaline phosphatase	15
Group unknown		5 units *reg-5*	15
reg-3,4	UV	Regulator of alkaline phosphatase	15
Group G		23 units *ad-3*, 5 units *acu-3*	15
reg-5	UV	Regulator of alkaline phosphatase	15
Group unknown		5 units *reg-2*	15
gp^s*-1*	wild	Inhibition by β-glycerophosphate	15
Group VI		15 units *gp*^{s}*-4*, 18.0 units *pab-2*	15
gp^s*-4*	spon	Inhibition by β-glycerophosphate	15
Group VI		3 units *pab-2*	15
Suppressors			
sup-3	spon,UV	Suppression of *met-1*	5
Group III		28 units *met-1*	5,17
sup-4	spon,UV	Suppression of *met-1*	5
Group III		1.5 units *met-1*	5,17
Color			
pur-3	β-PL	Secretion of purple quinone-type pigment	20,44
Group I		Close to mating type A	20
Resistance to Amino Acid and other Analogues			
can^r*-1*	wild	Resistance to 10^{-4}M L-canavanine	29
Group I		39 units *eth*^{r}*-1*, 50 units *ad-8*	29
can^r*-2*	wild	Resistance to 10^{-4}M L-canavanine	29
Group I		26.5 units from *ad-8*; unlinked *me-5*	29
can^r*-3*	wild	Resistance to 10^{-4}M L-canavanine	29
Group I		14.0 units *can*^{r}*-1*	29
eth^r*-1*	UV,spon	Resistance to 10^{-3}M L-ethionine	
	wild	(methionyl tRNA synthetase)	4,54
Group I		10 units *met-2*, 9-18 units mating type A	4
eth^r*-1*	UV	Resistance to 10^{-3}M L-ethionine	54
Group I		24 units *ad-8*, 50 units *eth*^{r}*-1*	54
eth^r*-3*	spon	Resistance to 10^{-3}M L-ethionine	54
Group I		6 units mating type A, 4 units *pfp*^{r}*-5*,	
		7 units *eth*^{r}*-1*	53,54

Mutant Linkage Group		Phenotypic Expression Relevant Map Distances	Reference
eth^r-*4*	UV	Resistance to 10^{-3}M L-ethionine	54
Group I		32 units mating type A, 0.5 units *met-5*, 8 units pfp^r-*6*	53,54
eth^r-*5*	UV	Resistance to 10^{-3}M L-ethionine	54
Group G		22 units *ad-3*, 13.5 units pfp^r-*7*	53,54
eth^r-*6*	spon	Resistance to 10^{-3}M L-ethionine	54
Group V		13 units pfp^r-*8*	53,54
eth^r-*8*	UV	Resistance to 10^{-3}M L-ethionine	58
Group unknown			
eth^r-*9*	UV	Resistance to 10^{-3}M L-ethionine	58
Group II		12.5 units distal to *L-2*	58
eth-x	UV	Complementary gene to eth^r-*8*	58
Group II		26.5 units distal to *adhi-1*	58
pfp^r-*1*	UV	Resistance to 10^{-4}M DL-*para*-fluorophenylalanine (permease)	54
Group I		2 units mating type A, 16 units eth^r-*1*	54
pfp^r-*1*	UV,wild	Resistance to 10^{-4}M PFP (permease)	52,55
Group I		4 units mating type A, 6 units pfp^r-*1*	52,54
pfp^r-*3*		Resistance to 10^{-4}M PFP, semi-dominant,	51,52
Group possibly VI		overproduction of aromatic compounds	
pfp^r-*4*	spon	Resistance to 10^{-4}M PFP (phenylalanine	
Group unknown		tRNA synthetase)	52,55
pfp^r-*5*	spon	Resistance to 10^{-4}M PFP	54
Group I		9 units mating type A, 3.4 units eth^r-*3*	53,54
pfp^r-*6*	UV	Resistance to 10^{-4}M PFP	54
Group I		8-16 units eth^r-*4*	53,54
pfp^r-*7*	UV	Resistance to 10^{-4}M PFP	54
Group G		12 units *ad-3*	53,54
pfp^r-*8*	spon	Resistance to 10^{-4}M PFP	54
Group V		13 units eth^r-*6*, 18 units *ad-6*	53,54
pfp^r-*10*	UV	Resistance to 10^{-4}M PFP	29
Group II		4.1 units proximal to mating type B	58
pfp^r-*11*	UV	Resistance to 10^{-4}M PFP	29
Group II		33.7 units distal to mating type B	29
pfp^r-*12*	UV	Resistance to 10^{-4}M PFP	58
Group II		Very close to pfp^r-*10*	58
pfp^r-*13*	UV	Resistance to 10^{-4}M PFP	58
Group II		11.0 units mating type B on *cho-1* arm	58
mod-10	wild	Modifies dominance of pfp^r-*10*	29
Group unknown			
mod-11	wild	Modifies dominance of pfp^r-*11*	29
Group II		20 units pfp^r*11*	29,58
L-1	UV	Lethal gene associated with dominant PFP resistance (pfp^r-*12)*	58
Group II		32.0 units distal to mating type B	58
L-2	UV	Lethal gene associated with dominant PFP resistance (pfp^r-*13)*	58
Group II		35.0 units distal to mating type B	58
L-3	UV	Lethal gene associated with dominant PFP resistance (pfp^r-*10*)	29
Group I		6.6 units distal to mating type A	56
iar-2	UV,spon	Indole analogue resistance	51,57
Group H		5.8 units *trp-4*; 7.8 units *acu-7*	51,57
iar-3	UV,spon	Indole analogue resistance	51,57
Group H		34.9 *trp-4* away from *acu-7*	51,57
iar-4	UV,spon	Indole analogue resistance	51,57
Group III		12.6 units *me-1*; 4.2 units *ad-2*	51,57
Resistance to Antibiotics			
cy-1	wild	Resistance to 3.6 μM cycloheximide	30
Group VII		36.5 units *ftr*	30
cy-2	UV	Resistance to 3.6 μM cycloheximide	30
Group II		27.9 units distal to *cho-1*	30
modcy	UV	Modifies dominance of *cy-2*	30
Group II		27.0 units distal to *cy-2*	30
Mating type			
A	wild	Mating type A	
Group I		15.0 units centromere	17
B	wild	Mating type B	
Group II		8.5 units centromere	17
Ribosomal RNA			
RDN-1	spon	Ribosomal RNA gene cluster	39
Group unknown (J) not I,II,III,IV or G		3.0 units centromere	39

Abbreviations: spon = spontaneous; UV = ultraviolet light; NTG = nitrosoguanidine; PL = propiolactate.

REFERENCES

Published Data

1. Lange, J.E., *Flora Agaricina Danica, Copenhagen, 1-5* (1935-1940).
2. Orton, P.D., *Trans. Br. Mycol. Soc.*, *40*, 263 (1957).
3. Day, P.R., and Anderson, G.E., *Genet. Res.*, *2*, 414 (1961).
4. Lewis, D., *Nature 200*, 151 (1963).
5. Lewis, D., *Genet. Res.*, *2*, 141 (1961).
6. Lu, B.C., and Raju, N.B., *Chromosoma*, *29*, 305-316 (1970).
7. Cowan, J.W., *Nature*, *204*, 1113 (1964).
8. Casselton, L.A. and Economou, A., p.213-219 *in* Developmental Biology of Higher Fungi, eds. Moore, D. *et al*. Cambridge University Press, Cambridge, UK (1985).
9. Amirkhanian, J.D., and Cowan, J.W., *Journal of Heredity 76*, 348-354 (1985).
10. Swiezynski, K.M., and Day, P.R., *Genet. Res.*, *1*, 114 (1960).
11. Casselton, L.A., *Genet. Res.*, *6*, 190 (1965).
12. Casselton, L.A., and Lewis, D., *Genet. Res.*, *9*, 63 (1967).
13. Casselton, L.A., and Casselton, P.J., *Trans. Br. Mycol. Soc.*, *49*, 569 (1966).
14. Kaplan, P.J., and Walls, D., *Genet. Res.*, *17*, 279 (1971).
15. North, J., and Lewis, D., *Genet. Res.*, *18*, 153 (1971).
16. Anderson, G.E., *Heredity*, *13*, 411 (1959).
17. Moore, D., *Genet. Res.*, *9*, 331 (1967).
18. Moore, D., *J. Gen. Microbiol.*, *74*, 163 (1967).
19. Moore, D., and Stewart, G.R., *Genet. Res.*, *18*, 341 (1971).
20. Morgan, D.H., *Genet. Res.*, *7*, 195 (1966).
21. Chang-Ho, Y., and Yee, N.T., *Trans. Br. Mycol. Soc.*, *68*, 167-172 (1977).
22. North, J., *Trans. Br. Mycol. Soc.*, *75*, 161-162 (1980).
23. Morimoto, N., Suda, S., and Sagara, N., *Plant and Cell Physiology*, *22*, 247-252 (1981).
24. Takemaru, T., p.356, *in* Experimental Methods in Microbial Genetics (ed. Ishikawa, T.); Kyoritsu Publishing Co. Ltd., Tokyo, Japan (1982).
25. Kemp, R.F.O., *Trans. Br. Mycol. Soc.*, *65*, 375-388 (1975).
26. Moore, D., Elhiti, M.M.Y., and Butler, R.D., *New Phytologist*, *83*, 695-722 (1979).
27. Waters, H., Moore, D., and Butler, R.D., *New Phytologist*, *74*, 207-213 (1979).
28. Hereward, F.V., and Moore, D., *J. Gen. Microbiol. 113*, 13-18 (1979).
29. SenathiRajah, S., and Lewis, D., *Genet. Res.*, *25*, 95-107 (1975).
30. North, J., *J. Gen. Microbiol.*, *128*, 2747-2753 (1982).
31. Wu, M.M.J., Cassidy, J.R., and Pukkila, P.J., *Current Genetics*, *7*, 385-392 (1983).
32. Rao, P.S., and Niederpruem, D.J., *J. of Bacteriology*, *100*, 1222-1228 (1969).
33. Raju, N.B., and Lu, B.C., *Canadian Journal of Botany*, *48*, 2183-2186 (1970).
34. Pukkila, P.J., Binninger, D.M., Cassidy, J.R., Yashar, B.M. and Zolan, M.E., p.499-512 *in* Developmental Biology of Higher Fungi, Eds. Moore, D. *et al*. Cambridge University Press, Cambridge, UK. (1985).
35. North, J., *J. Gen. Microbiol.*, *98*, 529-534 (1977).
36. Takemaru, T., Kamada, T., and Murakami, S., *Rept. Tottori. Mycol. Inst. (Japan)*, *10*, 377-382 (1973).
37. Takemaru, T., and Kamada, T., *Rept. Tottori. Mycol. Inst. (Japan)*, *9*, 21-35 (1971).
38. Lewis, D., and North, J., p.691-699, *in* Handbook of Microbiology *4* (ed. Laskin, A.I., and Lechevalier, H.A.) CRC Press, Cleveland, Ohio (1974).
39. Cassidy, J.R., Moore, D., Lu, B.C., and Pukkila, P.J., *Current Genetics*, *8*, 607-613 (1984).
40. Lu, B.C., *Canadian Journal of Genetics and Cytology*, *11*, 834-847 (1969).
41. Tilby, M.J., *J. Gen. Microbiol.*, *93*, 126-132 (1976).
42. King, H.R., and Casselton, L.A., *Mol. and Gen. Genetics*, *157*, 319-325 (1977).
43. Kimura, K., *Botanical Magazine*, *65*, 232-235 (1952).

Unpublished Data

44. Day, P.R., Plant Breeding Institute, Cambridge, UK.
45. Morgan, D.H., John Innes Institute, Norwich, Norfolk, UK.
46. North, J., Department of Biological Sciences, City of London Polytechnic, London, UK.
47. Cowan, J.W., Department of Botany, King's College, London, UK.
48. Greenaway, W., Department of Botany, King's College, London, UK.
49. Rahman, M.A., Department of Botany, King's College, London, UK.
50. Moore, D., Department of Botany, University of Manchester, Manchester, UK.
51. Casselton, L.A., School of Biological Sciences, Queen Mary College, London, UK.
52. Barker, C., University College, London, UK.
53. Talmud, P.J., Department of Biochemistry, Queen Mary's Hospital Medical School, London, UK.
54. Lewis, D., School of Biological Sciences, Queen Mary College, London, UK.
55. Maggs, G., University College, London, UK.
56. SenathiRajah, S., Thesis, University of London, UK (1973).
57. Veal, D., School of Biological Sciences, Queen Mary College, London, UK.
58. Jehan, M., Thesis, University of London, UK (1979).

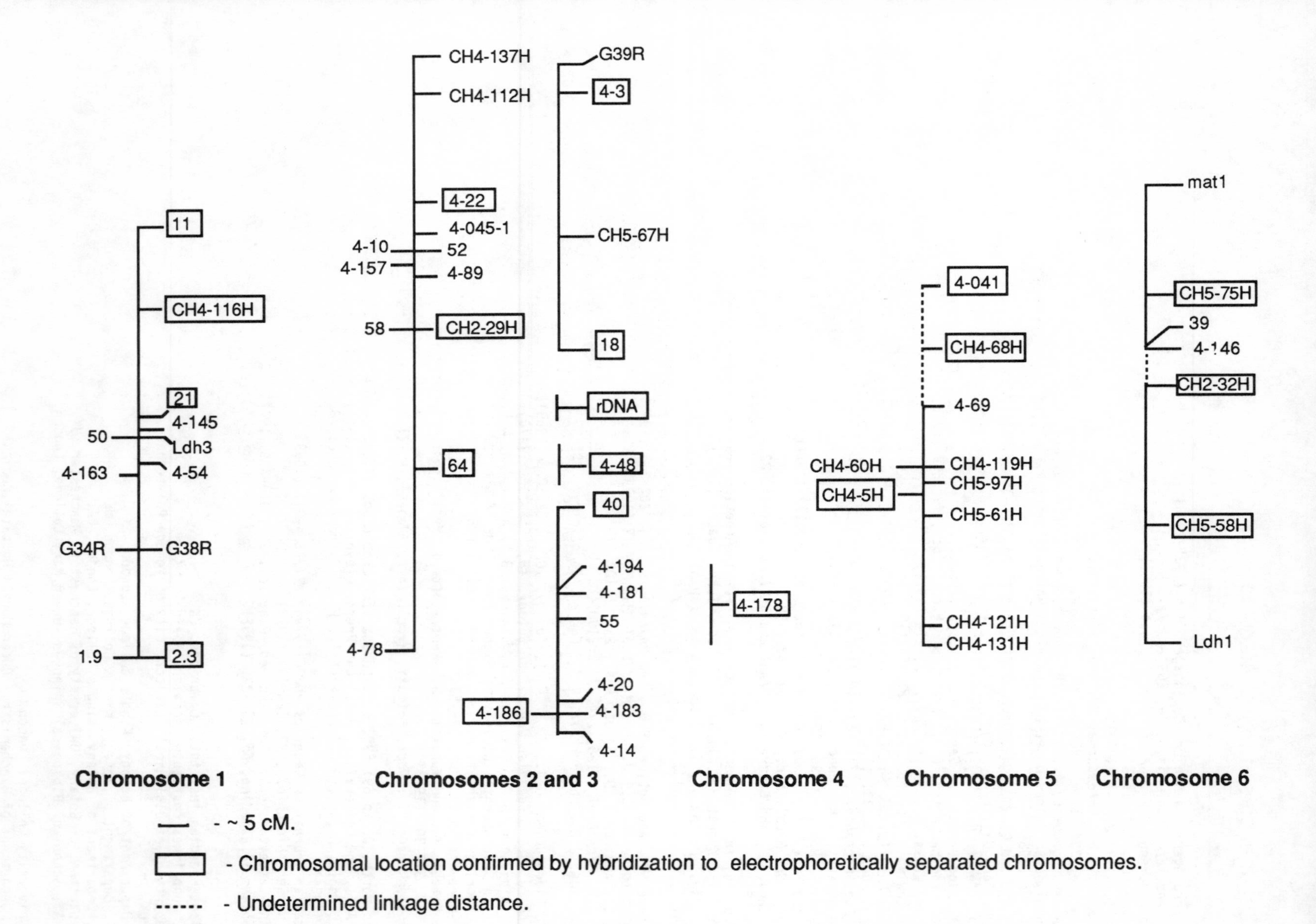

Fig. 1. Genetic map of *Magnaporthe grisea* based on isolates Guy 11 and 2539.

Genetic map of the blast fungus *Magnaporthe grisea* (n=6)

D. Z. Skinner, H. Leung, and S. A . Leong. Dept. of Plant Pathology and USDA-ARS, University of Wisconsin, Madison, WI 53706. USA.

August, 1989.

Magnaporthe grisea (anamorph = *Pyricularia oryzae* and *P. grisea*), causal agent of rice blast disease, is a haploid ascomycete-pyrenomycete with six chromosomes (2). We have constructed a preliminary genetic map of *M. grisea* primarily based on RFLP's (Fig. 1). Lactate dehydrogenase loci ldh1 and ldh3, the mating type locus (mat1), and the ribosomal DNA repeat (rDNA) were also mapped. The map spans approximately 400 cM. Fungus isolates used were Guy11 (mat1-2), and 2539 (mat1-1).

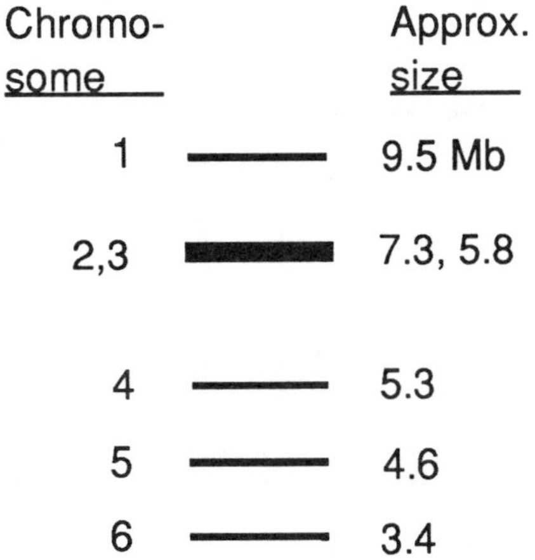

Chromosome		Approx. size
1	——	9.5 Mb
2,3	▬▬	7.3, 5.8
4	——	5.3
5	——	4.6
6	——	3.4

Fig. 2. Electrophoretic karyotype of *M. grisea*.

Linkage groups were assigned to chromosomes by hybridization of selected markers to electrophoretically separated chromosomes (Fig. 2). We have not yet separated chromosome 2 from chromosome 3, even though they apparently are quite different in size. The size of chromosome 5 was estimated to be 4.6 Mb because it co-migrated with chromosome II of *Schizosaccharomyces pombe*, currently thought to be that size (3). All other chromosome size estimates were calculated from published physical length measurements (2), using chromosome 5 as a molecular size standard.

References

1. Leung, H., and Williams, P. H. 1985. Can. J. Genet. Cytol. 27:697-704.
2. Leung, H., and Williams, P. H. 1987. Can. J. Bot. 65:112-123.
3. Steele, P. E., Carle, G. F., Kobayashi, G. S., and Medoff, G. 1989. Mol. Cell. Biol. 9:983-987.

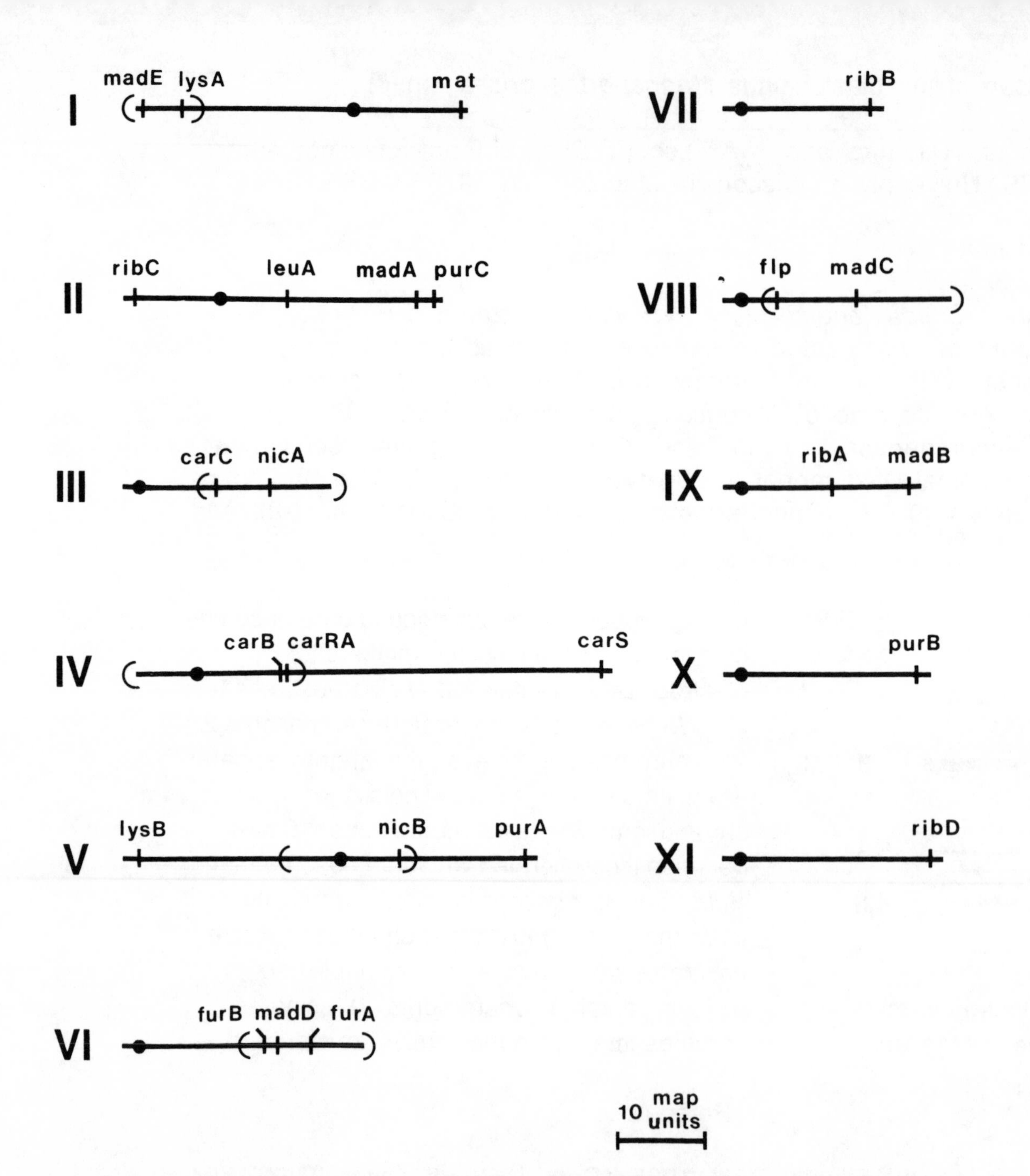

Linkage groups found in Phycomyces blakesleeanus among the 25 markers used. The ordening of the genes within parentheses has not been determined relative to outside markers.

GENETIC LOCI OF <u>PHYCOMYCES BLAKESLEEANUS</u>

Margarita OREJAS, José María DIAZ-MINGUEZ, María Isabel ALVAREZ and Arturo P. ESLAVA

Department of Microbiology, Genetics and Public Health, Faculty of Biology, University of Salamanca, E-37008, Salamanca, Spain

July, 1989

Chromosome number: unknown

In *Phycomyces*, hundreds or thousands of haploid nuclei of both sexes enter and form part of the zygospores, which after a dormancy of 2-3 months, germinate and give rise to a germsporangium containing some 15.000 multinucleate germspores. Nearly all nuclei of both sexes that enter the zygospore degenerate, either after fusion or in their initial haploid state. In general, only one diploid nucleus undergoes meiosis; the four meiotic products later undergo successive mitotic divisions (9, 16, 17). The germsporangium is partitioned into cellular masses, often containing a single nucleus. Further mitotic divisions give rise to the two three or four nuclei most often found in mature germspores that are homokaryotic. Since a very small proportion of germspores are heterokaryotic, it is assumed that some germspore primordia contain more than one nucleus (3).

Once the zygospores have germinated, the progeny may be analyzed by two methods. In the first method, a sample of germspores from each germsporangium is characterized phenotypically, so as to reveal separately the products of each meiosis. This is an example of unordered and amplified tetrad analysis. The second method consists in mixing the germspores from at least 100 germsporangia and analyzing a sufficiently large sample of germspores. For background and references on mapping methods, linkage groups, genetic nomenclature, etc, see references 8, 10, 14, 15, 35. Stocks of most mutants are available from the following collections: Departamento de Genética y Biotecnia, Facultad de Biología, Universidad de Sevilla, Sevilla, Spain (Prof. E. Cerdá-Olmedo); Department of Physics, Syracuse University, Syracuse, New York, USA (Prof. E.D. Lipson); Institute of Genetic Ecology, Tohoku University, Katahira, Sendai, Japan (Prof. T. Ootaki); Departamento de Microbiología, Genética y Salud Pública, Facultad de Biología, Universidad de Salamanca, Salamanca, Spain (Prof. A.P. Eslava). The genetic nomenclature and strain catalogue of *Phycomyces* is available from Prof. T. Ootaki (17).

TABLE I

Phycomyces blakesleeanus loci

Linkage Group	Gene Symbol	Mutant Phenotype	References
I[a]	lysA	Lysine requirement	35
	madE	Abnormal phototropism	2,18,27,30,34,35
	mat	Mating-type locus, (+) and (-)	1,2,4,9,16,17,35
II	leuA	Leucine requirement	2,16,17,35
	madA	Abnormal phototropism	2,18,27,30,34,35
	purC	Purine requirement (adenine or hypoxanthine)	35
	ribC	Riboflavin requirement	35
III	carC	Abnormal carotene biosynthesis	38
	nicA	Nicotinic acid requirement	1,2,17,18,23,35
IV[b]	carB	Abnormal carotene biosynthesis	9,33,39
	carRA	Abnormal carotene biosynthesis	9,17,33,39,41
	carS	Abnormal carotene biosynthesis	31,39,40
V	lysB	Lysine requirement	35
	nicB	Nicotinic acid requirement	35
	purA	Purine requirement (adenine or hypoxanthine)	2,35
VI	furA	5-Fluorouracil resistance	2,35
	furB	5-Fluorouracil resistance	2,35
	madD	Abnormal phototropism	2,18,27,30,34,35
VII	ribB	Riboflavine requirement	2,35
VIII	flp	Flavoprotein (previously FP_2)	21,36
	madC	Abnormal phototropism	18,27,30,34,35
IX	madB	Abnormal phototropism	18,27,30,34,35
	ribA	Riboflavine requirement	22,35
X	purB	Purine requirement (adenine or hypoxanthine)	35
XI	ribD	Riboflavine requirement	35

[a] The mating type locus has been redefined as mat to distinguish it from those mutants disturbed in some steps leading to the formation of zygospores, which are called sex.

[b] The carR mutants are red and accumulate lycopene. The carA mutants are white and they do not accumulate any carotenoids, except for traces of ß-carotene. Genetic evidence points to the existence of a single carRA gene containing two segments responsible for the two biochemical functions (41). The carS marker in linkage group IV must be considered tentative until further confirmatory likage studies have been conducted.

TABLE 2

Unmapped markers of Phycomyces blakesleeanus

Markers	Mutant Phenotype	References
adh	Little or no alcohol dehydrogenase	19,20
amp	Resistant to amphotericine	32
arb	Arbusculus, a type of colonial mutant	3,7,32
arg	Arginine requirement	32
can	Canavanine resistance	32
carE	Extranuclear genetic factor	12
carI	Defective in the chemical stimulation of ß-carotene biosynthesis	14,39
carD	Superyellow. They complete the sexual reaction and respond to retinol, while carS do not	40
chx	Resistant to cycloheximide	32
col	Colonial growth in mycelia	3,7,32
dar	Resistant to 5-deazariboflavin	13,14
dwf	Dwarf thin sporangiophore (microsporangiophore)	3,32
fol	Folic acid requirement	32
gal	Resistant to 2-deoxigalactose in sorbitol medium	11
ger	The spores germinate spontaneously without heat or chemical activation	42
ilv	Isoleucine and valine requirement	32
imb	Abnormal macrophorogenesis (unable to produce sporangiophores under certain conditions); latin imberbis or beardless	23
leuB	Leucine requirement	Eslava, A.P. (unpublished)
liv	Leucine, isoleucine and valine requirement	32
madF	Abnormal phototropism	14,18,27
madG	Abnormal phototropism	14,18,27
madH	Abnormal phototropism	14,28,30
met	Methionine requirement	32
mic	Abundant microphores	23
nan	The nan mutants (latin nanus or dwarf) develop short sporangiophores but should not be confused with the dwf mutants which have thinner sporangiophores	3,7
pan	Pantothenic acid requirement	32
pde	Little or no phosphodiesterase activity	37
picA	Abnormal photoinduction of carotenogenesis	24,29
picB	Abnormal photoinduction of carotenogenesis	24,29
pil	They have a bulge in the sporangiophore growing zone, reminding one of the related fungus Pilobolus	3,5,6,25,26
sex	Disturbed in some steps leading to the formation of zygospores	14,32
sgm	Temperature sensitive for sporangium formation	32
sul	Sulphur requirement (cysteine, methionine or S_2O_3)	32
xtv	Resistant to crystal violet	9,32

REFERENCES

1. ALVAREZ, M.I., and A.P. ESLAVA. 1983. Genetics 105:873-879.
2. ALVAREZ, M.I., M.I. PELAEZ, and A.P. ESLAVA. 1980. Mol. Gen. Genet. 179: 447-452.
3. BERGMAN, K. et al. 1969. Bacteriol. Rev. 33: 99-157.
4. BLAKESLEE, A.F. 1904. Proc. Natl. Acad. Arts. Sci. 40: 205-319.
5. BURGEFF, H. 1912. Ber. Dtsch. Bot. Ges. 30: 679-685.
6. BURGEFF, H. 1915. Flora 108: 353-448.
7. BURGEFF, H. 1928. Z. Vererbungsl. 49: 26-94.
8. CERDA-OLMEDO, E. 1974. In "Handbook of Genetics", R.C. King ed., pp 343-357. Plenum Press, NY.
9. CERDA-OLMEDO, E. 1975. Genet. Res. 25: 285-296.
10. CERDA-OLMEDO, E. 1985. In "Methods Enzymol." 110: 220-243.
11. CERDA-OLMEDO, E. 1987. In "Phycomyces", E. Cerdá-Olmedo and E.D. Lipson eds., pp 341-344. Cold Spring Harbor, NY.
12. DE LA CONCHA, A., and F.J. MURILLO. 1984. Planta 161: 233-239.
13. DELBRUCK, M., and T. OOTAKI. 1979. Genetics 92: 27-48.
14. ESLAVA, A.P. 1987. In "Phycomyces", E. Cerdá-Olmedo and E.D. Lipson eds., pp 27-48. Cold Spring Harbor, NY.
15. ESLAVA, A.P., and M.I. ALVAREZ. 1987. In "Phycomyces", E. Cerdá-Olmedo and E.D. Lipson eds., pp 361-365. Cold Spring Harbor, NY.
16. ESLAVA, A.P., M.I. ALVAREZ, P.V. BURKE, and M. DELBRUCK. 1975. Genetics 80: 445-462.
17. ESLAVA, A.P., M.I. ALVAREZ, and M. DELBRUCK. 1975. Proc. Natl. Acad. Sci. USA 72: 4076-4080.
18. ESLAVA, A.P., M.I. ALVAREZ, E.D. LIPSON, D. PRESTI, and K. KONG. 1976. Mol. Gen. Genet. 147: 235-241.
19. GARCES, R., E. SANTERO, and J.R. MEDINA. 1984. Genet. Iber. 36: 133.145.
20. GARCES, R., M. TORTOLERO, and J.R. MEDINA. 1986. Exp. Mycol. 9: 356-358.
21. GARCES, R., C.L. CHMIELEWICZ, and E.D. LIPSON. 1986. Mol. Gen. Genet. 203: 341-345.
22. GAUGER, W., M.I. PELAEZ, M.I. ALVAREZ, and A.P. ESLAVA. 1980. Exp. Mycol. 4: 56-64.
23. GUTIERREZ-CORONA, F., and E. CERDA-OLMEDO. 1988. Devel. Genet. 9: 733-741.
24. JAYARAM, M., D. PRESTI, and M. DELBRUCK. 1979. Exp. Mycol. 3: 42-52.
25. KOGA, K., and T. OOTAKI. 1983. Exp. Mycol. 7: 148-160.
26. KOGA, K., and T. OOTAKI. 1983. Exp. Mycol. 7: 161-169.
27. LIPSON, E.D., D.T. TERASAKA, and P.S. SILVERSTEIN. 1980. Mol. Gen. Genet. 179: 155-162.
28. LIPSON, E.D., I. LOPEZ-DIAZ, and J.A. POLLOCK. 1983. Exp. Mycol. 7: 241-252.

29. LOPEZ-DIAZ, I., and E. CERDA-OLMEDO. 1980. Planta 150: 134-139.
30. LOPEZ-DIAZ, I., and E.D. LIPSON. 1983. Mol. Gen. Genet. 190: 318-325.
31. MURILLO, F.J., and E. CERDA-OLMEDO. 1976. Mol. Gen. Genet. 148: 19-24.
32. OOTAKI, T., and K. KOGA. 1982. "Genetic nomenclature and strain catalogue of Phycomyces", pp 1-131. Yamagata University Press. Japan.
33. OOTAKI, T., A.C. LIGHTY, M. DELBRUCK, and W.J. HSU. 1973. Mol. Gen. Genet. 121: 57-70.
34. OOTAKI, T., E.P. FISHER, and P. LOCKHART. 1974. Mol. Gen. Genet. 131: 233-246.
35. OREJAS, M., M.I. PELAEZ, M.I. ALVAREZ, and A.P. ESLAVA. 1987. Mol. Gen. Genet. 210: 69-76.
36. POLLOCK, J.A., E.D. LIPSON, and T.D. SULLIVAN. 1985. Biochem. Genet. 23: 379-390.
37. REDDY, V.M., P. GALLAND, and E.D. LIPSON. 1985. Mol. Gen. Genet. 201: 124-125.
38. REVUELTA, J.L., and A.P. ESLAVA. 1983. Mol. Gen. Genet. 192: 225-229.
39. RONCERO, M.I.G., and E. CERDA-OLMEDO. 1982. Curr. Genet. 5: 5-8.
40. SALGADO, L.M., E.R. BEJARANO, and E. CERDA-OLMEDO (submitted for publication).
41. TORRES-MARTINEZ, S., F.J. MURILLO, and E. CERDA-OLMEDO. 1980. Genet. Res. 36: 299-309.
42. VAN LAERE, A.J., and F. RIVERO. 1986. Arch. Microbiol. 145: 290-295.

ACKNOWLEDGEMENTS

We (M.I.A. and A.P.E.) are indebted to Prof. E. Cerdá-Olmedo and the late Prof. Max Delbrück for introducing us in the field. Financial support was provided at different times by the Commission for Educational Exchange between Spain and the United States, the Alexander von Humboldt Foundation, Junta de Castilla-León, and Dirección General de Investigación Científica y Técnica of Spain (grants PB86-0209 and PB88-0376).

Gene Map of the Basidiomycete Schizophyllum commune. 1N=11
August, 1989
Carlene A. Raper
Department of Microbiology and Molecular Genetics
University of Vermont
Burlington, Vermont 05405

Previous maps published are referenced as follows: #17 was derived by meiotic recombination and gives distances between loci; #4 was derived by mitotic recombination; and #13 combines data from both meiotic and mitotic studies.

LINK. GROUP	GENE/LOCUS IN ORDER	DESCRIPTION	REFERENCES
I	cen	centromere	14,12,13,17
I	blu2	late indigo producer (prev. l.blu)	13,17,18
I	pig	pigmy morph.	13,17,18
I	str	streak morph.	12,13,17
I	ade11*	adenine	4
I	dom9	dome colonial morph.	1,13,17
I	com2,3, 6,9,4	cluster, compact colonial morph.	1,13,17
I	dom1	dome colonial morph.	13,17
I	<u>Aα</u>	9 alleles Aα mating-type locus (MAT)	4,7,13,16,17
I	<u>pab</u>	paramino benzoic acid	4,5,13,17
I	<u>ade5</u>	adenine	4,5,13,17
I	Aβ	32 alleles Aβ MAT	4,13,16,17
I	su[Aβ1(1)]	supressor constitutive mut. Aβ1(1)	13,17
I	com1	compact colonial morph.	13,17
I	pdx1	pyridoxine (prev. ade-pyr,x15)	3,4,13,17
I	rst	restricted colonial morph.	13,17
II	ade6	adenine	13,17
II	mod8-21	modifier "flat" phenotype B MAT funct.	13,14,17
II	Bα	9 alleles Bα MAT	4,13,16,17
II	su(mod4-34)	supressor of mod4-34 mut.	7,13,17
II	Bβ	9 alleles Bβ MAT	4,13,16,17
II	def1(Bβ2)	blocks control acceptance of migr. nuclei by Bβ2 (prev. Bβf)	13,15,17
II	dom2	dome colonial morph.	1,13,17
II	rec1(Bα-β)	recomb. freq. between Bα-Bβ	13,17,19
II	mod4-56,59 61,34,80,86 11,23,33,22	cluster, block acceptance migr.nuclei	2,13,17
II	mod4-4	blocks acceptance of migrating nuclei	13,14,17
II	ura2	uracil	13,14,17
II	thn	thin colonial morph.,corkscrew hyphae	13,17
II	dom7	dome colonial morph.	13,17
II	pdx2*	pyridoxine	4,17
II	mod9-1	blocks dikaryosis,induces nucl.fusion (prev. dik$^-$)	13,17
II	leu*	leucine	4,17
II	thn2	thin colonial morph.	13,17
II	blu1	early indigo producer (prev.e.blu-1)	13,17,18
II	thy*	thymine	4,13
II	nic2	nicotinic acid	3,4,13,17
II	puf1	hyphae branch as puffs	13,17,18
II	ade3	adenine	3,4,13,17
II	met*	methionine	4,13
II	<u>ura1</u>	oratidine-5'-phosphate decarboxylase	3,4,5,13,17

LINK. GROUP	GENE/LOCUS IN ORDER	DESCRIPTION	REFERENCES
III	gua	guanine	13,17
III	puf2	hyphae branch as puffs	13,17,18
III	arg1	arginine	3,4,13,17
III	spi*	colockwise spiral morph.	4,13
III	ade4	adenine, pink	4,13,17
III	dom10*	dome colonial morph.	1,13,17
III	min*	minute colonial morph.	4,13
III	pol^r*	polymixin resistance	13,17
III	ade7	adenine	13,17
IV	rib	riboflavin	4,13
IV	ade2	adenine	3,4,5,13,17
IV	unk11	unknown nutritional requirement	4,13,17
IV	cnc	concentric rings, colonial morph.	13,17
IV	su(arg2)*	supressor arg2 mutation	4,10,13
IV	arg2	arginine	3,4,13,17
IV	nic1	nicotinic acid	3,4,13,17
IV	dom5	dome colonial morph.	13,17
IV	chc	chocolate color, septateless	3,4,13,17
IV	mnd*	mound colonial morph.	6,13
V	trp1	indole-3-glycerol phospate synthetase	11
V	arg6	arginine	3,4,13
V	ino1	inositol	4,13
V	pan	pantothenate	4,13
VI	aro	aromatic amino acids	4,13
VI	nic3	nicotinic acid	3,4,13
VI	com	compact colonial morph	4,13

Asterisk indicates uncertain location relative to adjacent loci. A locus underlined signifies that gene has been cloned--three alleles of Aα MAT (1,3 & 4) have been cloned; 1 and 4 have been sequenced (7).

The emphasis on mapping of gene loci in this organism has been to identify genes relevant to control of mating and development (16,13). Stocks are currently available in this laboratory, including strains containing 119 different A MAT factors and 63 different B MAT factors from the world-wide population. Many mutants of the MAT loci and mutants modifying sexual development are also available.

REFERENCES

1. CHANG,S.T. 1972. Aust. J. Biol. Sci. 25:757-764.
2. DUBOVOY,C. 1975. Genetics 82:423-428.
3. ELLINGBOE,A.H. and J.R.RAPER. 1962. Genetics 47:85-98.
4. FRANKEL,C. and A.H.ELLINGBOE. 1976. Genetics 85:417-425.
5. FROELIGER,E.H. et al. 1987. Curr. Genet. 12:547-554.
6. GABER,R.F. and T.J.LEONARD. 1981. Nature 291:342-344.
7. GIASSON,L. et al. 1989. Mol. Gen. Genet. 218:72-77.
8. KOLTIN,Y. and J.STAMBERG. 1972. Jour. Bacteriol. 109:594-598.
9. MIDDLETON,R.B., 1964. Genetics 50:701-710.
10 MILLS,D.I. and A.H.ELLINGBOE. 1969. Mol. Gen. Genet. 104:313-320.
11. MUNOZ-RIVAS,A.,et al. 1986. Mol. Gen. Genet. 205:103-106.
12. PAPAZIAN,H.P. 1950. Genetics 36:441-459.
13. RAPER,C.A. 1988. Advances in Plant Path. (Sidhu,ed.) 6:511-521.
14. RAPER,C.A. and J.R.RAPER. 1966. Genetics 54:1151-1168.
15. RAPER,C.A. and J.R.RAPER. 1973. Proc.Natl.Acad.Sci. 70:1427-1431.
16. RAPER,J.R. 1966. Genet.of Sex.in Higher Fungi. Ronald Press, 283pp.
17. RAPER,J.R.and R.H.HOFFMAN. 1974. Handbk.Genet.(King,ed.) 1:597-626.
18. RAPER,J.R. and P.G.MILES. 1958. Genetics 43:530-546.
19. STAMBERG,J. and Y.KOLTIN. 1973. Genet.Res.Camb. 22:101-111.

Ustilago maydis , causal agent of corn smut, (n=20)

August 1989

Allen D. Budde and Sally A. Leong
USDA/ARS and Dept of Plant Pathology
University of Wisconsin-Madison
Madison, WI 53706

Gene assignments to the electrophoretic karyotype (Fig. 1) of *Ustilago maydis* (strain #518 from R. Holliday) were obtained by lysing spheroplasts(1) in agarose, subjecting the DNA to a contour-clamped homogeneous electric field(2) and probing Southern transfers of gels with the genes listed below (Table 1). Twenty chromosome-sized DNA bands(3) and six defined linkage groups(4) have been reported in *U. maydis*.

Table 1

Gene	Probe source	Karyotype band
*a*2(5)	*U. maydis*	XIX-XX[a]
actin(6)	*Saccharomyces cerevisiae*	II
*b*1(7)	*U. maydis*	XVIII
*b*2(7)	*U. maydis*	XVIII
GAPDH(8)	*U. maydis*	XVIII
hsp 70(9)	*U. maydis*	IX,XII-XIII,XIX-XX
*leu*2(10)	*U. maydis*	XIX-XX
metallothionein(11)	*Mus musculus*	XIX-XX
oxygenase(12)	*U. maydis*	XIX-XX
*pan*1(5)	*U. maydis*	XIX-XX
*pyr*6(10)	*U. maydis*	XI
rDNA(13)	*Neurospora crassa*	XIX-XX
*tpi*1(10)	*U. maydis*	XV

[a] hyphen indicates non-resolved bands

Electrophoretic karyotype of *U. maydis*

Conditions
140 Volts
60 sec pulse 15 hr
100 sec pulse 9.5 hr

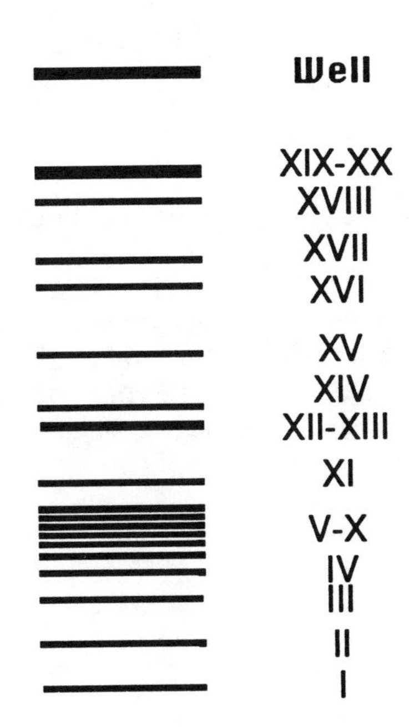

References

(1) Wang, J., et. al. 1988. PNAS 85:865-869.
(2) Chu, G., et. al. 1986. Science 234:1582-1585.
(3) Kinscherf, T. G. and S. A. Leong. 1988. Chromosoma 96:427-433.
(4) Holliday, R. 1974. In: Handbook of Genetics (R. C. King, ed.). Plenum Press, N.Y. pp. 575-595.
(5) Froeliger, E. and S. A. Leong. Unpublished.
(6) Ng, R. and J. Abelson. 1980. PNAS 77:3912-3916.
(7) Kronstad, J. W. and S. A. Leong. 1989. PNAS 86:978-982.
(8) Smith, T. and S. A. Leong. Unpublished.
(9) Holden, D. W., et. al. 1989. EMBO 8(7):1927-1934.
(10) Kronstad, J. W., et. al. 1989. Gene 79:97-106.
(11) Durnam, D. M., et. al. 1980. PNAS 77:6511-6515.
(12) Wang, J., et. al. 1989. J. Bacteriol. 171(5):2811-2818.
(13) Free, S. J., et. al. 1979. J. Bacteriol. 137:1219-1226.

Genetic loci of the anther smut fungus Ustilago violacea

September, 1986

Dr. Alan W. Day
Department of Plant Sciences
University of Western Ontario
London, Ontario N6A 5B7
Canada

Dr. Edward D. Garber
Department of Biology
University of Chicago
Chicago, Illinois
U.S.A.

Genetic maps of this phytopathogenic basidiomycete begun in 1965, have been developed using methods based on both sexual and parasexual recombination. Reviews giving details of mapping methods have recently been published 4,5. The haploid chromosome number deduced by these genetic means has been shown to be at least 15 more probably about 20. Recently electrophoretic chromosome sorting has shown a probable chromosome number of 20-21 for the related species U. maydis (Dr. Sally Leong, personal communication). Stocks of many of the mutants described here are available from Dr. E.D. Garber (address above).

Linkage Group[1]	Gene Symbol	Phenotype	CLV[2]	Other Linkage Data[1]	Reference
01	arg-2	arginine requirement	--	arg-2/orn-1 - 10 cm	1
	arg-3	arginine requirement	--	arg-3/orn-1 - 6 cm	1
	orn-1	arginine, ornithine or citrulline requirement	--	--	1
	orn-2	arginine, ornithine or citrulline requirement	--	--	1
	ts-17c	temperature sensitive	03	--	11
02	his-1	histidine requirement	22	his-1/lys-1 - 41 cm	1,3
	met-2	methionine cysteine, or $S_2O_3^{2-}$ requirement	--	--	1
	lys-1	lysine requirement	--	--	1
	ade-2	adenine requirement	--	ad-2/lys-1? - 3 cm	1
	cxr-1	cycloheximide resistance	--		3
03	lys-2	lysine requirement	07	--	1,11
	glu-1	glutamate requirement	09	--	1,11
04	pdx-1	pyridoxin requirement	03	--	1,9
	lys-c	lysine requirement	05	--	9
	arg-b	arginine requirement	05	--	9
	Bot-2	botran resistance	05	--	9
	ph	psuedohyphal colonial - mutant	08	--	7
	tm	temperature sensitive morphology mutant	09	--	12
	lys-a	lysine requirement	10	--	9
	arg-a	arginine requirement	16	--	9
	cf	cauliflower - colonial morphology mutant	31	--	10
	mph	modified psuedohyphal colonial mutant	31	--	10
	inos-1	inositol requirement	--	--	1
05	met-3	methionine cysteine, $S_2O_3^{2-}$, or SO_3^{2-} requirement	03	--	1,11
06	MT	mating type locus, allele a_1 and a_2	00	--	1,13
07	y	yellow colonies } dominant alleles = pink colonies	00	--	1,5
	w	white colonies	00	--	5
	P	pumpkin color colonies	00	--	6
	O	orange colonies	00	--	1,5

Linkage Group[1]	Gene Symbol	Phenotype	CLV[2]	Other Linkage Data[1]	Reference
07	mor10.2	morphology mutant	24	--	13
	car	carboxin resistant	27	--	9
08	his-3	histidine requirement	--	--	1
09	met-1	methionine requirement	12	met-1/cit-1 - 2.4 cm	1,11
	lys-3	lysine requirement	--	met-1/lys-3 - 44 cm	1
	cit-1	arginine or citrulline requirement	--	cit-1/lys-3 - 44 cm	1
	pro-1	proline requirement	--	pro-1/lys-3 - 15 cm pro-1/met-1 - 19 cm pro-1/cit-1 - 30 cm	1
10	his-2	histidine requirement	--	--	1
11	his-4	histidine requirement	05	--	1,11
12	arg-1	arginine requirement	06	--	1,unpub
13	ure-1	urease (structure or permease)	--	--	8
14	ure-2	urease (structure or permease)	--	--	8
15?	Tb2-1	thiobendazole resistance	02	--	9
	pro-2	proline requirement	08	--	9
unassigned markers					
	ade-a	adenine requirement	--	--	
	arg-a	arginine requirement	16	--	9
	arg-e		02	--	14
	arg-f		01	--	14
	arg-j		03	--	14
	arg-n		03	--	14
	arg-p		10	--	14
	Bot-1	botran resistance	10	--	9
	Chl-1	chloramphenicol resistance	06	--	9
	Chl-2		14	--	9
	Chl-3		19	--	9
	cit-k	arginine or citrulline requiement	22	--	9
	cit-m		14	--	9
	cit-o		31	--	9
	ch	chain forming	16	--	12
	cxr-2	cycloheximide resistance	--	--	
	dl	dull colonies	16	--	10
	glu-1	glutamic acid	09	--	11
	ile-1		25	--	unpub.
	lys-a	lysine requirement	10	--	9
	lys-b		09	--	9
	lys-c		05	--	9
	met-1	methionine requirement	12	--	11
	met-3		03	--	11
	mor14.4	morphology mutation	29	--	13
	mor13.1		21	--	13
	mor12.3		28	--	13
	mor12.1		15	--	13
	nic-1	nicotinic acid requirement	21	--	
	nir	NH_4^+m asparagine, arginine	--	--	
	orn-1	arginine ornithine	16	--	14
	orn-g	or citrulline requirement	23	--	14
	orn-i		05	--	14
	pab-1	para-aminobenzoic acid requirement	21	--	9
	pdx-1	pyridoxin requirement	05	--	9
	ser-a	serine requirement	--	--	
	sul-1	SO_3^{2-}	--	--	
	tm9.34	temperature sensitive morphology mutant	25	--	13
	ts-1	temperature sensitive mutant	05	--	11
	ts-3		03	--	11
	ts-6		08	--	11
	ts-8		32	--	11
	ts-9		09	--	11
	ts-10.9		33	--	11
	ts-33		04	--	11
	ts-1046		11	--	11
	tyr-1	tyrosine requirement	19	--	11

Linkage Group[1]	Gene Symbol	Phenotype	CLV[2]	Other Linkage Data[1]	Reference
unassigned markers					
	uvs1	ultraviolet light sensitive	--	--	2
	uvs2		--	--	2
	uvs3		--	--	2
	uvs4		--	--	2

1 See reference #1
2 Centromere linkage value (distance from centromere in cM)

REFERENCES

1. DAY, A. W. and J. K. JONES. 1969. Genet. Res. 14:195-221.
2. DAY, A. W. and L. L. DAY. 1970. Canad. J. Genet. Cytol. 12:891-904.
3. DAY, A. W. 1978. Nature 273:753-755.
4. DAY, A. W. and E. D. GARBER. 1987. In G. S. Sidhu (ed.) Genetics of Pathogenic Fungi. Academic Press.
5. GARBER, E. D., et al. 1975. Bot. Gaz. 136:341-346.
6. GARBER, E. D., et al. 1980. Bot. Gaz. 141:220-222.
7. GARBER, E. D., et al. 1981. Bot. Gaz. 142:589-591.
8. BAIRD, M. L. and E. D. GARBER. 1981. Biochem. Genet. 19: 1101-1114.
9. GARBER, E. D., et al. 1982. Bot. Gaz. 143:524-529.
10. ZIELINSKI, A. B. and E. D. GARBER. 1982. Bot. Gaz. 143:530-533.
11. GARBER, E. D., et al. 1983. Bot. Gaz. 144:584-588.
12. SALTIEL, F. S. and E. D. GARBER. 1984a. Bot. Gaz. 145:140-144.
13. SALTIEL, F. S. and E. D. GARBER. 1984b. Bot. Gaz. 145:565-568.
14. STEVENS, D. M. and E. D. GARBER. 1987. Bot. Gaz. 147:in press.

ASPERGILLUS NIDULANS

A.J. Clutterbuck, Department of Genetics, University of Glasgow
Glasgow G11 5JS, Scotland, U.K.

Chromosomes: n=8 Compiled June 1989

Aspergillus genetics and methods are outlined in references 261 and 80.

LINKAGE MAP

Note that the accompanying map and list are summaries only: **please do not quote them** without checking with the references to individual markers. The sources of linkage data are mostly to be found in the references to markers in the region concerned, but in some cases may be unpublished.

The map is an updated version of those previously published in refs 158, 106, 80 & 83. Linkages are given as uncorrected recombination percentages and in most cases are given to the nearest marker only; more extensive data is usually available in the references.
Since the map refers only to Glasgow strains derived from Pontecorvo's original single wild isolate, unbiased map distances are usually consistent with each other. On the other hand the data come from a wide variety of sources so this map should be taken only as a guide; some attempts to combine diverse map distances have required considerable graphic gymnastics and not a little guesswork: consult the references for authentic information.
Where the distance, but not the orientation, of a linkage is known, it is shown as a vertical or diagonal line. Centromeres are shown as closed circles. The position and orientation of the isolated *cysD-E* fragment of linkage group II is unknown. Where linkages are disturbed by the presence of known translocations, the recombination percentages are given below the line.

LOCUS LIST

Gene symbols given in parenthesis after the standard one are alternatives or obsolete. Symbols of extranuclear genes are given in square brackets. Linkage groups in parenthesis have been determined by haploidization, but the marker has not been located meiotically.

ABBREVIATIONS

def.= deficiency	inhib.= inhibited by	regl.= regulation
req.= requiring	res.= resistant	sup.= suppressor
upt.= uptake	ut.= utilization	deH = dehydrogenase

* = Cloned gene and reference to cloning # = mutant strains available from Glasgow or Fungal Genetics Stock Centre

Linkage map of

```
                            10
                        uvsF----                                                                                    .2
                                                          16                                         uaZ      --fpaA
                                                      rA----        pyroB--.4                                  12
         20       6       5       3       2       4      3      3     5       .6          3          9            35         19     |9     ----fpaI
I   aroC-----acuJ---acuM---trpB--fpaB--medA--areB--adA--rD---creA--galD--pyrG----suAadE------sulA----riboA----anA--->
                                                   |                                                               18         9
                   T1(I;VII)--                 T1(I;III)                            3|17           4|   <1
                          4                    -areB406                           T2(I;VIII)            pacJ--
                                                                                   -sD50          clB

                          36           7        17          22                                        8
                  iodA-----        ---etbA----        -----tsA                                     ---fpaP
        37       14        20         9          17       1        20         9       1        6       16        3         1       30        28
II  cysB-----adH----fpaQ----acrA---apnA----wA--methA----palcA-o-drkB--gatB---riboE----creD---creC--glnA-----pdeAB----->
                                                |35                  |
                                               saA             T1(II;VIII)-nis-5

  -brlA12
T2(III;VIII)                                                    13                fmdS
                               16           8               ----penB
                         spsA----        --sorB                                    |2
      |         28            1          18          22         3     4       1       27         14         11          7          3            7        4        .6
III   ----galE----meaB--pantoC----cnxH----sA--sC---adI--moC-----ivoA----alX----palG---chlA--methH---argB---dilA-->
                           35                    .5       20                                         1            16
                     suApro-----       /   |       -----palcB                              pycA--        ----sG
                            Inv1(III)      |            \T1(III;VIII)
                              -xprD    T1(III;VII)
                                       -lysD20

       iclA                                                                                                         penC
                   23                14
         2|      -----bncA       ----cadA                                                                            |?
         >45            22         1       6         4          34          3         5         14         11          14          10          25
IV  uvsH- - - -methG-----hisA--gdhB---meaA--acuK-----palC--pimA---bzuA----inoB----mauA----pabaB----pyroA-----apsA
         ?                  ------o--|<1          |4                                                  |<.1                  40
  ssbA---                     13                                                                                         -------acuL
                       T1(IV;VIII)-frA1       T1(III;IV)                                                                     .5
                                              -areA18                                           suApabaB              --ornA
                            suCadE---  |?
                                  4    |
                                     uvsB

                                                                                                                          12
          20        24                                                                                                ----sltA
    glcC----cnxG----                  11          2        .4        29            >50           10       <1       <1         22          16        6       11
VI    - - o - - - - - -nicC----apsB--methB--tsB-----lacA- - - -bwA----gabI--gabA---molA-----tamA-----argA---uaX---->
                                             |0.1       .2
                                                     --hfaB                   ----T1(V;VI)                                          .3|
                                             cadB                              20
                                                                              -----T1(VI;VII)                                       agaA
                                                                               22

                   drkA     acoC
                                                        <1                  20              4
                    \18     /24                     nimO--               ----prnCBDA---cnxJ
          2          15         24             37                                 4       44                                23          5          12
VII  actB--suaD----alcAR---o-phenB-------oliC---------pantoB--sF-----palF-----lysD----malA-----aauB---wetA---->
                                  |<1   7 |            32 | 15                     23          /     18
                                 telA     T1(I;VII)    T1(VI;VII)        T1(III;VII)        13         15
                                                                          -lysD50       ----hisJ----melB

                                           16                                                                   <1
         choC                  uvsC----                                                                     crnA--
          >50   <1|    33          5       3          12                            16       1        12         10         6        26         <1        10         3
VIII  pdhC- - o - -tsD-----sD---coA--ornB----qutCDBGEAR----galG-gamB----fwA----veA---facC-----niiA--niaD----brlA-->
                22                            5
       fluF----             |     benA---                                                                           |4          |                |
                    T2(I;VIII)                                                                                   arpB  T1(II;VIII)    T2(III;VIII)
                       -sD50                                                                                          -nis-5          -brlA12
```

Revised June 1989

ASPERGILLUS NIDULANS

```
                                                                              binD                           20
                                                                                                             ----fpaR
                                                            3                  |6                            3
                                           2          .5    ---mecA          ?      5        ?               ---acuE
                  .6     .4             --ornD  proB--      9            1   --binB--lacD--   .1             26          14
                  -davA-                                    ----hisB--                        --adE          -----sorA----pdhA
9        6        9           4        21         21        6         5      11           4       6        1   29         >50        >50
---pyrE---adG------pyrF---luA-----stuA-----proA-●-lysF---pabaA-------uvsA---yA---fpaM--biA-----niiC- - -uapA- - -camC
         18                   16            30              3         7      5 |              |<1          40          12
         ----fluC      pimB-----       fpaO-----            ---moA----                                     -----sgpD  ----hfaA
                                                            6         4      T1(I;II)         binG
                                        33                  --suDpro--
                                        mecB-----                     .3
                                                                  adF--

                                                                                                      24
                                                                                 25        29     aauC----
                 ?    ?                                 3                        ------rB------           .1      2       2
                 -inoA-                           imaC---                        18        21             --ygA---acoB---benB
                                                                                 -----tsC------                                |.5
28          44            14           25       .7       .4    .7       .1         12      30        23      .1              29             .3
----intA-----gmdA------thiA-----anB--mauB--puA--ileA--ssuA----cnxE-----trpA-----adC--adD---------------acrB--creB
                                       16                 |                                          4              24
                                 senA----                                                20          ---punA        -----lacB
                                               T1(II;VIII)-nis-5                         ----mauC
                                                                                                                                          11
                                                                                                                                  cysD----cysE

                                                                                                                    Inv(III)
                                             sgpC                32                                                 -xprD1
                                                                 -----actC
                                             |39                                                                     |
.6        14          14         5      7         5            14         5          8          1         8                 >50
--gdhA----palA----galA---smA---actA-●-phenA----cbxA---suaB---amdS--amdI---suBpro - - - - - areA
                                                                                                               34
                                           2                                                                   ornC-----
                                    suaA--                                                                         12|11
                                                                                                                T1(I;III)-areB406

                                                     21
                                                     -----cysC
                                                       30           18          20
                                                     -------pdhB-----riboG------
                                    11                                                                    21
                              cysA----               32         3       19                          -----smsA             26
                                                     -----gcnA---mnrA----                                                 -----acuN
                                                                  40                                                      36             20
V  uvsJ-----uvsD-----nicA---lysB-------●-----pA-----facA---hxA---acuG---lysE-----riboD-----uvsE
        25           27          4                                     27          14          7         3
                                 |14                      |10      <1         31
                                                                acuD--        -----fpaU
                             suCpalF              T1(V;VI)  fpaG----
                                                                 20

                              10
                              ----fpaJ
                              ?     ?
                              --hetB--
              7         31
              ---fpaH-----
11            19         18          41           36
----lysA----fpaF----sB--------sbA-----npeA
     |        5     2
     |        ---glrA--pacC
     |
Inv(VI)-lysA1

                                                                                    4
                                                                              nimQ--
                                                          2            14          15
                                                          flA--uapB----palD----
12            42          2      4         28          9          9          12         43
----imaA-----benC--acoA--choA----methE---gatA---alcC----cbxB-----nicB
                                                       5|                  |11           |?
                                                       amdA              sumA          binF
                                                       11|                                                       uvsG
                                                        cnxF
                                                                                                     manA  ?|   T1(VI;VIII)-mipB
                                                                                                      .8\   |  /7
                                                                                                         \  | /
                                                                                           5                |
                                                                                     aroA---                |   29
                                                             ?         ?                              yB    |   -----fluA
                                  1      20                  -hisDEI-          19       12                  |   20
                                  --adB-----                                   ----fluB----   10/   \10     |   ----fpaN
3          7        .4      .1   5            4         10               12               15         10         4         6         >50
--cnxABC---cbxC--oxpA--uaY---fpaD---pyrD---palcC--------argC----hisC----facB---riboB---trpC- - - ->
                                  <1/    \20         9                13             3                   .3       37
                                                     ----araA   fluD---           ---suApalB      mipA--
                                  lamAB    aauD                                                              aldA-----

                                                                    T1(III;VIII)
                                                                        |    16              9
                                                            ivoB   fluE   ----------pacB-------
               <1                     2                                       ?              ?
               --ureA              rC--                     |<1    15|    ------penA-----------
    <50        2          20          10           1         1      3        8               12        23
< - -ureB--abaA----moB----palB--pppA--ureD---chaA---pantoA----sE-----nirA
               23                                            12             3                         6   8
               -----ahrA                               tubA----          --imaD                      |
                                                              30            16
                                                       fpaK-----         ----palE      T1(IV;VIII)-frA
```

Locus	Linkage gp.	Phenotype, enzymes etc.	References
aaX #	(VI)	Allantoic acid ut. allantoinase	98,201,285,183
aauA	(VII)	Glutamate & aspartate upt.	245,183
aauB	VII	Amino acid upt.	180,183
aauC	II	" " "	180,183
aauD	VIII	" " "	180,183
abaA* (aba) #	VIII	"abacus" aconidial	76,81,213*
acl	-	Acleistothecial	248,350
aciA*	?	Unknown - regl. by amdA	32*
acnA*	?	Actin	115*
acoA	VII	Aconidial	66,50
acoB	II	"	66
acoC	VII	"	66
acrA (Acr1) #	II	Acriflavine res.	275
acrB (acr2) #	II	" "	275
acrC,D	?	" "	312
actA (=imaG?) #	III	Actidione res.	328,341
actB #	VII	" "	328
actC (=imaH)	III	" "	328
actD	(II)	" "	328
actE	(II)	" "	328
acuA,B,C see facA,B,C			
acuD* (icl) #	V	Acetate ut. isocitrate lyase	10,192,195,48*
acuE* (mas) #	I	" " malate synthase	9,10,195,102*
acuF #	(VII)	" " PEP carboxykinase	10,195
acuG #	V	" " fructose-1,6-diphosphatase	10,195
acuH #	(V)	" "	10,195
acuJ #	I	" "	10,195
acuK #	IV	" " malic enzyme.	10,192,195
acuL #	IV	" "	10,195
acuM #	I	" " malic enzyme	10,192,195
acuN #	V	" "	10,195
adA #	I	Adenine req., adenylosuccinase	116
adB #	VIII	" " AMP succinate synthetase	116
adC (ad1) #	II	" " AIR carboxylase	116,261
adD (ad3) #	II	" " AIR carboxylase	261
adE* (ad8) #	I	" "	158,262,223*
adF (ad9) #	I	" "	71,158
adG (ad14) #	I	" "	158
adH (ad23) #	II	" "	158
adI (ad50) #	III	" "	108
agaA	VI	Arginase	38,51,183
ahrA #	VIII	Aspartic hydroxamate res., asparaginase	110,183
alX #	III	Allantoin non-ut., allantoinase	100,283,285
alcA*	VII	Ethanol non-ut., alcohol deH I	188*,232,195
alcB	?	Alcohol deH II	290
alcC*	VII	Alcohol deH III	157*
alcR*	VII	Ethanol ut. regl.	114*,188*290
aldA*	VIII	" " " aldehyde deH	290,232,251*,188*
alP	?	Allantoin non-ut., allantoin permease	98,285,183
amaA	?	Ammonium ut. (with gdhA)	183
amdA #	VII	Acetamidase regl.	145,19,241
amdI*	III	" "	90*,145,241
amdR* (=intA)	II	Acrylamide non-ut. acetamidase regl.	3*,146,147,241
amdS*	III	Amide non-ut., acetamidase	90*,146,147,195

Locus	Linkage gp.	Phenotype, enzymes etc.	References
amdT*(= areA)	III	Acrylamide ut., acetamidase regl.	68*,146,147,241
ammE (see amrA)			
amrA (ammE)	(II)	Ammonium derepressed	246,241
amrB (see tamA)			
anA (an1) #	I	Aneurine req.	158
anB (an2) #	II	" "	117
aniA	?	Arginase repression	52,52
anthA-D	?	Anthranillate req.	222,183
apA	(VIII)	Aminopterin res.	6
aplA (allp)	(VI)	Allopurinol res., xanthine deH regl.	140,283,285
apnA #	II	Asparagine ut. asparaginase II	20
apsA	IV	Anucleate primary sterigmata	81
apsB	VI	" " "	81
ar	-	Actidione res.	337,338
araA (galB?) #	VIII	L-arabinose ut.	82
arcA (suGpro)	(VIII)	Arginine catabolism regl.	51,52
arcs	-	Actidione res., cold sensitivity	337,338
areA* (amdT, xprD)	III	Ammonium repression	68*,241,285,19,183,17,31
areB	I	" "	26,31,320
argA (arg1) #	VI	Arginine req., arginosuccinase	44,261,95,183
argB* #	III	" " OTCase	58*,95,117,261,183
argC #	VIII	" " AS synthetase	97,117,261,183
argD #	?	" "	298
aroA* (arom)	VIII	Aromatic metabolite req.	73*,270,183
aroB	(V)	" " "	270,183
aroC #	I	" " "	270,183
arpA	(VII)	Altered ribosomal profile	338
arpB	VIII	" " "	338
arpC	(VIII)	" " "	338
as	-	Actidione ultrasensitivity	337,338
asnA	(III)	Asparagine req.	183
asuA #	?	Antisuppressor of suaC	207
asuB #	(II)	" " "	207
asuC #	(I)	" " "	207
asuD #	(V)	" " "	207
asuE #	?	" " "	207
azgA	?	8-Azaguanine res., purine upt.	98,285,183
azgB	?	" " "	312
benA* #	VIII	Benlate res. β-tubulin	134,210*,217,292,328
benB #	II	" "	328
benC #	VII	" "	328
bgaA	?	Lactose non-ut.(germination) β-galactosidase	113,195
bgaB	?	" " " "	113,195
bgaC	?	" " " "	113,195
bguA	(II)	Benzoate & quinate ut.	150
biA (bi1) #	I	Biotin req.	273
bimA	(I)	Blocked in mitosis	215
bimB	(III)	" " "	215
bimC	(VI)	" " "	215
bimD	(IV)	" " "	215
bimE	(VI)	" " "	215,216
bimF	?	" " "	215
binA	(VI)	Inhibition by benzamide	144

Locus	Linkage gp.	Phenotype, enzymes etc.	References
binB	I	Inhibition by benzamide	144
binC	(I)	" " "	144
binD	I	" " "	144
binE	(VIII)	" " "	144
binF	VII	" " "	144
binG (bin-4)	I	" " "	144
bioA see biA			7
blA (bl1) #	(II)	Blue ascospores	4
bncA	IV	Binucleate conidia	258
brlA* (brl,br) #	VIII	"Bristle" aconidial	76,77,81,156*,213*
bwA (Bw) #	VI	Brown conidia	158
bzuA	IV	Benzamide ut.	143
cadA	IV	Cadmium res.	88
cadB	VI	" "	88
[camA]	cytoplasmic	Chloramphenicol res.	324,128
camB	(II)	" "	18,128
camC	I	" "	128
camD	(V)	" "	128
carA,B,C see cbxB,C,A			
casA	(VIII)	Chloramphenicol sensitivity	186
cbxA (=carC) #	III	Carboxin res.	127,327,328
cbxB (=carA) #	VII	" "	127,327,328
cbxC (=carB) #	VIII	" "	127,327,328
chaA (cha) #	VIII	Chartreuse coloured conidia	160
chlA (=dafA, pcnbA) #	III	Chloroneb res.	328,36
choA	VII	Choline req. PEA methyltransferase	11,158
choB	(VI?)	" "	337
choC #	VIII	" "	206
clA (cl4) #	(IV)	Colourless ascospores	4
clB (cl6) #	I	" "	4
cnxA, B, C (ni50) #	VIII	Nitrate and hypoxanthine ut.	91,92,243,285
cnxE (ni3) #	II	" "	91,92,158,243,285
cnxF #	VII	" " "	91,92,243,285
cnxG #	VI	" " "	91,92,243,285
cnxH #	III	" " "	91,92,243,285
cnxJ	VII	" " "	28
coA (co) #	VIII	Compact morphology	261
[cobA]	cytoplasmic	Apocytochrome b	200
[comA]	cytoplasmic	Compact morphology	276
comB?	(V)	" " mitotic instability	277
creA*	I	Carbon repression	26,37,19,102*
creB (molB) #	II	" "	148
creC #	II	" "	26,148
creD (cre-34)	II	" "	26,175,30
crnA*	VIII	Nitrate upt., chlorate resistance	65,155*,321
[csA] [cs-67]	cytoplasmic	Cold sensitivity	324,337
csuA #	(V)	Choline-O-sulphate ut.	12
cysA #	V	Sup. of meth, serine transacetylase	237,253,96
cysB (cysE1) #	II	" " cysteine synthesis	237,253,96
cysC #	V	" " "	237,253,96
cysD #	II	" " homocysteine synthase	253,96
cysE #	II	" " cysteine synthase	253,96
dafA (=chlA)	III	Iprodione res.	53
davA	I	delta-aminovaleric acid ut.	18

Locus	Linkage gp.	Phenotype, enzymes etc.	References
dcl	-	Dense cleistotheciation	248,350
dilA (dil) #	III	Dilute conidial colour	154
dilB	(I)	" " "	323
drkA (drk) #	VII	Dark conidial colour	76
drkB #	II	" " " , sup. of brlA12	77
etbA	II	Ethidium bromide res.	281
facA* (acuA) #	V	Fluoroacetate res., acetate ut. acetyl-coA synthetase	5,10,102*,195
facB* (acuB) #	VIII	" " " "	5,10,102*,195
facC (acuC) #	VIII	" " " "	5,10,195
fanA #	(V)	" "	5
fanB #	(VII)	" "	5
fanC	(VI)	" "	5
fanD #	(VIII)	" "	5
fanE #	(VI)	" "	5
flA (fl)	VII	Fluffy morphology	47
fluA (fl2)	VIII	" "	107
fluB (flu3)	VIII	" "	107
fluC (flu7)	I	" "	107
fluD (flu11)	VIII	" "	107
fluE (flu16)	VIII	" "	322
fluF (f)	VIII	" "	311
fmdS	III	Formamide ut., formamidase	146,147
fpaA (tyrA) #	I	p-fluorophenylalanine res., tyrosine req.	295
fpaB #	I	" " "	297
fpaD #	VIII	" " aromatic amino acid upt.	296,183,317
fpaE (=trpA)	II	" " tryptophan req.	295
fpaF	VI	" "	308
fpaG	V	" "	308
fpaH	VI	" "	308
fpaI #	I	" "	308
fpaJ	VI	" "	308
fpaK	VIII	" " aromatic amino acid upt.	308,317
fpaL	(VI)	" " tryptophan ut.	300
fpaM	I	" "	300
fpaN	VIII	" "	300
fpaO	I	" " acid, neutral, basic amino acid upt.	312,317
fpaP	II	" "	317
fpaQ	II	" " acid, neutral, aromatic amino acid upt.	317
fpaR	I	" "	317
fpaS	(II)	" "	317
fpaT	(I)	" "	317
fpaU	V	" " phenylalanine tRNA synthetase	316
frA (fr, suc) #	IV,VIII	Fructose ut.	267,195
fulA	(III)	Fluoropyrimidine res.	235
fulB	(I)	" "	235
fulC (=pyrC,D?)	(VIII)	" "	235
fulD (=argC)	VIII	" " arginosuccinate synthetase?	235
fulE	?	" "	235
fulF	?	" "	235
furA	?	" "	235
fwA* (fw,bge) #	VIII	Fawn conidia	75,81,326*
gaaA	(VI)	D-galacturonate ut.	331
gaaB	(I)	" " "	331

Locus	Linkage gp.	Phenotype, enzymes etc.	References
gabA	VI	GABA permease	13,38
gabI	VI	" " regl.	38
galA (gal1) #	III	Galactose ut., galactokinase and galactose-1-P uridyl transferase regl.	19,267,195
galB (gal3 =araA?) #	(II)	Galactose ut.	267,195
galC (gal4) #	(VIII,IV?)	" " molybdate res.	22,267,195
galD (gal5) #	I	" galactose-1-P uridyl transferase	267,195
galE (gal9) #	III	" " galactokinase	267,195
galF (gal2) #	(VIII)	" "	267,195
galG #	VIII	" " sup. of brlA12	77,195
galH (galC7) #	(IV?)	" " molybdate res.	22,267,195
gamA #	(II)	" " molybdate res.	22
gamB #	VIII	" " " "	22
gamC #	?	" " " "	22
gatA*	VII	GABA transaminase	13,102*157
gatB	II	GABA ut.	27
gcnA (glcA) #	V	Glucosamine req.	46
gdhA #	III	Ammonium sensitive NADP-glutamate deH	241,181,183
gdhB #	IV	Glutamate ut., NAD-glutamate deH	182,195,183
gdhC	(III)	" " NAD-glutamate deH regl.	177,195,183
ghrA,B	-	Glutamic hydroxamate res.	183
glcA (see also gcnA)	(V)	Glycerol ut.	195,330
glcB	(I)	" "	195,330
glcC	VI	" " glycerol upt.	336
glnA	II	Glutamine synthetase	196,30,183
glrA	VI	D-glucuronate ut.	38
glrB	(IV)	" "	150
glrC	(V)	" "	150
glrD,E	(VII)	" "	150
glrF,G	(VIII)	" "	150
gmdA	II	Benzamide ut., General amidase	143
gpdA*	?	Glyceraldehyde 3-phosphate deH	263*
hetA	(V)	Heterokaryon incompatibility	93,94,97
hetB	(VI)	" "	93,94,97
hetC	(V)	" "	93
hetD,E,F	(III)	" "	93
hetG	(VII)	" "	93
hetH	(II)	" "	93
hfaA	I	High frequency of aneuploids	332
hfaB	VI	" " " "	332
hfaC,F,H	(III)	" " " "	332
hfaD,I	(VIII)	" " " "	332
hfaE,L	(I)	" " " "	332
hfaG,K	(VI)	" " " "	332
hisA #	IV	Histidine req.	57,250,183
hisB #	I	" " IGP dehydrase	57,250,183
hisC #	VIII	" " IAP transaminase	57,250,183
hisD	VIII	" " histidinol deH	57,250,183
hisE	VIII	" " PR-ATP pyrophosphorylase	57,250,183
hisF	(VII)	" "	57,250,183
hisG #	(II)	" " PRP-ATP pyrophosphorylase	57,250,183
hisH #	(VIII)	" "	57,250,183
hisI	VIII	" " PR-AMP cyclohydrase	57,250,183
hisJ (hisEL,his122) #	VII	" "	250,183

Locus	Linkage gp.	Phenotype, enzymes etc.	References
hxA* #	V	Hypoxanthine non-ut.,xanthine deH I	100,102*,140,285
hxB	(VII)	" " " " I & II	100,140,285
hxnC see xanA			
icl see acuD and iclA,B			
iclA	IV	Isocitrate lyase regl.	193
iclB	(I)	" " "	193
ileA (abA) #	II	Isoleucine req. Threonine dehydratase	15,197,250
imaA #	VII	Imazalil res.	328
imaB (=camD) #	(V)	" "	328
imaC #	II	" "	328
imaD #	VIII	" "	328
imaE	(II)	" "	328
imaF	(I)	" "	328
imaG (=actA?) #	III	" "	328
imaH (=actC) #	III	" "	328
indA (ind)	?	Indole req.	183,222
inoA #	II	Inositol req.	120
inoB #	IV	" "	80
intA* (=amdR) #	II	Integration of GABA metabolism	13,26,19,3*,241
Inv1(VI)-lysA1 #	VI	Inversion	162
Inv1(III)-xprD1 #	III	"	17
iodA (Iod) #	II	Iodoacetate res.	341
ipsA*	?	Isopenicillin-N-synthetase	266*
ivoA #	III	Ivory conidiophores	76,81
ivoB* #	VIII	" " phenol oxidase	76,81,61*
lacA (lac1) #	VI	Lactose ut.	267
lacB (lac3) #	II	" "	267
lacC	(VII)	" " β-galactosidase	113,124
lacD	I	" "	113,124
lacE	?	" " lactose upt.	113,124
lacF	?	" " lactose upt.	113,124
lacG,I	(VI)	" " β-galactosidase	113,124
lacH	?	" " " "	113,124
lamA*	VIII	Lactam ut.	26,173*
lamB*	VIII	" "	173*
lcl	-	Low cleistotheciation	248,350
luA (lu) #	I	Leucine req.	117,297,183
lysA (lys1, lysAL) #	VI	Lysine req.	250,261,183
lysB (lys5, lysCL) #	V	" "	158,250,183
lysC (lys6, lysDL) #	(VII)	" "	250,183
lysD (lys7, lysBL) #	VII	" "	250,183
lysE (lys10, lysEL) #	V	" "	250,183
lysF (lys51, lysFL) #	I	" "	250,183
malA (mal) #	VII	Maltose ut.	267
manA #	VIII	Mannose ut., phosphomannose isomerase	195,46
mapA (suAmeth, su25)	?	Methionine alternative pathway	237
mapB	(I)	" " "	240
masA see acuD			
mauA (mau) #	IV	Monoamine ut., monoamine oxidase	21,232,183
mauB #	II	" " " "	232,183
mauC	II	Monoamine ut. in puA strains	232,183
meaA (mea8) #	IV	Methylammonium res., NH_4 derepressed	21,183,241
meaB (mea6) #	III	" "	21,183,241
mecA (meth-i)	I	Inhib. methionine, cystathionine β-synthase	252,183

Locus	Linkage gp.	Phenotype, enzymes etc.	References
mecB	I	Inhib. methionine, γ-cystathionase	252,183
mecC #	?	" " methionine adenosyltransferase	252,183
medA (med) #	I	Medusa morphology	76,350,81
melA #	(VII)	Excess melanin formation	208
melB #	VII	Reduced melanin formation	208
met see meth			7,96
methA	II	Methionine req. cystathionine γ-synthase	123,183
methB (meth3) #	VI	" " cystathionine γ-synthase	123,183
methC	(I)	" "	123,183
methD (=methH) #	III	" "	123,183
methE #	VII	" " homoserine transacetylase	123,183
methG (meth1,methF) #	IV	" " cystathionine β-lyase	123,158,183
methH (meth2,methD) #	III	" " THPTG methyl transferase	117,123,183
mipA*	VIII	γ-tubulin	226,344
mipB	(VIII,VI?)	Microtubule associated protein,	344
mnrA	V	Mannose relief, phosphomannose mutase	195,334
moA (mo1)	I	Morphologically abnormal	40
moB (mo9)	VIII	" "	42
moC (mo96) #	III	" "	40,42
modAmeaA	?	Modifier of meaA	183,246
molA #	VI	Molybdate res.	22
molB (=creB) #	II	" "	22
morA (mor-34)	?	Morphologically abnormal	2
nadA	(I)	Adenine ut., adenine deaminase	286
napA	?	Neutral amino acid permease	256
niaD* (ni7) #	VIII	Nitrate ut. nitrate reductase	91,92,155*,203*,242
nicA (nic2) #	V	Nicotinic or anthranilic acid req.	158,261,183
nicB (nic8) #	VII	" " " or tryptophan req.	158,183
nicC (nic10) #	VI	Nicotinic acid req.	158,183
niiA*	VIII	Nitrite ut., nitrite reductase	91,92,155*,243
niiB (see nirA)			
niiC #	I	Nitrite ut.	92
nimA*	(III)	Never in mitosis, protein kinase	215,231*
nimB, C, J	(V)	" " "	215
nimD, H, U	(VII)	" " "	215
nimE, G, R, T	(II)	" " "	215
nimF, W	?	" " "	215
nimI, K, M, V	(I)	" " "	215
nimL	(IV)	" " "	215
nimN	(VIII)	" " "	215
nimO	VII	" " "	23,215
nimP, S	(VI)	" " "	215
nimQ	VII	" " "	215,67
ninA	(VII)	Nitrite inhibited, sup. of areA[r]	318
nirA* (ni51,niiB, nir,am2) #	VIII	Nitrite ut., nitrate and nitrite reductase regl.	91,92,102*,241,243
nis -T1(II,VIII)	II,VIII	Nitrite ut. regl.	27
npeA	VI	Reduced penicillin production	87,94
npeB	(II)	" " "	87,202
npeC	(II)	" " "	87,202
npeD	VII	" " "	87,202
nudA, D, E	?	Nuclear distribution	215
nudB	(VIII)	" "	215
nudC	(I)	" "	215

Locus	Linkage gp.	Phenotype, enzymes etc.	References
[oliA]* #	cytoplasmic	Oligomycin res., ATP synthetase	64*,205,278,324
[oliB]*	cytoplasmic	" "	64*,278,280,324
oliC* (oliA) #	VII	" "	278,280,328,339*
ornA (orn4) #	IV	Ornithine req.	95,261
ornB #	VIII	" " N-acetyl ornithine transaminase	95,117
ornC #	III	" " N-acetyl ornithine deacetylase	95,17
ornD	I	" "	18,95
otaA (ota)	(VII)	Arginine ut., ornithine transaminase	255,183
[oxiA]*	cytoplasmic	Cytochrome oxidase I	64*,200
[oxiB]*	cytoplasmic	" " II	64*,200
[oxiC]*	cytoplasmic	" " III	64*,200
oxpA (oxp)	VIII	Oxallopurine res.	140,286
pA (p) #	V	Pale conidia	8
pabaA* (paba1) #	I	p-aminobenzoic acid req.	261,315*
pabaB (paba22) #	IV	" " "	190
pacA #	(IV)	Acid phosphatase def. phosphatase PII	22,67,105
pacB	VIII	" " "	22,105
pacC #	VI	" " " pH regl.	22,69,105
pacG	(V)	" " " phosphatase regl.	67
pacJ	I	" " "	70
palA #	III	Alkaline phosphatase def. pH regl.	22,69,105
palB #	VIII	" " " " "	22,69,105
palC #	IV	" " " " "	22,69,105
palD #	VII	" " " phosphatase PI	22,67,105
palE #	VIII	" " " pH regl.	22,69,105
palF #	VII	" " " " "	22,69,105
palG #	III	" " " phosphatase PII	67
palcA #	II	Acid and alkaline phosphatase def.	22,70,105
palcB	III	" " " "	22,105
palcC #	VIII	" " " "	22,105
pantoA #	VIII	Pantothenate req. pantothenate synthetase	16,158
pantoB #	VII	" "	16
pantoC #	III	" "	16
pcnbA (pcnb, =chlA?)	III	Pentachloronitrobenzene res.	313,328
pdeA	II	Piperidine ut.	27
pdeB	II	" "	26,30
pdhA #	I	Acetate req. pyruvate deH E1	330,195,248,272,19
pdhB #	V	" " pyruvate deH E2	330,195,248,62
pdhC #	VIII	" " " "	330,62
penA (pen)	VIII	Increased penicillin production	198
penB	III	" " "	198
penC	IV	" " "	198
pgkA?*	?	Phosphoglycerate kinase	74*
phenA (phen2) #	III	Phenylalanine req.	158,295
phenB #	VII	" "	295
pimA	IV	Pimaricin res.	328
pimB #	I	" "	328
pkiA*	(V)	Pyruvate kinase	103*,247,195,330
plnA	(III)	Pyrrolnitrin res.	130
plnB	?	" "	130
popB (=uvsB)	IV	High mitotic recombination	236
pppA #	VIII	Pentose-phosphate path, transaldolase	92,132,330
pppB #	(III)	" "	92,132,195
prnA* #	VII	Proline ut. regl.	23,111*,183

Locus	Linkage gp.	Phenotype, enzymes etc.	References
prnB*	VII	Proline ut. proline uptake	23,111*,183
prnC*	VII	" " pyrroline-5-carboxylate deH	23,111*,183
prnD*	VII	" " proline oxidase	23,111*,183
proA* (pro1) #	I	Proline req.	315*,118,343,183
proB (pro3) #	I	" "	118,343,183
puA #	II	Putrescine req. ornithine decarboxylase	15,303,310
punA #	II	Putrescine ut.	183,306
purA	?	Purine res.	25
pxnA see pyroA			7
pycA	III	Glutamate req. pyruvate carboxylase	330,301,195
pycB	?	" " " "	330
pyrA	(VIII)	Pyrimidine req. carbamyl phosphate synthetase	233,183
pyrB	(VIII)	" " aspartate transcarbamylase	233,183
pyrC	(VIII)	" " " "	233,183
pyrD #	VIII	" " dehydro-orotase	233,183
pyrE	I	" " dehydro-orotate deH	233,183
pyrF #	I	" " orotidine MP-pyrophosphorylase	233,183
pyrG* #	I	" " orotidine MP decarboxylase	227*,233,167
pyroA (pyro4) #	IV	Pyridoxine req.	158
pyroB #	I	" "	187,30
qutA* (qua)	VIII	Quinate ut. activator	6,138*,56*,125
qutB*	VIII	" " quinate dehydrogenase	35,136,137*,138*
qutC*	VIII	" dehydroshikimate dehydratase	135,136,137*,138*
qutD*	VIII	" " permease	136,138*,346
qutE*	VIII	" " dehydroquinase	136,137*,138*
qutG*	VIII	Quinate cluster, function unknown	138*
qutR*	VIII	Quinate ut. repressor	125*,138*
rA #	I	Enhanced phosphatase	105
rB	II	" "	108
rC	VIII	" "	108
rD (=rA?)	I	" "	70
rE	(VII)	" "	70
rec #	-	Low mitotic recombination	236
riboA (ribo1, rinB) #	I	Riboflavin req.	92,265
riboB*(ribo2,rinA) #	VIII	" "	92,158,228*
riboC (ribo3) #	(V)	" "	265
riboD (ribo5) #	V	" " for nitrate ut.	92,265
riboE (ribo6) #	II	" "	158,265
riboF (ribo8) #	(I)	" "	104
riboG (rinC)	V	" " for nitrate ut.	92
rin see ribo			
sA (s1) #	III	Sulphate ut., PAPS reductase	11,126,158
sB (s3) #	VI	" " , sulphate upt	11,126,158
sC (s0,s12,cys2) #	III	" " , ATP sulphurylase	11,126,261
sD (s50) #	VIII	" " , APS kinase	11,108,126
sE #	VIII	" " , PAPS reductase	126
sF #	VII	" "	126
sG	III	sulphite regulation	220
saA (sa)	II	Sage coloured conidia	176
sarA	(VII)	Sup. of areA, histidine ut.	183,260,241
sarB	(VII)	" " " " "	183,260,241
sbA (sb) #	VI	Sorbitol ut.	267
senA	II	Increased initial growth	27
sepA,C	(I)	Septation deficient	215

Locus	Linkage gp.	Phenotype, enzymes etc.	References
sepB	(V)	Septation deficient	215
sepD	?	"	215
sgpA (sgp1)	(VII)	Slow growing, noncleistothecial	141
sgpB (sgp2)	(V)	" " "	141
sgpC (sgp3)	III	" " "	141
sgpD (sgp4)	I	" " "	141
sgpE (sgp5)	(IV)	" " "	141
sltA #	VI	Salt sensitivity	305
smA (sm) #	III	Small colony	158
smsA	V	Suppressor of mutagen sensitivity	165
smsB	(III)	" " " "	165
sodA (=tsB) #	VI	Stabilisation of III disomics	333
sodB	?	" " VI "	55
sorA #	I	Sorbose res., sorbose upt.	63,112,195
sorB #	III	" " , phosphoglucomutase	112,195
spd	?	Spermidine supplementable (for puA)	307
SpoC1*	(VIII)	Spore-expressed gene cluster	131*
spsA #	III	Spermidine sensitivity	183,307
ssbA #	IV	Sup. of sbA	170
ssuA	II	GABA ut. SSA dehydrogenase	13,15,183
stfA	(VIII)	Duplication stability factor	33
stuA* (stu) #	I	Stunted conidiophores	81,76,212*,350
su-V	-	Sup. of deteriorated variants	211
suAadE20 (su1ad20) #	I	" " adE20	262
suCadE #	IV	" " adE	162
suAbenA33	?	" " benA33	225
suAmeth (mapA)	?	" " meth	123
suApabaB22 (supaba22)#	IV	" " pabaB22	190
suApalB7 #	VIII	" " palB7	22,105
suBpalB7 #	(VI)	" " "	22,105
suCpalF15 #	V	" " palF15	106
suDpalA1 #	(I)	" " palA1	106
suXpalC4	?	" " palC4	22
suApro (Su1pro)	III	" " pro	52,158,343
suBpro (Su4pro) #	III	" " "	158,343
suCpro (su6pro)	(III)	" " "	343
suDpro (su19pro)	I	" " " arginine regl.	52,343
suEpro (su11pro)	(V)	" " " " "	52,343
suFpro	?	" " " carbon repression	52,343
suH,J,Lpro	-	" " "	52
suaA #	III	Allele specific sup.	271,287
suaB	III	" " "	271,287
suaC #	(VII)	" " "	271,207
suaD	VII	" " "	271
sucA #	?	Succinate ut.	191
sulA (sul1) #	I	Sulfanilamide res.	152
sul-regA	?	Arylsulphatase derepressed	238,293
sul-regB	?	" " low sulphite reductase	238
sul-regC	?	" "	238
sul-regD	?	" " low ATP sulphurylase	238
sul-regE	?	" "	238
sumA	VII	Sup. of mitochondrial [cs-67]	340
sumB	(III)	" " " "	340
sumC	(IV)	" " " "	340

Locus	Linkage gp.	Phenotype, enzymes etc.	References
[sumD]	Cytoplasmic	Sup. of mitochondrial [cs-67]	340
sumE	(I)	" " " "	340
T1(I;II)	I	Translocation breakpoint	60,221
T1(I;III)-areB406	I,III	" "	31
T1(I;VII) #	I,VII	" "	161,162
T2(I;VIII)-sD50 #	I,VIII	" "	161,162
T1(II;VIII)-nis-5	II,VIII	" "	27
T1(III;IV)-areA18	III,IV	" "	68,31
T1(III;VII)-lysD20 #	III,VII	" "	162
T1(III;VIII)	III,VIII	" "	43,60
T2(III;VIII)-brlA12	III,VIII	" "	77
T1(IV;VIII)-frA1 #	IV,VIII	" "	162
T1(V;VI) #	V,VI	" "	162
T1(VI;VII) #	VI,VII	" "	162
T1(VI;VIII)-mipB	VIII	" "	344
tamA #	VI	Ammonium regulation	241,178,183,30
teA(te6)	(III)	Teoquil res.	341
telA #	VII	Mound (tel)-shaped colony	80
tgsA	(II)	Threonine, glycine & serine ut.	150
tguA	(I)	Threonine & glycine ut.	150
tguB	(VIII)	" " "	150
tguC	(IV)	" " "	150
thiA (thi4) #	II	Thiazole req.	158
trpA (trypA,fpaE) #	II	Tryptophan req. anthranilate synthetase	142,268,295
trpB (trypB) #	I	" " tryptophan synthetase	142,268
trpC* (trypC,ind2) #	VIII	" " IGP synthetase, PRA isomerase, anthranilate synthetase	142,163,218*,268
trpD (trypD) #	(II)	" " PR transferase	142,268
trpE (trypE)	(VI)	" "	142,268
tsA	II	Temperature sensitive	121
tsB (=sodA) #	VI	" "	121
tsC #	II	" "	121
tsD (ts) #	VIII	" "	121
tsE (ts6) #	(VI)	" " , glucosamine req.,	86,172
tsl	-	Temperature sensitive lethal	230
tubA	VIII	Suppressor of benA, alpha-tubulins 1 & 2	217
tubB	?	Beta-tubulin 2	342
tubC*	(I)	Beta-tubulin 3	210*,342
tutA	(III)	Threonine ut.	150
tutB	(II)	" "	150
tutC	(VII)	" "	150
tutD	(V)	" "	150
tutE	(VI)	" "	150
tyrB	(III)	Tyrosine req.(if fpaA^{-}) phenylalanine hydroxylase	295
uX,Y,Z see ureB,C,D			
uaX #	VI	Uric acid ut.	201,283,285
uaY #	VIII	" " " xanthine dehydrogenase and urate oxidase regl.	283,285,286,18
uaZ	I	" " " urate oxidase	283,285,286
uapA*	I	Uric acid and xanthine uptake	98,102*,283,183
uapB	VII	" " " " "	98,283,183
uapC	(I)	" " " " "	98,283,183
ufaA	?	Unsaturated fatty acid req. $^{a}\Delta$-desaturase	101
ureA (uruA) #	VIII	Thiourea res., urea uptake	285,183,201,241
ureB (uX) #	VIII	Urea ut., urease	285,183,201,241
ureC (uY) #	(VII)	" " "	285,183,201,241
ureD* (uZ) #	VIII	" " "	285,183,201,241,61*
ureE	(VIII)	" " "	285,183,241
uvr	?	UV repair defective	171
uvs	-	UV sensitive	168,347
uvsA (uvs1) #	I	" "	168,151
uvsB (popB,uvs12) #	IV	" "	168,347,236,153
uvsC #	VIII	" "	168,153
uvsD (uvs8) #	V	" "	168,347,122
uvsE #	V	" "	168,122
uvsF (uvsB) #	I	" "	168,291
uvsG #	VIII	" "	168
uvsH (uvs4,11,17) #	IV	" "	168,133,347
uvsJ #	V	" "	185
uvsK? (uvsH)	(III)	" "	72
v	-	Deteriorated variants of duplication strains	35
veA (ve, velA) #	VIII	Velvet morphology	75,166
velA see veA			7
vitA	(VII)	Vitavax res.	36
wA* (w) #	II	White conidia	261,282*,81
wetA* (wet) #	VII	Wet-white conidia	76,81,213*
xanA (hxnC)	(VIII)	Xanthine ut. (with hxB)	288
xprA	(VIII)	Extracellular protease deficient	84
xprB	(VII)	" " "	84
xprC	(VIII)	" " "	84
xprD* (=areA) #	III	Extracellular protease regl.	68*,84,85,17
yA* (y, yelA) #	I	Yellow conidia, laccase I	223*,261,78
yB	VIII	" " copper deficiency	184
yelA see yA			7
ygA (yg) #	II	Yellow-green conidia, copper deficiency	78

REFERENCES

1. Adams,T.H., M.T.Boylan & W.E.Timberlake 1988 Cell 54; 353-362
2. Almeida, C.G. & Y.B. Rosato 1984 Trans. Brit. Mycol. Soc. 83;139-143
3. Andreanopoulos,A. & M.J.Hynes 1988 Mol. Cell. Biol. 8; 3532-3541
4. Apirion, D. 1963 Genet. Res. 4, 276
5. Apirion, D. 1965 Genet. Res. 6, 317
6. Apirion, D., G.L. Dorn & E. Forbes 1963 Aspergillus Newsletter 4;15-16
7. Aramayo, R., T.H.Adams & W.E.Timberlake 1989 Genetics 122;65-71
8. Arkel, G.A. van 1962 Aspergillus Newsletter 3;4
9. Armitt, S., C.F. Roberts, & H.L. Kornberg 1971 FEBS Lett. 12, 276
10. Armitt, S., W. McCullough & C.F. Roberts 1976 J. Gen. Microbiol. 92;263-282
11. Arst, H.N.jr. 1968 Nature, 219, 268
12. Arst, H.N.jr. 1971 Genet. Res. 17, 273.
13. Arst, H.N.jr. 1976 Nature 262, 231-234.
15. Arst, H.N.jr. 1977 Molec. Gen. Genet. 151, 105:110.
16. Arst, H.N.jr. 1978 Molec. Gen. Genet. 163, 23-27.
17. Arst, H.N.jr. 1982 Molec. Gen. Genet. 188, 490-493
18 Arst, H.N.jr. 1988 Molec. Gen. Genet. 213; 545-547
19. Arst, H.N.jr. & C.R. Bailey 1977 In ref. 302. pp.131-146.
20. Arst, H.N.jr. & C.R. Bailey 1980 J. Gen Microbiol. 121, 243-247.

21. Arst, H.N.jr. & D.J. Cove. 1969 J. Bacteriol., 98, 1284
22. Arst, H.N.jr. & D.J. Cove. 1970 Mol. Gen. Genet. 108,146
23. Arst, H.N.jr. & D.W. MacDonald 1978 Molec. Gen. Genet. 163,17-22.
25. Arst, H.N.jr., D.W. MacDonald & D.J. Cove. 1970 Mol. Gen. Genet., 108, 129.
26. Arst, H.N.jr., H.A. Penfold & C.R. Bailey 1978 Molec.Gen.Genet. 166, 321-327.
27. Arst, H.N.jr., K.N. Rand & C.R. Bailey 1979 Molec. Gen. Genet. 174, 89-100
28. Arst, H.N.jr., D.W.Tollervey & H.M.Sealy-Lewis 1982 J.Gen.Microbiol. 128,1083
30. Arst, H.N.jr., A.G.Brownlee & S.A.Cousen 1982 Curr. Genet. 6, 245-257.
31 Arst, H.N.jr., D.W.Tollervey & M.X.Caddick 1989 Mol. Gen. Genet. 215;364-367
25. Atkinson P.W., J.A. King & M.J. Hynes 1985 Curr. Genet. 10, 133-138.
33. Azevedo, J.L. 1975 Genet. Res. Camb., 26, 55-61.
35. Azevedo, J.L. & J.A. Roper 1970 Genet. Res., 16, 79.
36. Azevedo, J.L., E.P. Santana & R. Bonatelli 1977 Mut. Res. 48, 163-172.
37. Bailey, C. & H.N. Arst jr. 1975 Eur. J. Biochem. 51, 573-577.
38. Bailey, C.R., H.A. Penfold & H.N. Arst jr. 1979 Molec. Gen. Genet. 169, 79-83
40. Bainbridge, B.W. 1966 Aspergillus Newsletter 7, 19.
42. Bainbridge, B.W. 1970 Genet. Res. Camb. 15, 317.
43. Bainbridge, B.W. & J.A. Roper 1966 J. Gen. Microbiol. 42, 417.
44. Bainbridge, B.W., H. Dalton & J.H. Walpole 1966 Aspergillus Newsletter 7, 18.
46. Bainbridge, B.W., P.Markham & P.B.Valentine 1979 Fungal Walls & Hyphal Growth. Symp. Soc. Brit. Mycol. Ed. J.H.Burnet & A.P.J.Trinci. Cambridge Univ. Press. pp 71-91
47. Ball, C. & J.L. Azevedo 1964 Aspergillus Newsletter 5, 9.
48. Ballance,D.J. & G.TURNER 1986 Molec. Gen. Genet. 202;271-275
50. Barbata, G.L., L. Valdes & G. Sermonti 1973 Molec. Gen. Genet. 126, 227-232.
51. Bartnik, E. & P. Weglenski 1974 Nature 250, 590-592.
52. Bartnik, E., J. Guzewska & P. Weglenski 1973 Molec. Gen. Genet. 126, 85-92.
52. Bartnik, E., J. Guzewska, J. Klimczuk, M. Piotrowska & P. Weglenski 1977 In ref. 302 pp243-254.
54. Beever,R. in ref. 67
55. Bergen,L.G., A.Upshall & N.R.Morris 1984 J. Bacteriol. 159;114-119
56. Beri,R.K., H.Whittington, C.F.Roberts & A.R.Hawkins 1987 Nucleic Acids Res. 15; 7991-8001
57. Berlyn, M., 1967 Genetics, 57, 561.
58. Berse,B., A.Dmochowska, M.Skrzypek, P.Weglenski, M.A.Bates & R.L.Weiss 1983 Gene 25; 109-117
60. Birkett, J.A. & J.A. Roper 1977 In ref. 302 pp.293-303.
61. Birse, C.E. & A.J.Clutterbuck, unpublished
62. Bos, C.J., M.Slakhorst, J.Visser & C.F.Roberts 1981 J.Bacteriol. 148,594-599.
63. Boschloo, J.G. 1985 Ph.D. thesis, Wageningen
64. Brown, T.A. 1990 Genetic Maps 5 (this volume)
65. Brownlee, A.G. & H.N. Arst jr. 1983 J. Bacteriol. 155, 1138-1146.
66. Butnik, N.Z., L.N.Yager, M.B.Kurtz & S.P.Champe 1984 J.Bacteriol.160, 541-545
67. Caddick, M.X. & H.N. Arst, jr. 1986 Genet. Res. Camb. 47,83-91.
68. Caddick,M.X. H.N.Arst, L.H.Taylor, R.I.Johnson & A.G.Brownlee 1986 EMBO J. 5,1087-1090
69. Caddick, M.X., A.G.Brownlee & H.N.Arst jr. 1986 Mol.Gen.Genet. 203,346-353
70. Caddick, M.X., A.G. Brownlee & H.N. Arst jr. 1986 Genet.Res.Camb. 47,93-102.
71. Calef, E., 1957 Heredity 11, 265.
72. Chae, S-K., D-M.King & H-S.Kang 1986 Kor. J. Microbiol. 241(3);221-227
73. Charles,I.G., J.W.Keyte, W.J.Brammer & A.R.Hawkins 1985 Nucleic Acids Research 13; 8119-8128
74. Clements, J.M. & C.F.Roberts 1985 Curr. Genet. 9; 293-298
75. Clutterbuck, A.J. 1965 Aspergillus Newsletter 6, 12.
76. Clutterbuck, A.J. 1969 Genetics, 63, 317.
77. Clutterbuck, A.J. 1970 Genet. Res. 16, 303.
78. Clutterbuck, A.J. 1972 J. Gen. Microbiol. 70, 423-435.
80. Clutterbuck, A.J. 1974 Handbook of Genetics I,447-510.Ed. R.C.King, Plenum NY
81. Clutterbuck, A.J. 1977 In ref. 302 pp.305-317.
82. Clutterbuck, A.J. 1981 Aspergillus News Letter 15, 21.
83. Clutterbuck, A.J. 1987 Genetic Maps 4, 325-335
84. Cohen, B.L. 1972 Heredity 29;131.
85. Cohen, B.L. 1972 J. Gen. Microbiol. 71;293-299
86. Cohen, J., D. Katz & R.F. Rosenberger 1969 Nature 224, 713.
87. Cole, D.S., G. Holt & K.D. MacDonald 1977 J. Gen. Microbiol. 96, 423-426.
88. Cooley, N.R., H.R.Haslock & A.B.Tomsett 1986 Curr. Microbiol. 13; 265-268
90. Corrick,C.M., A.P.Twomey & M.J.Hynes 1987 Gene 53; 63-71
91. Cove, D.J. 1977 In ref. 302 pp. 81-95.
92. Cove, D.J. 1979 Biol. Rev. 54, 291-327.
93. Croft, J.H. 1983 In Fungal protoplasts: their uses in biochemistry, genetics and physiology. Ed. J.F.Peberdy & L.Ferenczy, Cambridge Univ. Press.
94. Croft, J.H. & J.L. Jinks 1977 In ref. 302 pp. 339-360.
95. Cybis, J., M. Piotrowska & P. Weglenski 1972 Acta Microbiol. Pol, A4, 163-169
96. Cybis, J., R.Natorff, I.Lewandowska, W.Prazmo & A.Paszewski 1988 Genet. Res. Camb. 51;85-88
97. Dales, R.G.B.,J. Moorhouse & J.H.Croft 1983 J. Gen. Microbiol. 129; 3637-3642
98. Darlington, A.J. & C. Scazzocchio 1967 J. Bacteriol., 93, 937.
100. Darlington, A.J., C. Scazzocchio & J.A. Pateman 1965 Nature 206, 599.
101. Das, T.K. & K. Sen 1983 Ind. J. Exp. Biol. 21, 339-342.
102. Davis,M.A. & M.J.Hynes 1989 Trends Genet. 5; 14-19
103. de Graaf, L., H.van den Broek & J.Visser 1988 Curr. Genet. 13; 315-321
104. Devi, C.S.S. & E.R.B.Shanmugasundaram 1969 Curr.Sci.(Bangalore) 38, 193-195.
105. Dorn, G.L. 1965 Genet. Res. 6, 13.
106. Dorn, G.L. 1967 Genetics, 56, 619.
107. Dorn, G.L. 1970 Genetics, 66, 267.
108. Dorn, G.L. & W. Rivera 1965 Aspergillus Newsletter 6, 13.
110. Drainas, C., J.R.Kinghorn & J.A.Pateman 1977 J. Gen. Microbiol. 98,493-501.
111. Durrens,P., P.M.Green, H.N.jr.Arst & C.Scazzocchio 1986 Mol. Gen. Genet. 203; 544-549
112. Elorza, M.V. & H.N. Arst jr. 1971 Mol. Gen. Genet. 111, 185.
113. Fantes, P.A. & C.F. Roberts 1973 J. Gen. Microbiol. 77, 471-486.
114. Felenbok,B., D.Sequeval, M.Mathieu, S.Sibley, D.I.Gwynne & R.W.Davies 1988 Gene 73; 385-396
115. Fidel,S., J.H.Doonan & N.R.Morris 1988 Gene 70; 283-293
116. Foley, J.M., N.H. Giles & C.F. Roberts 1965 Genetics 52, 1247.
117. Forbes, E. 1959 Heredity 13, 67.
118. Forbes, E. 1959 Microb. Genet. Bull. 13, 9.
120. Forbes, E. unpublished
121. Forbes, E & U. Sinha 1966 Aspergillus Newsletter 7, 17.
122. Fortuin, J.J.H. 1971 Mutation Res. 11, 149-162.
123. Gajewski, W. & J. Litwinska 1968 Mol. Gen. Genet. 102,210.
124. Gajewski, W., J. Litwinska, A. Paszewski & T. Chojnacki 1972 Molec. Gen. Genet. 116,99-106
125. Grant, S., C.F.Roberts, H.Lamb, M.Stout & A.R.Hawkins 1988 J. Gen. Microbiol. 134;345-358
126. Gravel, R.A., E.Kafer, A.Niklewicz-Borkenhagen, & P.Zambryski 1970 Can. J. Genet. Cytol. 12, 831.
127. Gunatilleke, I.A.U.N., H.N.Arst & C.Scazzocchio 1975 Genet. Res. Camb. 26, 297-305.
128. Gunatilleke, I.A.U.N., C.Scazzocchio & H.N.Arst jr. 1975 Molec. Gen. Genet. 137, 269-276.
130. Gunatilleke, A.I.U.N., C. Scazzocchio & H. N. Arst. jr. Unpublished.

131. Gwynne,D.I., B.L.Miller, K.Y.Miller & W.E.Timberlake 1984 J. Mol. Biol. 180; 91-109
132. Hankinson, O. 1974 J. Bacteriol. 117, 1121-1130.
133. Hansen, P.A., L.E. Harvey & J.E. Lennox 1971 Genetics 68, s27.
134. Hastie, A.C. & S.G. Georgopoulos 1971 J. Gen. Microbiol. 67, 371-373.
135. Hawkins, A.R., N.H. Giles & J.R. Kinghorn 1982 Biochem. Genet. 20,271-286.
136. Hawkins, A.R., A.J.Francisco da Silva & C.F.Roberts 1984 J. Gen. Microbiol. 130, 567-574
137. Hawkins,A.R., A.J.Francisco da Silva & C.F.Roberts 1985 Curr.Genet.9;305-311
138. Hawkins, A.R., H.K.Lamb, M.Smith, J.W.Keyte & C.F.Roberts 1988 Mol. Gen. Genet. 214;224-231
140. Holl, F.B. & C. Scazzocchio 1970 FEBS Letters 12, 51.
141. Houghton, J.A. 1970 Genet. Res. 16,285.
142. Hutter, R. & J.A. deMoss 1967 Genetics, 55, 241.
143. Hynes, M.J. 1975 J. Gen. Microbiol. 91, 99-109
144. Hynes, M.J. 1975 Aspergillus Newsletter 12, 16-17.
145. Hynes, M.J. 1977 J. Bacteriol. 131, 770-775.
146. Hynes, M.J. & J.A. Pateman 1970 Mol. Gen. Genet. 108, 97.
147. Hynes, M.J. & J.A. Pateman 1970 Mol. Gen. Genet., 108, 107.
148. Hynes, M.J. & J.M. Kelly 1977 Mol. Gen. Genet., 150, 193-204.
150. Hynes, M.J. & J.M. Kelly 1981 Aspergillus News Letter 15, 18-21.
151. Jansen, G.J.O. 1967 Aspergillus Newsletter 8, 20-21.
152. Jansen, G.J.O. Unpublished.
153. Jansen, G.J.O. 1970 Mutat. Res. 10, 21.
154. Jansen, G.J.O. 1975 Aspergillus Newsletter 12, 20.
155. Johnstone, I.L. & J.R.kinghorn unpublished
156. Johnstone, I.L. S.G.Hughes & A.J.Clutterbuck 1985 EMBO J 4;1307-1311
157. Jones, I.G. & H.M. Sealy-Lewis 1989 Curr. Genet. 15; 135-142
158. Kafer, E. 1958 Adv. Genet., 9, 105.
160. Kafer, E. 1961 Genetics 46, 1581.
161. Kafer, E. 1976 Genetics 82, 605-627.
162. Kafer, E. 1977 Adv. Genet. 19, 33-131.
163. Kafer, E. 1977 Can. J. Genet. Cytol. 19, 723-738.
165. Kafer, E. 1987 Genetics 115;671-676
166. Kafer, E. & T.L. Chen 1964 Can. J. Genet. Cytol. 6, 249-254.
168. Kafer, E. & O. Mayor 1986 Mutat. Res. 161; 119-134
170. Kafer, E., B.R. Scott & A. Kappas 1986 Mutat. Res. 167,9-34
171. Kameneva, S.V. & Y.M. Romanova 1969 Genetika 5, 196-198.
172. Katz, D. & R.F. Rosenberger 1970 Biochim. Biophys. Acta. 208, 452.
173. Katz, M.E. & M.J.Hynes 1989 Gene 78; 167-171
175. Kelly, J.M. & M.J. Hynes 1977 Molec. Gen. Genet. 156;87.
176. Kilbey, B.J. 1960 Nature 186, 906-907.
177. Kinghorn, J.R. & J.A. Pateman 1974 Genet. Res. Camb. 23, 119-124.
178. Kinghorn, J.R. & J.A. Pateman 1975 Molec. Gen. Genet. 140, 137-147.
180. Kinghorn, J.R. & J.A. Pateman 1975 J. Gen. Microbiol. 86,174-184.
181. Kinghorn, J.R. & J.A. Pateman 1975 J. Gen. Microbiol. 86, 295-300.
182. Kinghorn, J.R. & J.A. Pateman 1976 J. Bacteriol. 125, 42-47.
183. Kinghorn, J.R. & J.A. Pateman 1977 In ref. 302 pp 147-202.
184. Kurtz, B.M. & S.P. Champe 1981 J. Bacteriol. 148, 629-638.
185. Lanier, W.B., R.W. Tuveson & J.E.Lennox 1968 Mutat. Res. 5,23-31.
186. Lazarus, C.M. & G. Turner 1977 Molec. Gen. Genet. 156, 303-311.
187. Lieber, M.M. 1975 Aspergillus Newsletter 12, 26.
188. Lockington,R.A., H.M.Sealy-Lewis, C.Scazzocchio & R.W.Davies 1985 Gene 33; Gene 33; 137-149
190. Luig, N.H. 1962 Genet. Res. 3, 331.
191. Luig, N.H. unpublished
192. McCullough, W. & C.F. Roberts 1974 FEBS letters 41, 238-242.
193. McCullough, W. & C.F. Roberts 1980 J. Gen. Microbiol. 120, 67-84.
195. McCullough, W., M.A. Payton & C.F. Roberts 1977 In ref. 302, pp.97-129.
196. MacDonald, D.W. 1982 Curr. Genet. 6;203-208.
197. MacDonald,D.W., H.N.Arst,jr. & D.J.Cove 1974 Biochim.Biophys.Acta 362,60-65.
198. MacDonald, K.D. & G. Holt 1976 Sci. Prog. Oxford 63, 547-573.
200. Macino, G., C. Scazzocchio, R.B. Waring, M.McP. Berks & R.W. Davies 1980 Nature 288,404-406
201. Mackay, E.M. & J.A.Pateman 1983 Biochem. Genet. 20, 763-776.
202. Makins,J.F., G. Holt & K.D. MacDonald 1983 J.Gen.Microbiol. 129,3027-3033.
203. Malardier, L. M.J.Daboussi, J.Julien, F.Roussel, C.Scazzocchio & Y.Brygoo 1989 Gene 78;147-156
205. Marahiel, M.A., G. Imam, P. Nelson, N.J. Pieniazek, P. Stepien & H. Kuntzel 1977 Eur. J. Biochem. 76, 345-354.
206. Markham, P. & B.W. Bainbridge 1978 Genet. Res. Camb. 32, 303-310.
207. Martinelli, S.D. 1987 Genet. Res. Camb. 49; 191-200
208. Martinelli, S.D. & B.W. Bainbridge 1974 Trans.Brit.Mycol.Soc. 69, 361-370.
210. May,G.S., M.L.-S.Tsang, H.Smith, S.Fidel & N.R.Morris 1987 Gene 55; 231-243
211. Menezes, E.M. & J.L. Azevedo 1978 Molec. Gen. Genet. 164, 255-258.
212. Miller, B.M. unpublished
213. Mirabito, P.M., T.H.Adams & W.E.Timberlake 1989 Cell 57 859-868
215. Morris, N.R. 1976 Genet. Res. Camb. 26, 237-254.
216. Morris, N.R. 1976 Exp. Cell Res. 98, 204-210.
217. Morris, N.R., M.H. Levi & C.E. Oakley 1979 Cell 16, 437.
218. Mullaney,E.J., J.E.Hamer, K.A.Roberto, M.M.Yelton & W.E.Timberlake 1985 Molec. Gen. Genet. 199;37-45
220. Nadolska-lutyk, K. & A. Paszewska A. 1988 Genet. Res. Camb. 51;1-3
221. Nga, B.H. & J.A. Roper 1968 Genetics, 58, 193.
222. Noronha, L. 1970 Ind. J. Exp. Biol. 8, 298-301.
223. O'Hara,E.B. & W.E.Timberlake 1989 Genetics 121; 249-254
225. Oakley, B.R. & N.R.Morris 1981 Cell, 24, 837-845
226. Oakley, C.E. & B.R.Oakley 1989 Nature 338;662-664
227. Oakley,B.R., J.E.Rinehart, B.L.Mitchell, C.E.Oakley, C.Carmona, G.L.Gray & G.S.May 1987 Gene 61; 385-399
228. Oakley,C.E., C.F.Weil, P.L.Kretz & B.R.Oakely 1987 Gene 53; 293-298
230. Oliviera, P.C.de, & M.H.Andersen 1975 Mem. Ist. Oswaldo Cruz, Rio de Janeiro 73,65-70
231. Osmani, S.A., R.T.Pu & N.R.Morris 1988 Cell 53;237-244
232. Page, M.M. & D.J. Cove 1972 Biochem. J. 127, 17P.
233. Palmer, L.M. & D.J. Cove 1975 Molec. Gen. Genet. 138, 243-255.
235. Palmer, L.M., C. Scazzocchio & D.J. Cove 1975 Molec.Gen.Genet. 140,165-173.
236. Parag, Y. & G. Parag 1975 Molec. Gen. Genet. 137, 109-137.
237. Paszewski, A. & J. Grabski 1975 J. Bacteriol. 124, 893-904.
238. Paszewski, A., W. Prażmo, J. Nadolska & M. Regulski 1984 J. Gen. Microbiol. 130,1113-1121
240. Paszewski. A. Personal communication
241. Pateman, J.A. & J.R. Kinghorn 1977 In ref. 302 pp.203-241.
242. Pateman, J.A. D.J. Cove, B.M. Rever & D.B. Roberts 1964 Nature, 210, 58-60.
243. Pateman, J.A., B.M. Rever & D.J. Cove 1967 Biochem. J. 104, 103.
245. Pateman, J.A., J.R. Kinghorn & E. Dunn 1974 J. Bacteriol. 119, 534-542.
246. Pateman, J.A., E.Dunn, J.R.Kinghorn & E.C.Forbes 1974 Molec. Gen. Genet. 133,225-236.
247. Payton, M. & C.F. Roberts 1976 FEBS Letters 66, 73-76.
248. Payton, M.A., W. McCullough, C.F. Roberts & J.R. Guest 1977 J. Bacteriol. 129,1222-1226
250. Pees, E. 1966 Aspergillus News Letter 7, 11.

251. Pickett,M., D.I.Gwynne, F.P.Buxton, P.Elliott, R.W.Davies, R.A.Lockington, C.Scazzocchio & H.M.Sealy-Lewis 1987 Gene 51; 217-226
252. Pieniazek,N.J., I.M.Kowalski & P.P.Stepien 1973 Molec.Gen.Genet. 126,367-374
253. Pieniazek, J., J. Bal, E. Balbin & P.P. Stepien 1974 Molec. Gen. Genet. 132, 363-366.
255. Piotrowska, M., M. Sawacki and P. Weglenski 1969 J. Gen. Microbiol. 55,301.
256. Piotrowska, M., P.P. Stepien, E. Bartnik & E. Zakrewska 1976 J. Gen. Microbiol. 92,89-96.
258. Pizzirani-Kleiner & Azevedo, J.L. 1986 Trans. Brit. Mycol. Soc. 86,123-130
260. Polkinghorne, M. & M.J. Hynes 1975 Genet. Res. Camb. 25, 119-135.
261. Pontecorvo, G., J.A. Roper, L.M. Hemmons, K.D. MacDonald & A.W.J. Bufton 1953 Adv. Genet. 5, 141-238.
262. Pritchard, R.H. 1955 Heredity 9, 343.
263. Punt, P.J., M.A.Dingemanse, B.J.M.Jacobs-Meijsing, P.H.Pouwels & C.A.M.J.J.van den Hondel 1988 Gene 69; 49-57
265. Radha, K. & E.R.B. Shanmugasundaram 1962 Nature 193, 165.
266. Ramon,D., L.Carramolino, C.Patino, F.Sanchez & M.A.Penalva 1987 Gene 57; 171-181
267. Roberts, C.F. 1963 J. Gen. Microbiol. 31, 45.
268. Roberts, C.F. 1967 Genetics, 55, 233.
270. Roberts, C.F. 1969 Aspergillus Newsletter 10, 19.
271. Roberts, T.J., S.D.Martinelli & C.Scazzocchio 1979 Molec.Gen.Genet.177,57-64
272. Romano, A.H. & H.L. Kornberg 1968 Biochim. Biophys. Acta., 158, 491.
273. Roper, J.A. 1950 Nature, 166, 956.
275. Roper, J.A. & E. Kafer 1957 J. Gen. Microbiol. 16, 660.
276. Rosato, Y.B. & J.L. Azevedo 1980 Trans. Brit. Mycol. Soc. 75, 313-315.
277. Rosato, Y.B. & J.L. Azevedo 1987 Rev. Bras. Genet. 10;425-434
278. Rowlands, R.T. & G. Turner 1973 Molec. Gen. Genet. 126, 201-216.
195. Rowlands, R.T. & G. Turner 1977 Molec. Gen. Genet. 154, 311-318.
196. Scarazzatti,M.E., R.Bonatelli jr. & J.L.Azevedo 1979 Experientia 35,307-308.
282. Scazzocchio, C. unpublished
197. Scazzocchio, C. & A.J. Darlington 1967 Bull. Soc. Chim. Biol. 49, 1503.
198. Scazzocchio, C. & G. Gorton 1977 In ref. 302. pp255-265.
286. Scazzocchio, C., N. Sdrin & G. Ong 1982 Genetics 100, 185-208.
287. Sealy-Lewis, H.M. 1987 Curr. Genet. 12; 141-148
288. Sealy-Lewis, H.M., C. Scazzocchio & S. Lee 1978 Molec. Gen. Genet. 164, 303.
290. Sealy-Lewis, H.M. & R.A. Lockington 1984 Curr. Genet. 8,253-259
291. Shanfield, B. & E. Kafer 1969 Mutat. Res. 7, 485.
292. Sheir-Neiss, G., M.H. Lai & N.R. Morris 1978 Cell 15, 639-647.
293. Siddiqi, O.H., B.N. Apte & M.P.Pitale 1966 Cold Spr. Harb. Symp. 31,381-382.
295. Sinha, U. 1967 Genet. Res. 10, 261.
296. Sinha, U. 1969 Genetics, 62, 495.
297. Sinha, U. 1970 Arch. Mikrobiol. 72, 308.
298. Sinha, U. unpublished
300. Sinha, M. & U. Sinha 1979 Genet. Res. Camb. 34, 121-130.
301. Skinner, V.M. & S. Armitt 1972 FEBS Letters 20, 16-18.
302. Smith, J.E. & J.A. Pateman (eds.) 1977 Genetics and Physiology of Aspergillus. British Mycological Society Symposium I, Academic Press,London
303. Sneath, P.H.A. 1955 Nature 175, 818.
305. Spathas, D.H. 1978 Aspergillus News Letter 14, 28.
306. Spathas, D.H.,A.J.Clutterbuck & J.A.Pateman 1983 FEMS Microbiol.Lett.17,345
307. Spathas, D.H., J.A. Pateman & A.J. Clutterbuck 1983 J. Gen. Microbiol. 129,1865-1871.
308. Srivastava, S, & U. Sinha 1975 Genet. Res. Camb. 25, 29-38.
310. Stevens, L. 1975 FEBS Letters, 59, 80-82.
311. Tamame,M., F. Antequera, & E. Santos 1988 Molec. Cell Biol. 8; 3043-3050
312. Teow S.C. & A. Upshall 1983 Trans. Brit. Mycol. Soc. 81,513-521
313. Threlfall, R.J. 1968 J. Gen. Microbiol., 52, 35.
315. Timberlake, W.E. unpublished
316. Tiwary, B.N., P.S.Bisen & U.Sinha 1987 Mol. Gen. Genet. 209;164-169
317. Tiwary, B.N., P.S.Bisen & U.Sinha 1987 Curr. Microbiol. 15;305-311
318. Tollervey, D.W. & H.N.Arst jr. 1981 Curr. Genet. 4, 63-68.
320. Tollervey, D.W. & H.N.Arst jr. 1982 Curr. Genet. 6, 79-85.
321. Tomsett, A.B. & D.J. Cove 1979 Genet. Res. Camb. 34, 19-32.
322. Torre Melis, R.A. de la, 1976 Ciencias, Havana ser. 4, Ciencias Biologicas 56, 3-24.
323. Torre, R.A. de la, 1981 Aspergillus News Letter 15, 22-23.
324. Turner, G. & R.T. Rowlands 1977 In ref. 302 pp. 319-337.
326. Turner, G. unpublished
327. Tuyl, J.M.van 1975 Neth. J. Plant Pathol. 81, 122-133.
328. Tuyl, J.M.van 1977 Meded. Landbouwhogeschool Wageningen 77(2), 1-136.
330. Uitzetter, J.H.A.A., C.J.Bos & J.Visser 1982 Ant. van Leeuwenhoek 48,219-227
331. Uitzetter, J.H.A.A., C.J.Bos & J.Visser 1986 J.Gen.Microbiol. 132,1167-1172.
332. Upshall, A., & Mortimore, I.D. 1984 Genetics 108, 107-121.
333. Upshall, A., B. Giddings, S.C. Teow & I.D. Mortimore 1979 Genetics of Industrial Microorganisms: Proc. 3rd. Int. Cong., pp197-204. ed. O.K. Sebek & A. Laskin, Am. Soc. Microbiol., Washington.
334. Valentine, B.P. & B.W. Bainbridge 1978 J. Gen. Microbiol. 109, 155-168.
336. Visser, J., R.van Rooijen, C.Dijkema, K.Swart & H.M.Sealy-Lewis 1988 J. Gen. Microbiol. 134;655-659
337. Waldron, C. & C.F. Roberts 1974 Molec. Gen. Genet. 134, 99-113.
338. Waldron, C. & C.F. Roberts 1974 Molec. Gen. Genet. 134, 115-132.
339. Ward,M., B.Wilkinson & G.Turner 1986 Molec. Gen. Genet. 202;265-270
340. Waring, R.B. & C. Scazzocchio 1980 J. Gen. Microbiol. 119, 297-311.
341. Warr, J.R. & J.A. Roper 1965 J. Gen. Microbiol. 40, 273.
342. Weatherbee, J.A., G.S.May, J.Gambino & N.R.Morris 1985 J. Cell.Biol. 101,706-711
343. Weglenski, P. 1966 Genet. Res. 8, 311.
344. Weil, C.F., C.E.Oakley & B.R.Oakley 1986 Mol. Cell Biol. 6,2963-2968
346. Whittington, H.A., S.Grant, C.F.Roberts, H.Lamb & A.R.Hawkins 1987 Curr. Genet. 12; 135-139
347. Wright, P.J. and J.A. Pateman 1970 Mutat. Res. 9, 579.
348. Zonneveld, B.M.J. 1974 J. Gen. Microbiol. 81, 445-451.
350. Zonneveld, B.M.J. 1977 In ref. 302, pp. 59-80.

MITOCHONDRIAL GENOME OF ASPERGILLUS NIDULANS

Dr T.A.Brown
Dept of Biochemistry and Applied Molecular Biology
University of Manchester Institute of Science and Technology
Manchester M60 1QD
UK

August 1989

Primary reference: T.A.Brown et al (1985) Current Genetics 9:113-7

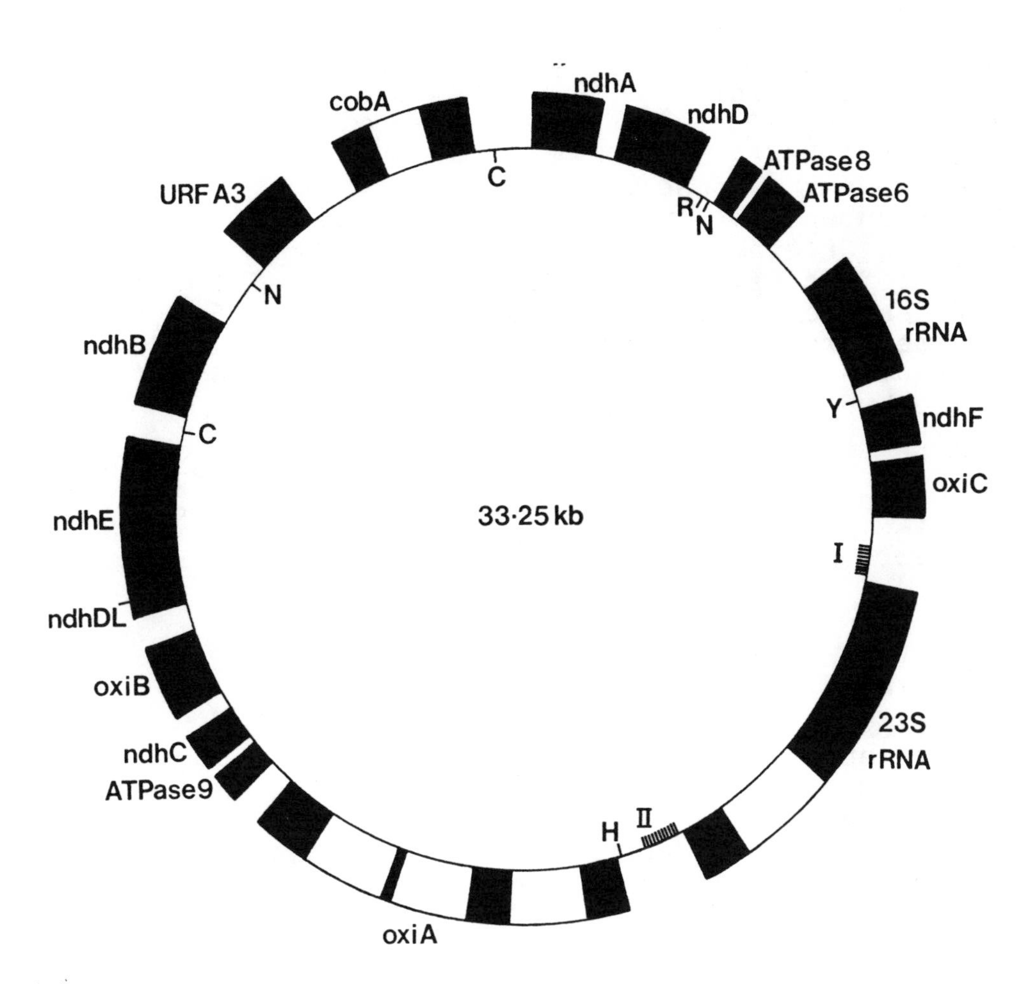

References
1 Brown TA et al (1983) EMBO J 2:427-435
2 Brown TA et al (1983) Nature 302:721-723
3 Brown TA et al (1984) Curr Genet 8:489-492
4 Brown TA et al (1989) Nucl Acids Res 17:4371
5 Brown TA et al (1989) Nucl Acids Res 17:5838
6 Brown TA unpublished
7 Dyson NJ et al (1989) Gene 75:109-118
8 Dyson NJ et al (1989) J Mol Biol, in press
9 Grisi E et al (1982) Nucl Acids Res 10:3531-3539
10 Gunatilleke IAUN et al (1975) Mol Gen Genet 137:269-276
11 Kochel HG & Kuntzel H (1981) Nucl Acids Res 9:5689-5696
12 Kochel HG & Kuntzel H (1982) Nucl Acids Res 19:4795-5001
13 Kochel HG et al (1981) Cell 23:625-633
14 Netzker R et al (1982) Nucl Acids Res 11:4783-4794
15 Rowlands RT & Turner G (1973) Mol Gen Genet 126:201-216

GENE	GENE PRODUCT	NOTES	REFERENCES
ndhA	NADH dehydrogenase subunit 1		1
ndhD	NADH dehydrogenase subunit 4		1, 14
R	arginyl-tRNA		9, 14
N	asparaginyl-tRNA	a	2, 9, 14
ATPase 8	ATPase subunit 8		2, 9, 14
ATPase 6	ATPase subunit 6	b	9, 14, 15, 16
16S rRNA	small rRNA subunit		7, 11
Y	tyrosinyl-tRNA		14
ndhF	NADH dehydrogenase subunit 6		14
oxiC	cytochrome c oxidase sub III		14
I	tRNA cluster I	c	13
23S rRNA	large rRNA subunit	d	7, 10, 12
II	tRNA cluster II	e	13
H	histidinyl-tRNA		14
oxiA	cytochrome c oxidase sub I		8, 20
ATPase 9	ATPase subunit 9	f	3, 8
ndhC	NADH dehydrogenase subunit 3		8
oxiB	cytochrome c oxidase sub II		8
ndhDL	NADH dehydrogenase subunit 4L		6
ndhE	NADH dehydrogenase subunit 5		4
C	cysteinyl-tRNA	a	6
ndhB	NADH dehydrogenase subunit 2		6
N	asparaginyl-tRNA	a	2
URF A3	?		2, 5
cobA	apocytochrome b	g	17, 18, 19
C	cysteinyl-tRNA	a	18

Notes

a. The asparaginyl and cysteinyl-tRNA genes are present as duplicated copies.
b. Probably corresponds to oliA and oli322.
c. tRNA cluster I contains genes for lys, gly1, gly2, asp, ser2, trp, ile, ser1, pro.
d. Probably corresponds to camA and cam51.
e. tRNA cluster II contains genes for thr, glu, val, met1, met3, leu1, ala, phe, leu2, gln, met2.
f. Possibly a pseudogene as there is a functional nuclear copy.
g. Probably corresponds to mucl.

16 Rowlands RT & Turner G (1977) Mol Gen Genet 154:311-318
17 Turner G, unpublished
18 Waring RB et al (1981) Cell 27:4-11
19 Waring RB et al (1982) Proc Natl Acad Sci USA 79:6332-6336
20 Waring RB et al (1984) EMBO J 3:2121-2128

The Nematode *Caenorhabditis elegans*

July 1989

Edgley, Mark L. and Donald L. Riddle, Caenorhabditis Genetics Center,
Division of Biological Sciences, Tucker Hall, University of Missouri, Columbia, MO 65211

This drawing of the *C. elegans* Genetic Map (N = 6) contains 830 genes. Each linkage group is divided into sections, and the gene cluster area of each chromosome (marked by a black box) is drawn at an enlarged scale on a separate page. Many genes that lie within the cluster appear only on the cluster enlargement. The scale of the main map is 1 map unit (1% recombination) to 1 centimeter, and the clusters are drawn at 1 map unit to 10 centimeters (except the LG V cluster, the LG II subcluster and the supplemental section of LG I). Each page of the map is identified by the appropriate Roman numeral or X, left and right chromosome arms are marked with "(L)" or "(R)" after the chromosome numeral, and a scale bar appears under each numeral. Linkage groups are displayed vertically, with the left (-) and right (+) arms extending from zero points defined by standard reference markers. Duplications (Dp) and deficiencies (Df) are drawn to the left of the main map line; duplications are indicated by a double bar, and deficiencies by a single bar. Endpoints are placed within the position of flanking markers on the main map line as determined by complementation testing. Dotted lines within the Df or Dp intervals correspond to positions of genes on the line that have not been tested for complementation. A few mapping markers positioned on the line are reproduced at the far right margin of the map at their correct left-right positions. These are included so a ruler can be placed across the map to facilitate reading endpoints of intervals off the line.

Conventions for gene placement and map symbols:

(1) Genes are placed on the line when they have been ordered with respect to their nearest neighbors on the line by three-factor (3F), Df, Dp or physical mapping data. Names of physically-mapped genes appear in italics. **(2)** Genes on the line that make up the "skeleton" for each linkage group (those that are placed by two-factor (2F) data in addition to ordering data) appear in boldface. If 3F data position a gene between two line markers and no 2F data exist, the gene is placed midway between the two markers, or the position is determined by interpolation from 3F data. **(3)** Genes are stacked at a single position on the line (within a brace if space allows, separated by commas if not) if they map to the same position by 3F, Df, Dp or physical data, but have not been ordered with respect to each other. **(4)** Genes that have been ordered with respect to each other, but for which no relevant 2F data exist, are shown separated in their correct order on a branched line pointing to a single position on the line. **(5)** Asterisks denote genes for which the available pieces of data conflict, or where a special ambiguity may exist. **(6)** Genes placed off the line to the right are indicated on bars representing intervals with endpoints of six types: (a) A solid single cross line is an endpoint determined by a 3F result with a gene on the line. The endpoint is placed *within* the position of the line marker. (b) A double cross line is an endpoint determined by complementation tests with deficiencies or duplications, placed *at* the position of the corresponding Df or Dp endpoint. (c) An arrow indicates that the gene in question was not separated from a marker on the line in a 3F cross. (d) A solid triangle is an endpoint determined from physical map data with respect to genes on the line. The endpoint is placed *within* the position of the line marker, and the names of genes so mapped appear in italics. When more than one physically mapped gene appear on a single bar, the genes are listed in correct order. (e) A solid dot represents a limit of a 2F 95% confidence interval. 2F dots are used both to represent genes placed by 2F in conjunction with very general 3F data, and to modify gene placement within 3F or Df/Dp bars. **(7)** Genes with 95% confidence intervals as large as half a linkage group are included in the Gene Map Positions list, but may be left off the map figure.

A uniform system of genetic nomenclature for *C. elegans* has been described (Horvitz *et al.*, MGG 175: 129 - 133). The accompanying gene list gives the general phenotype for each gene class, as well as linkage group, reference allele and bibliographic references for individual genes. Only genes appearing on the map figure are included in this list. An abbreviated bibliography follows. Detailed information on the mutations can be found in "The Nematode *Caenorhabditis elegans*" (Wood *et al.*, Cold Spring Harbor Laboratory, 1988, pp. 497 - 558).

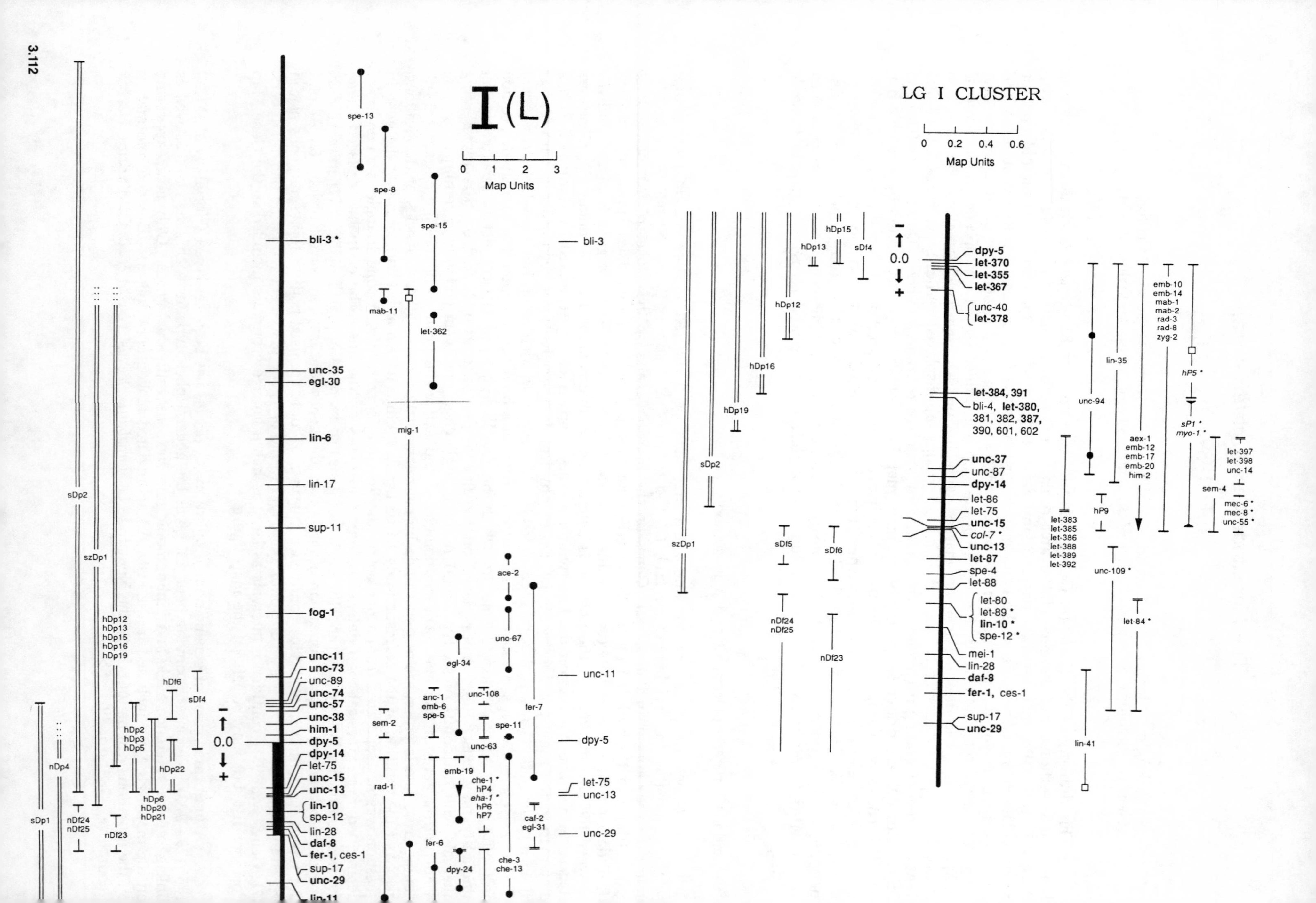

I (L)
0 1 2 3
Map Units
bli-3 *
unc-35
egl-30
lin-6
lin-17
sup-11
fog-1
unc-11
unc-73
unc-89
unc-74
unc-57
unc-38
him-1
dpy-5
dpy-14
let-75
unc-15
unc-13
lin-10
spe-12
lin-28
daf-8
fer-1, ces-1
sup-17
unc-29
lin-11
0.0
spe-13
spe-8
spe-15
mab-11
let-362
mig-1
bli-3
ace-2
unc-67
egl-34
unc-108
spe-11
unc-63
anc-1
emb-6
spe-5
sem-2
emb-19
che-1 *
hP4
eha-1 *
hP6
hP7
rad-1
fer-6
dpy-24
che-3
che-13
caf-2
egl-31
fer-7
unc-11
dpy-5
let-75
unc-13
unc-29
sDf4
hDf6
hDp22
hDp6
hDp20
hDp21
hDp2
hDp3
hDp5
hDp12
hDp13
hDp15
hDp16
hDp19
szDp1
sDp2
nDf23
nDf24
nDf25
nDp4
sDp1
LG I CLUSTER
0 0.2 0.4 0.6
Map Units
0.0
dpy-5
let-370
let-355
let-367
unc-40
let-378
let-384, 391
bli-4, let-380,
381, 382, 387,
390, 601, 602
unc-37
unc-87
dpy-14
let-86
let-75
unc-15
col-7 *
unc-13
let-87
spe-4
let-88
let-80
let-89 *
lin-10 *
spe-12 *
mei-1
lin-28
daf-8
fer-1, ces-1
sup-17
unc-29
sDf4
hDp15
hDp13
hDp12
hDp16
hDp19
sDp2
szDp1
sDf5
sDf6
nDf24
nDf25
nDf23
unc-94
lin-35
aex-1
emb-12
emb-17
emb-20
him-2
emb-10
emb-14
mab-1
mab-2
rad-3
rad-8
zyg-2
hP5 *
sP1 *
myo-1 *
let-397
let-398
unc-14
sem-4
mec-6 *
mec-8 *
unc-55 *
let-383
let-385
let-386
let-388
let-389
let-392
hP9
unc-109 *
let-84 *
lin-41

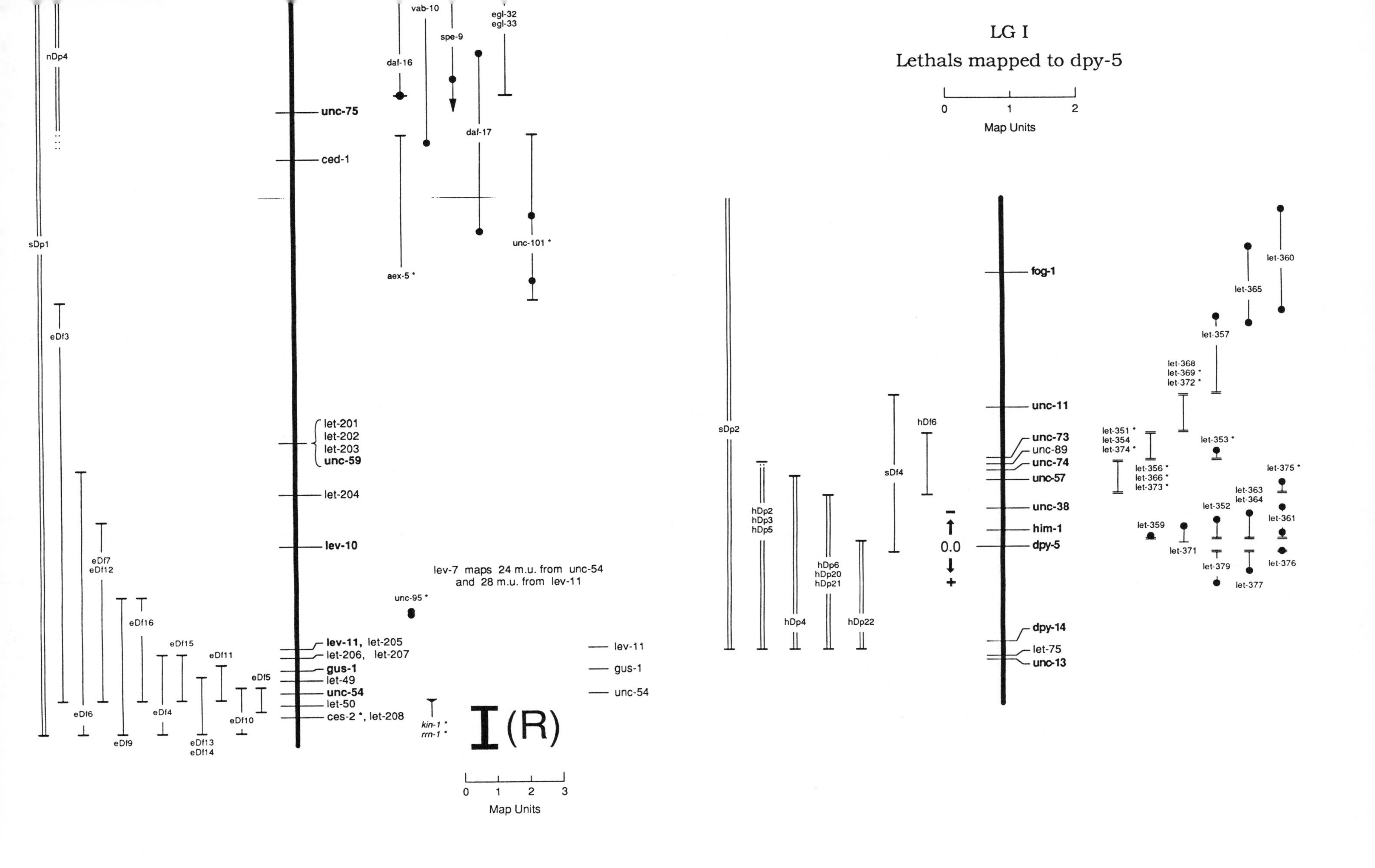

LG I
Lethals mapped to dpy-5
0
1
2
Map Units
fog-1
unc-11
unc-73
unc-89
unc-74
unc-57
unc-38
him-1
dpy-5
dpy-14
let-75
unc-13
0.0
let-360
let-365
let-357
let-368
let-369 *
let-372 *
let-353 *
let-351 *
let-354
let-374 *
let-356 *
let-366 *
let-373 *
let-375 *
let-363
let-364
let-352
let-361
let-359
let-371
let-379
let-377
let-376
sDp2
hDp2
hDp3
hDp5
hDp4
hDp6
hDp20
hDp21
hDp22
sDf4
hDf6
nDp4
sDp1
eDf3
eDf6
eDf7
eDf12
eDf9
eDf16
eDf4
eDf15
eDf13
eDf14
eDf11
eDf10
eDf5
unc-75
ced-1
let-201
let-202
let-203
unc-59
let-204
lev-10
lev-11, let-205
let-206, let-207
gus-1
let-49
unc-54
let-50
ces-2 *, let-208
daf-16
vab-10
spe-9
daf-17
egl-32
egl-33
aex-5 *
unc-101 *
unc-95 *
kin-1 *
rrn-1 *
lev-7 maps 24 m.u. from unc-54
and 28 m.u. from lev-11
lev-11
gus-1
unc-54
I (R)
0
1
2
3
Map Units

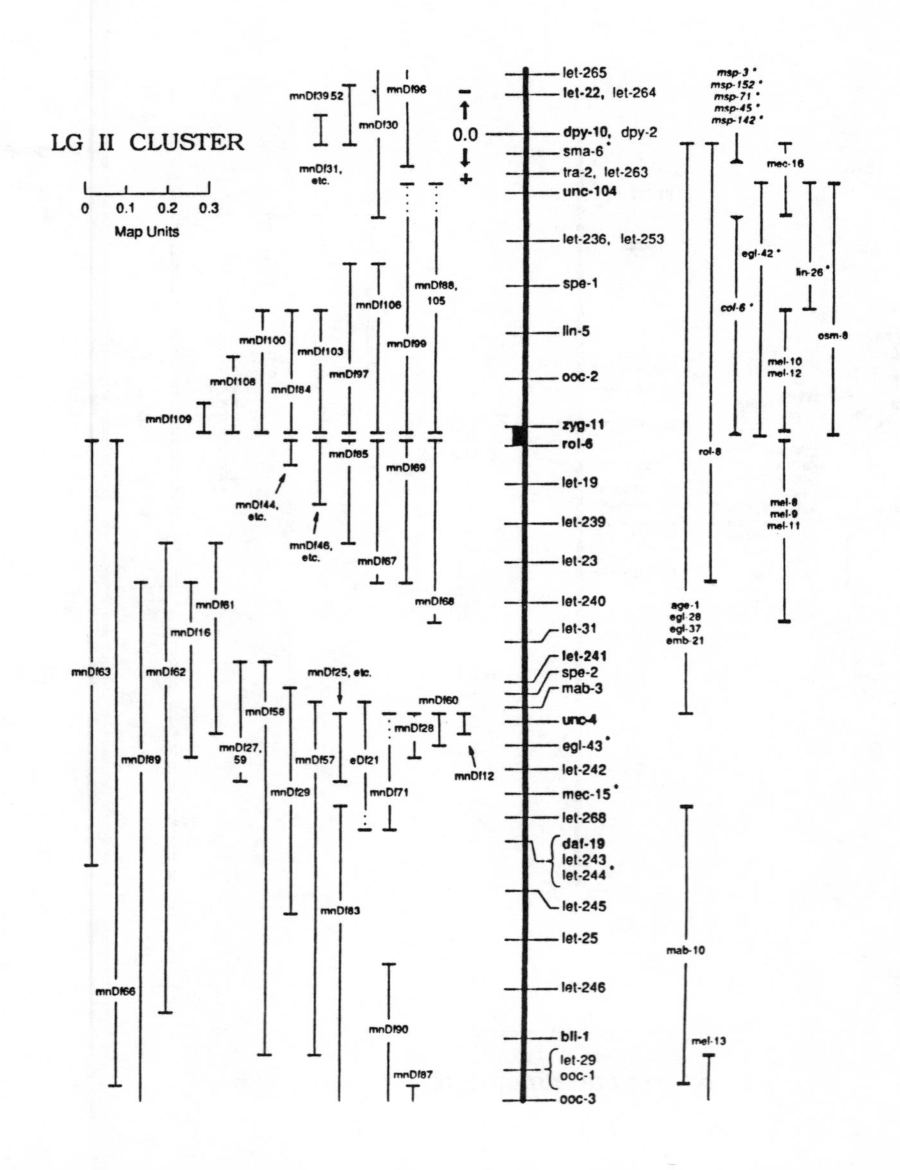
LG II CLUSTER
0
0.1
0.2
0.3
Map Units
let-265
let-22, let-264
0.0
dpy-10, dpy-2
sma-6*
tra-2, let-263
unc-104
let-236, let-253
spe-1
lin-5
ooc-2
zyg-11
rol-6
let-19
let-239
let-23
let-240
let-31
let-241
spe-2
mab-3
unc-4
egl-43*
let-242
mec-15*
let-268
daf-19
let-243
let-244*
let-245
let-25
let-246
bli-1
let-29
ooc-1
ooc-3
msp-3*
msp-152*
msp-71*
msp-45*
msp-142*
mec-16
egl-42*
lin-26*
col-6*
osm-8
mel-10
mel-12
rol-8
mel-8
mel-9
mel-11
age-1
egl-28
egl-37
emb-21
mab-10
mel-13
mnDf3952
mnDf96
mnDf30
mnDf31, etc.
mnDf88, 105
mnDf106
mnDf99
mnDf100
mnDf103
mnDf97
mnDf108
mnDf84
mnDf109
mnDf85
mnDf69
mnDf44, etc.
mnDf46, etc.
mnDf67
mnDf68
mnDf61
mnDf16
mnDf63
mnDf62
mnDf89
mnDf58
mnDf27, 59
mnDf25, etc.
mnDf57
eDf21
mnDf29
mnDf71
mnDf60
mnDf28
mnDf12
mnDf83
mnDf90
mnDf87
mnDf66

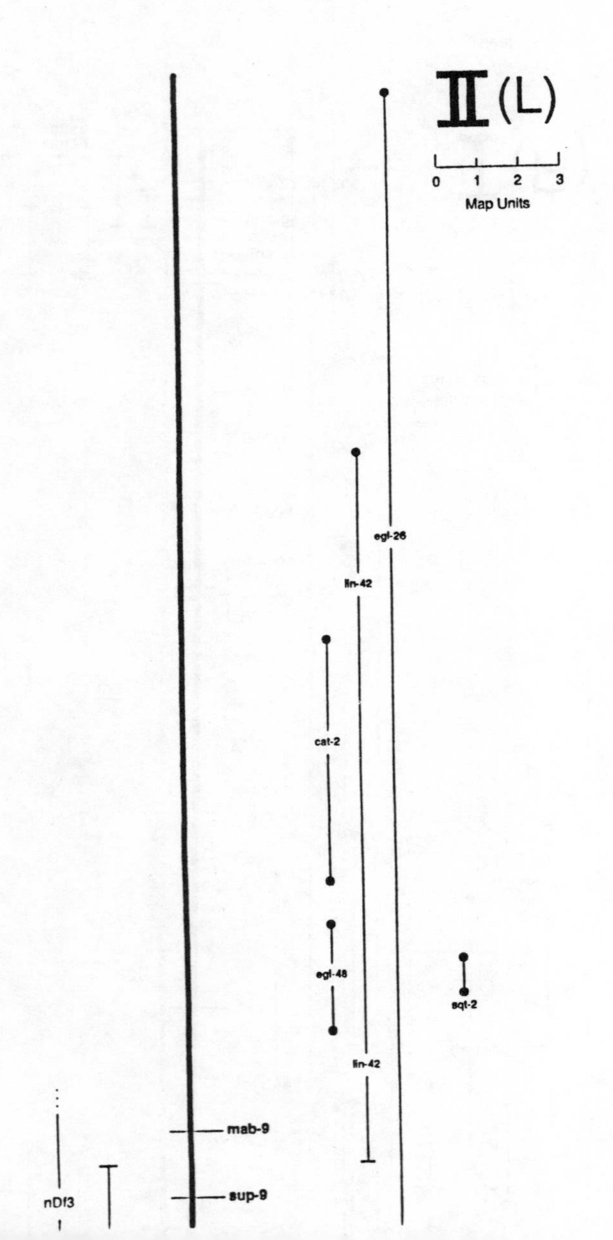
II (L)
0
1
2
3
Map Units
egl-26
lin-42
cat-2
egl-48
sqt-2
mab-9
sup-9
nDf3

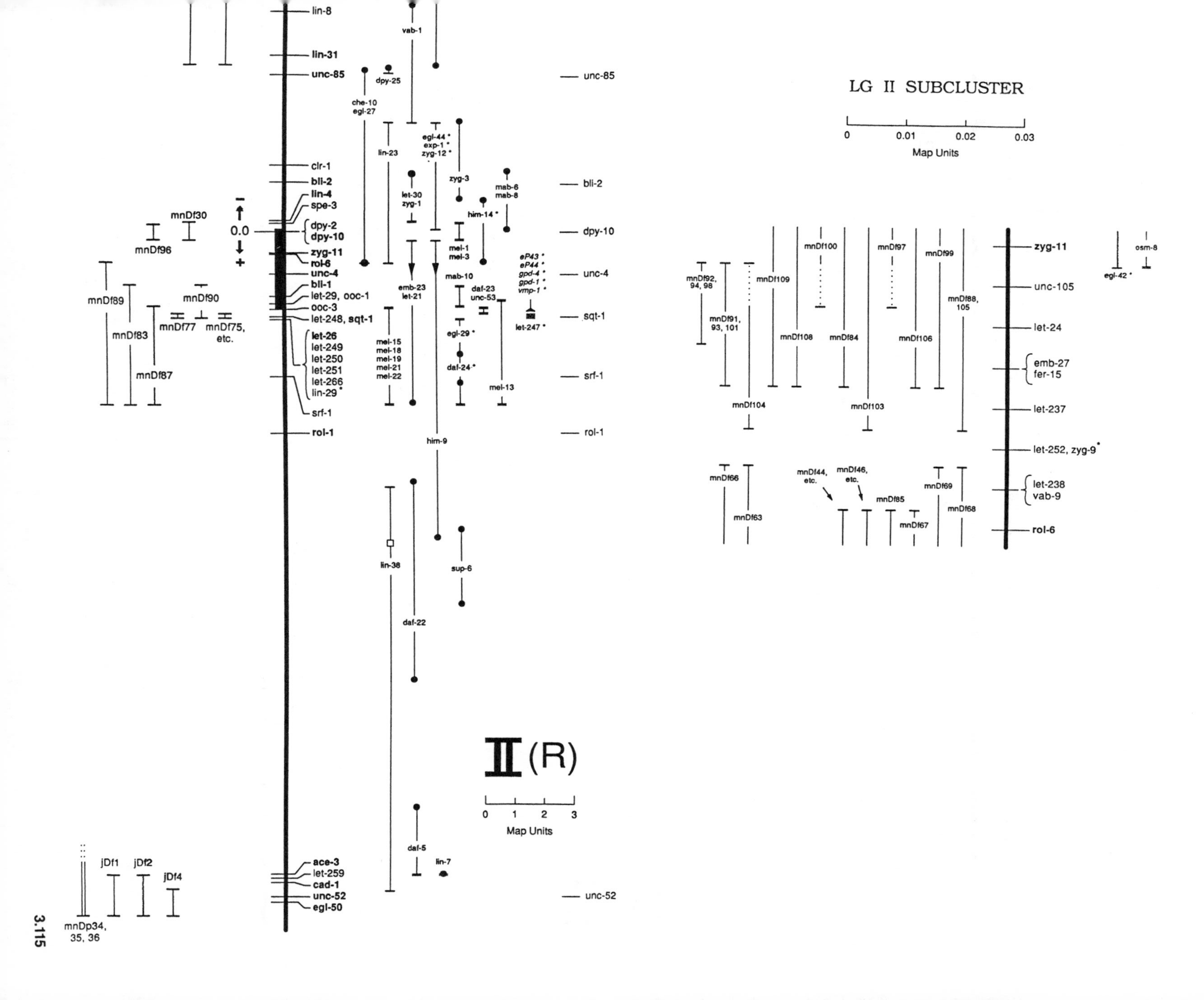

LG II SUBCLUSTER
0
0.01
0.02
0.03
Map Units
zyg-11
unc-105
let-24
emb-27
fer-15
let-237
let-252, zyg-9*
let-238
vab-9
rol-6
osm-8
egl-42*
mnDf88, 105
mnDf99
mnDf106
mnDf97
mnDf103
mnDf84
mnDf100
mnDf108
mnDf109
mnDf104
mnDf91, 93, 101
mnDf92, 94, 98
mnDf68
mnDf69
mnDf67
mnDf85
mnDf46, etc.
mnDf44, etc.
mnDf63
mnDf66
II (R)
0
1
2
3
Map Units
unc-85
bli-2
dpy-10
unc-4
sqt-1
srf-1
rol-1
unc-52
lin-8
lin-31
clr-1
lin-4
spe-3
dpy-2
zyg-11
rol-6
bli-1
let-29, ooc-1
ooc-3
let-248, sqt-1
let-26
let-249
let-250
let-251
let-266
lin-29*
ace-3
let-259
cad-1
egl-50
0.0
mnDf30
mnDf96
mnDf90
mnDf77
mnDf75, etc.
mnDf87
mnDf83
mnDf89
jDf1
jDf2
jDf4
mnDp34, 35, 36
vab-1
dpy-25
che-10
egl-27
lin-23
egl-44*
exp-1*
zyg-12*
let-30
zyg-1
zyg-3
mab-6
mab-8
him-14*
mel-1
mel-3
mab-10
daf-23
unc-53
emb-23
let-21
mel-15
mel-18
mel-19
mel-21
mel-22
egl-29*
daf-24*
mel-13
eP43*
eP44*
gpd-4*
gpd-1*
vmp-1*
let-247*
him-9
sup-6
daf-22
lin-38
daf-5
lin-7

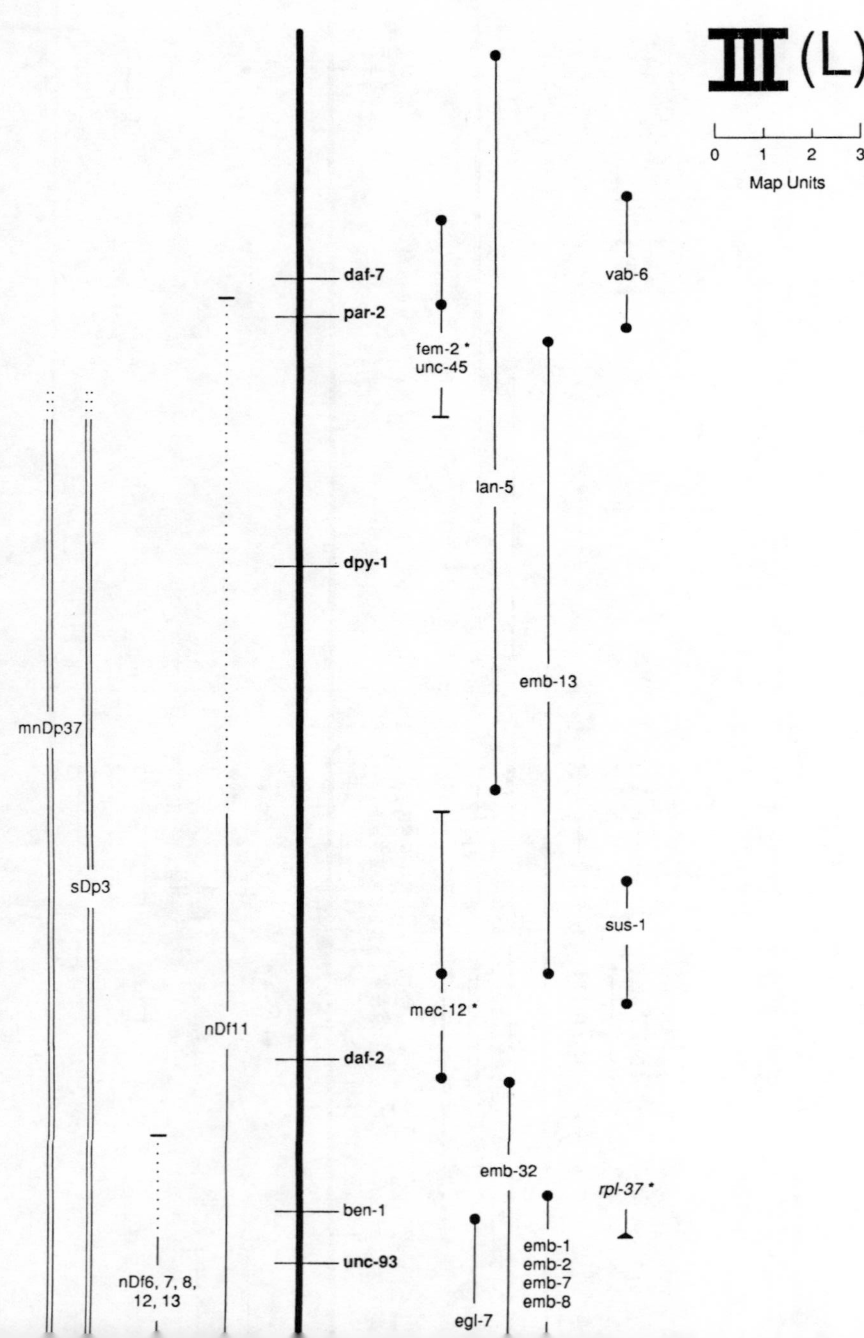

III (L)
0 1 2 3
Map Units
daf-7
par-2
dpy-1
daf-2
ben-1
unc-93
vab-6
fem-2 *
unc-45
lan-5
emb-13
sus-1
mec-12 *
emb-32
rpl-37 *
emb-1
emb-2
emb-7
emb-8
egl-7
nDf11
nDf6, 7, 8,
12, 13
sDp3
mnDp37

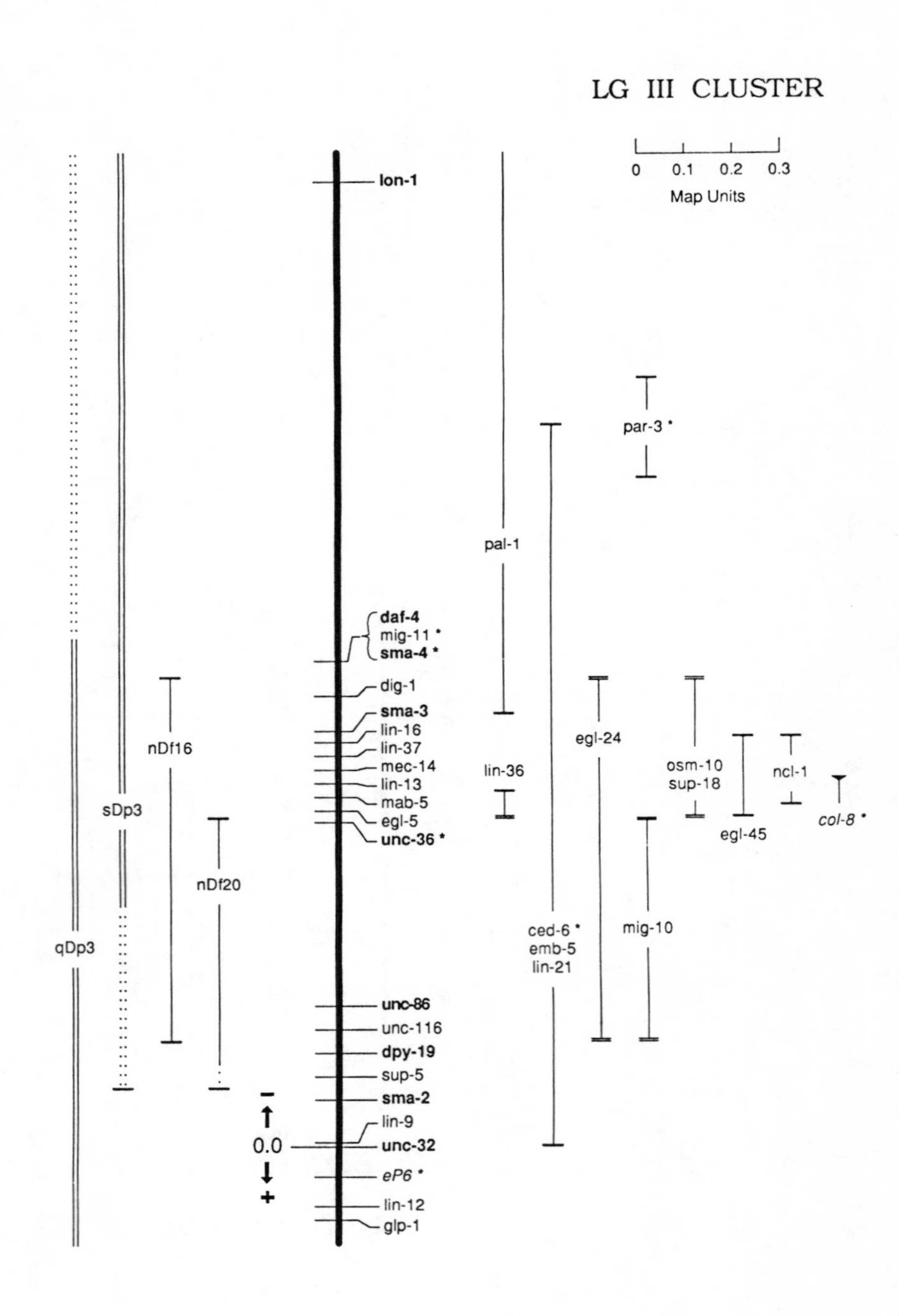

LG III CLUSTER
0 0.1 0.2 0.3
Map Units
lon-1
daf-4
mig-11 *
sma-4 *
dig-1
sma-3
lin-16
lin-37
mec-14
lin-13
mab-5
egl-5
unc-36 *
unc-86
unc-116
dpy-19
sup-5
sma-2
lin-9
unc-32
0.0
eP6 *
lin-12
glp-1
pal-1
lin-36
ced-6 *
emb-5
lin-21
egl-24
mig-10
par-3 *
osm-10
sup-18
egl-45
ncl-1
col-8 *
nDf16
nDf20
sDp3
qDp3

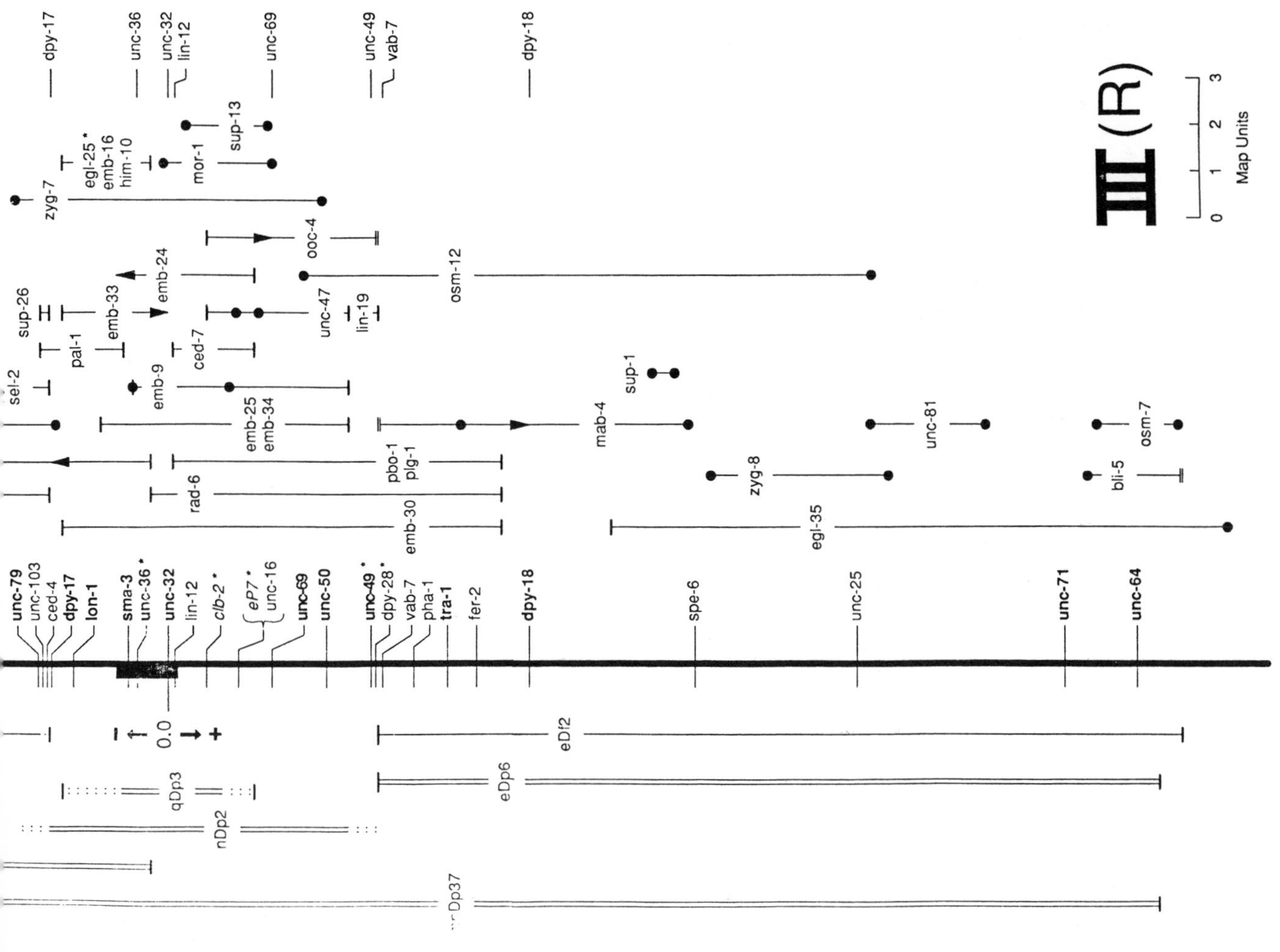

unc-79
unc-103
ced-4
dpy-17
lon-1
sma-3
unc-36 *
unc-32
lin-12
clb-2 *
eP7 *
unc-16
unc-69
unc-50
unc-49 *
dpy-28 *
vab-7
pha-1
tra-1
fer-2
dpy-18
spe-6
unc-25
unc-71
unc-64
0.0
eDf2
qDp3
eDp6
nDp2
Dp37
emb-30
egl-35
rad-6
zyg-8
bli-5
pbo-1
plg-1
emb-25
emb-34
mab-4
unc-81
osm-7
sel-2
emb-9
sup-1
pal-1
ced-7
sup-26
emb-33
unc-47
lin-19
emb-24
osm-12
ooc-4
zyg-7
egl-25 *
emb-16
him-10
mor-1
sup-13
dpy-17
unc-36
unc-32
lin-12
unc-69
unc-49
vab-7
dpy-18
III (R)
0
1
2
3
Map Units

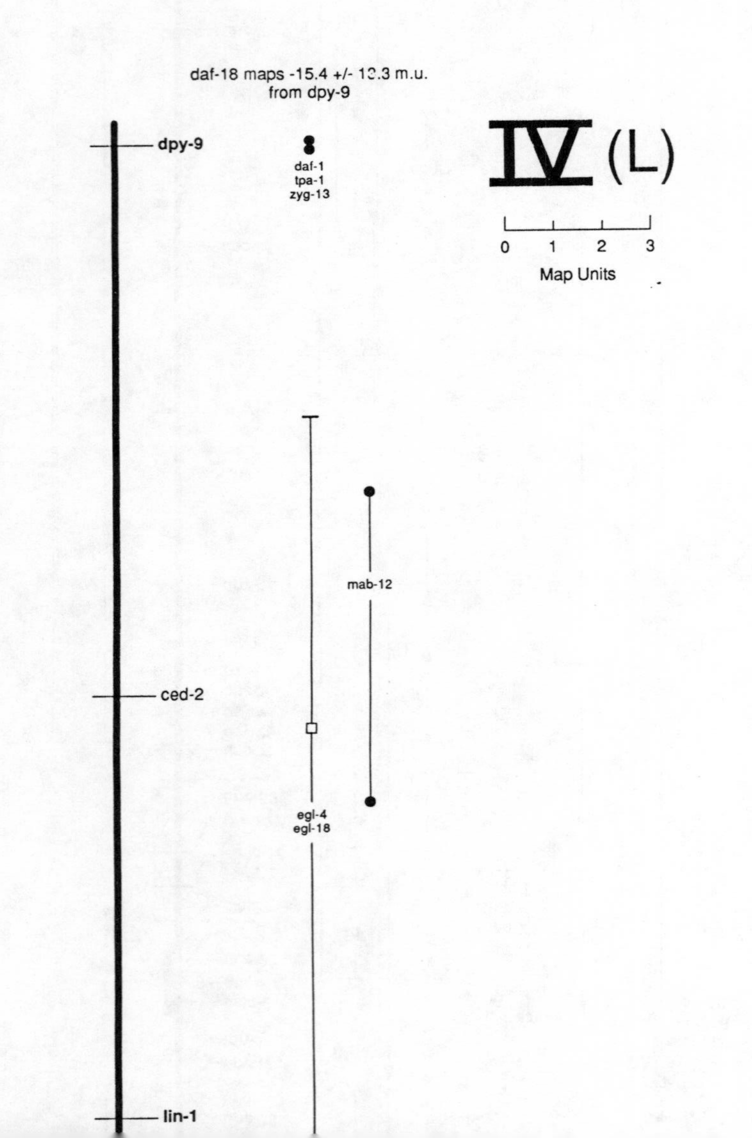
IV (L)
0 1 2 3
Map Units
daf-18 maps -15.4 +/- 12.3 m.u.
from dpy-9
dpy-9
daf-1
tpa-1
zyg-13
ced-2
lin-1
mab-12
egl-4
egl-18
mDp1

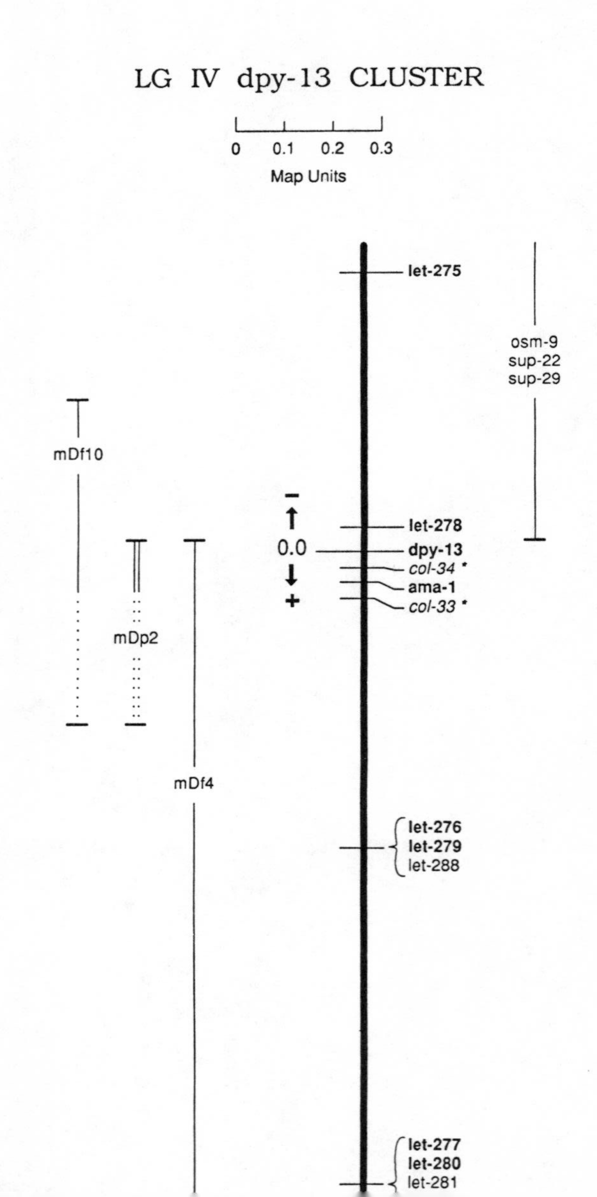
LG IV dpy-13 CLUSTER
0 0.1 0.2 0.3
Map Units
let-275
osm-9
sup-22
sup-29
mDf10
mDp2
mDf4
let-278
dpy-13
col-34 *
ama-1
col-33 *
0.0
let-276
let-279
let-288
let-277
let-280
let-281

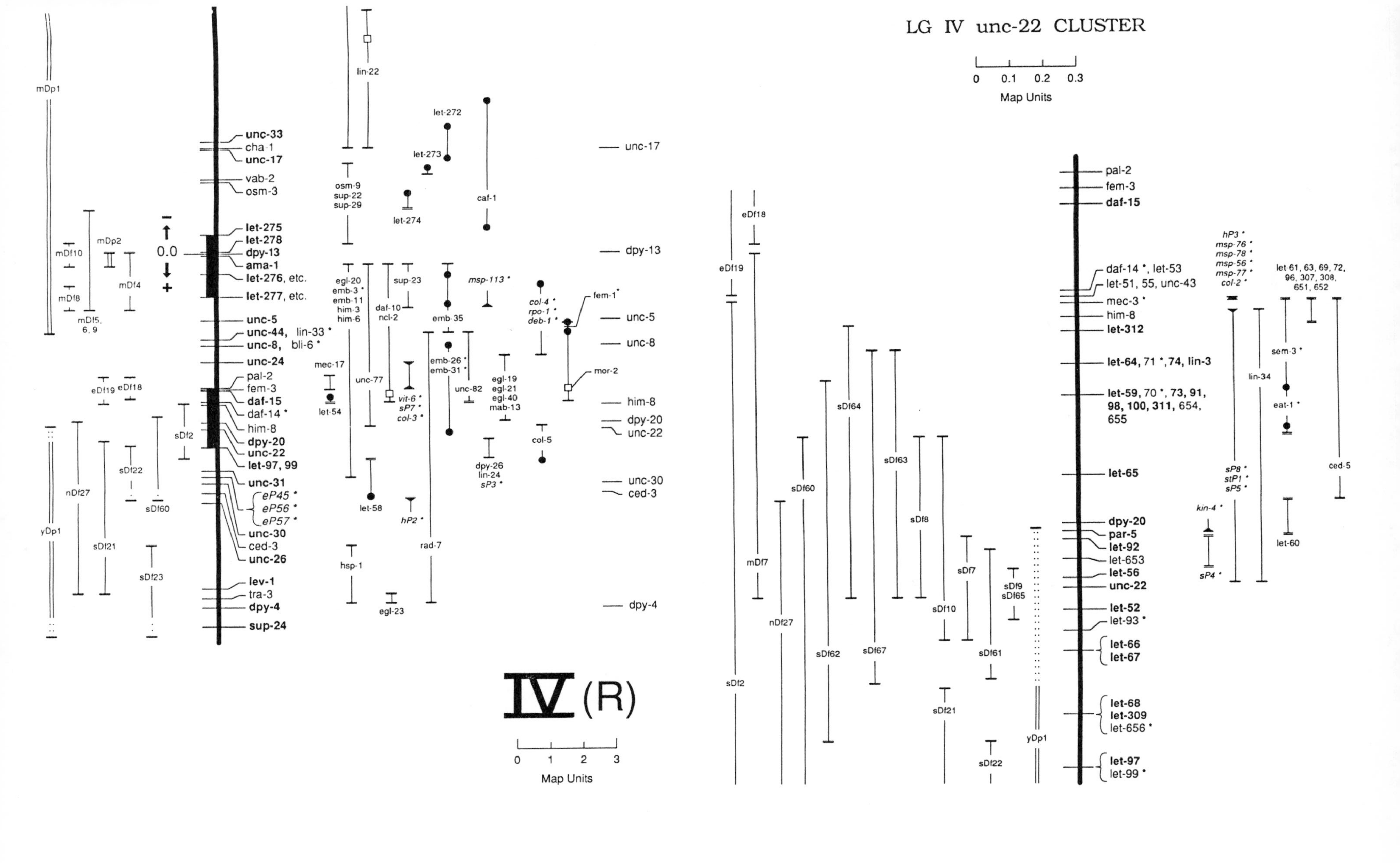

LG IV unc-22 CLUSTER
0 0.1 0.2 0.3
Map Units
mDp1
mDf10
mDp2
mDf4
mDf8
mDf5, 6, 9
– ↑ 0.0 ↓ +
unc-33
cha-1
unc-17
vab-2
osm-3
let-275
let-278
dpy-13
ama-1
let-276, etc.
let-277, etc.
unc-5
unc-44, lin-33 *
unc-8, bli-6 *
unc-24
pal-2
fem-3
daf-15
daf-14 *
him-8
dpy-20
unc-22
let-97, 99
unc-31
eP45 *
eP56 *
eP57 *
unc-30
ced-3
unc-26
lev-1
tra-3
dpy-4
sup-24
eDf19
eDf18
sDf2
nDf27
sDf22
sDf60
yDp1
sDf21
sDf23
lin-22
osm-9
sup-22
sup-29
let-272
let-273
let-274
caf-1
egl-20
emb-3 *
emb-11
him-3
him-6
unc-77
daf-10
ncl-2
sup-23
emb-35
msp-113 *
col-4 *
rpo-1 *
deb-1 *
fem-1 *
mor-2
mec-17
let-54
vit-6 *
sP7 *
col-3 *
emb-26 *
emb-31 *
unc-82
egl-19
egl-21
egl-40
mab-13
col-5
dpy-26
lin-24
sP3 *
let-58
hP2 *
rad-7
hsp-1
egl-23
unc-17
dpy-13
unc-5
unc-8
him-8
dpy-20
unc-22
unc-30
ced-3
dpy-4
IV (R)
0 1 2 3
Map Units
eDf18
eDf19
mDf7
sDf2
nDf27
sDf60
sDf62
sDf64
sDf67
sDf63
sDf8
sDf10
sDf21
sDf7
sDf61
sDf22
sDf9
sDf65
yDp1
pal-2
fem-3
daf-15
daf-14 *, let-53
let-51, 55, unc-43
mec-3 *
him-8
let-312
let-64, 71 *, 74, lin-3
let-59, 70 *, 73, 91, 98, 100, 311, 654, 655
let-65
dpy-20
par-5
let-92
let-653
let-56
unc-22
let-52
let-93 *
let-66
let-67
let-68
let-309
let-656 *
let-97
let-99 *
hP3 *
msp-76 *
msp-78 *
msp-56 *
msp-77 *
col-2 *
sP8 *
stP1 *
sP5 *
kin-4 *
sP4 *
lin-34
sem-3 *
eat-1 *
let-60
let-61, 63, 69, 72, 96, 307, 308, 651, 652
ced-5

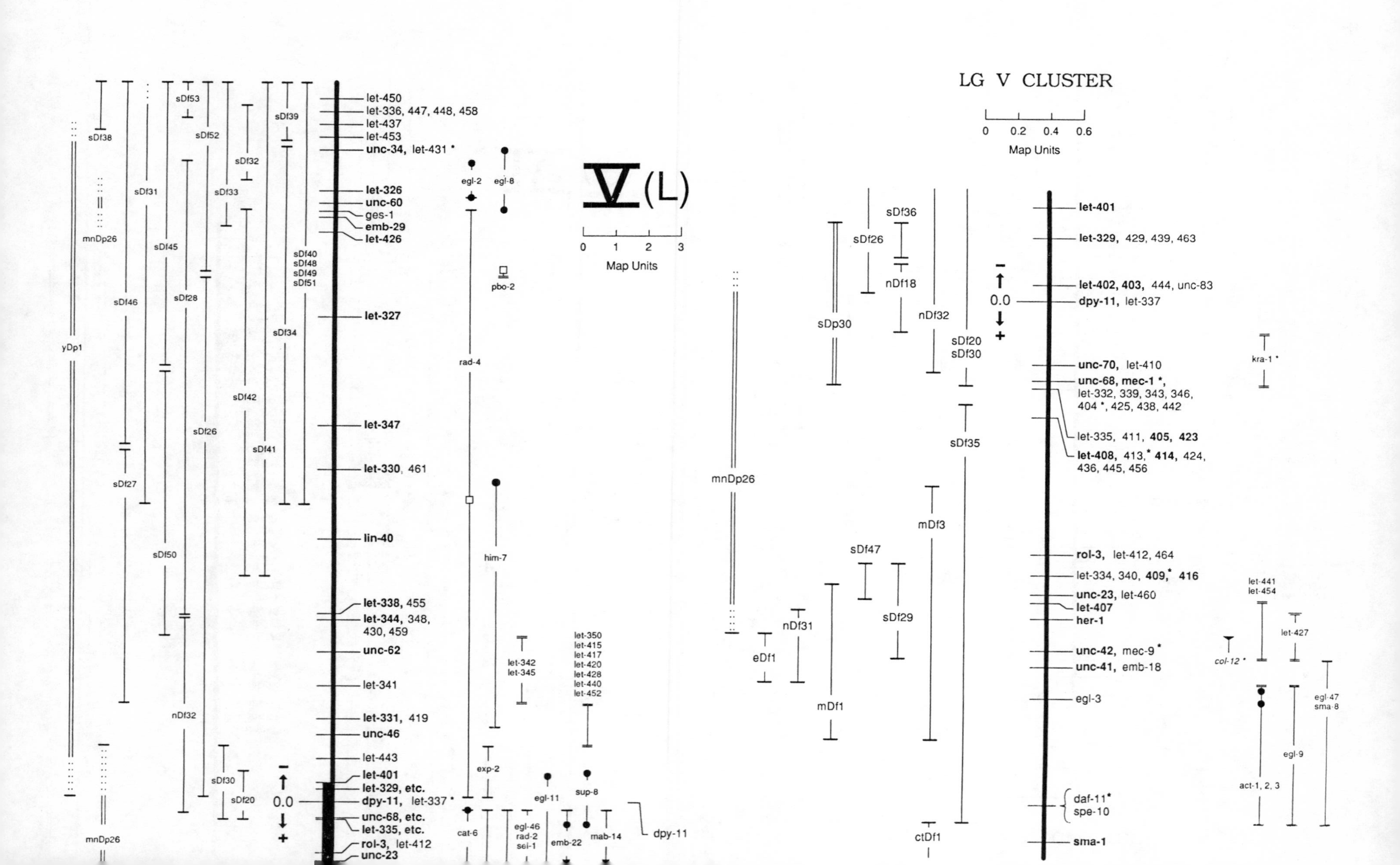

V (L)
Map Units
0
1
2
3
let-450
let-336, 447, 448, 458
let-437
let-453
unc-34, let-431 *
let-326
unc-60
ges-1
emb-29
let-426
let-327
let-347
let-330, 461
lin-40
let-338, 455
let-344, 348, 430, 459
unc-62
let-341
let-331, 419
unc-46
let-443
let-401
let-329, etc.
dpy-11, let-337 *
unc-68, etc.
let-335, etc.
rol-3, let-412
unc-23
egl-2
egl-8
pbo-2
rad-4
him-7
let-342
let-345
let-350
let-415
let-417
let-420
let-428
let-440
let-452
exp-2
egl-11
sup-8
cat-6
egl-46
rad-2
sel-1
emb-22
mab-14
dpy-11
0.0
sDf53
sDf38
sDf52
sDf39
sDf32
sDf31
sDf33
sDf45
sDf40
sDf48
sDf49
sDf51
sDf46
sDf28
sDf34
sDf42
sDf26
sDf41
sDf27
sDf50
nDf32
sDf30
sDf20
mnDp26
yDp1
LG V CLUSTER
0
0.2
0.4
0.6
Map Units
let-401
let-329, 429, 439, 463
let-402, 403, 444, unc-83
dpy-11, let-337
unc-70, let-410
unc-68, mec-1 *,
let-332, 339, 343, 346,
404 *, 425, 438, 442
let-335, 411, 405, 423
let-408, 413, * 414, 424,
436, 445, 456
rol-3, let-412, 464
let-334, 340, 409, * 416
unc-23, let-460
let-407
her-1
unc-42, mec-9 *
unc-41, emb-18
egl-3
daf-11 *
spe-10
sma-1
kra-1 *
let-441
let-454
let-427
col-12 *
egl-47
sma-8
egl-9
act-1, 2, 3
sDf36
sDf26
nDf18
sDp30
nDf32
sDf20
sDf30
sDf35
mnDp26
mDf3
sDf47
sDf29
nDf31
eDf1
mDf1
ctDf1

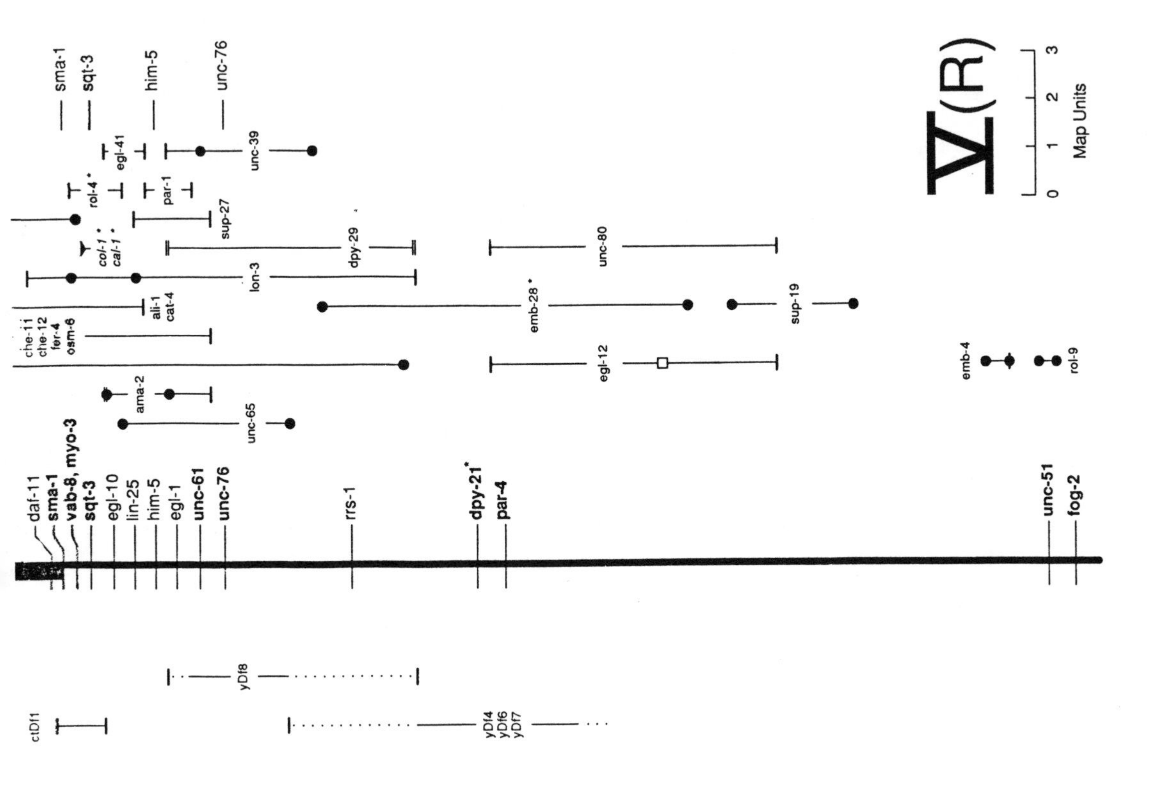

V(R)
0 1 2 3
Map Units
daf-11
sma-1
vab-8, myo-3
sqt-3
egl-10
lin-25
him-5
egl-1
unc-61
unc-76
rrs-1
dpy-21*
par-4
unc-51
fog-2
unc-65
ama-2
che-11
che-12
fer-4
osm-6
ali-1
cat-4
lon-3
col-1*
cal-1*
dpy-29
sup-27
rol-4*
par-1
egl-41
unc-39
egl-12
emb-28*
sup-19
unc-80
emb-4
rol-9
yDf8
ctDf1
yDf4
yDf6
yDf7

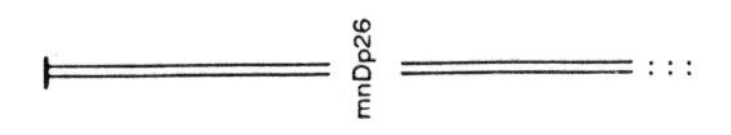

mnDp26

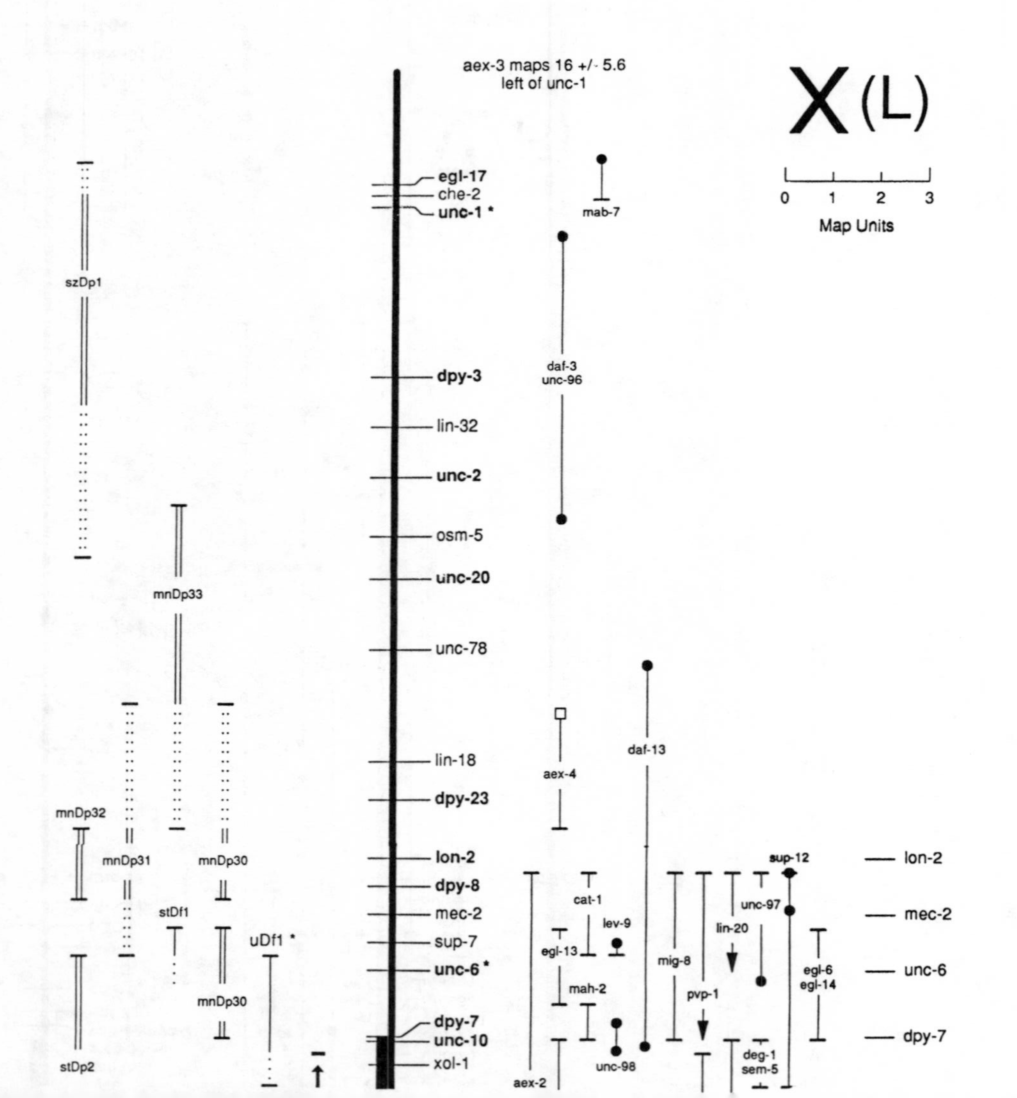

X (L)
0 1 2 3
Map Units
aex-3 maps 16 +/- 5.6
left of unc-1
egl-17
che-2
unc-1 *
dpy-3
lin-32
unc-2
osm-5
unc-20
unc-78
lin-18
dpy-23
lon-2
dpy-8
mec-2
sup-7
unc-6 *
dpy-7
unc-10
xol-1
mab-7
daf-3
unc-96
aex-4
daf-13
cat-1
lev-9
egl-13
mah-2
unc-98
aex-2
mig-8
pvp-1
lin-20
unc-97
sup-12
deg-1
sem-5
egl-6
egl-14
lon-2
mec-2
unc-6
dpy-7
szDp1
mnDp33
mnDp32
mnDp31
mnDp30
stDf1
mnDp30
uDf1 *
stDp2

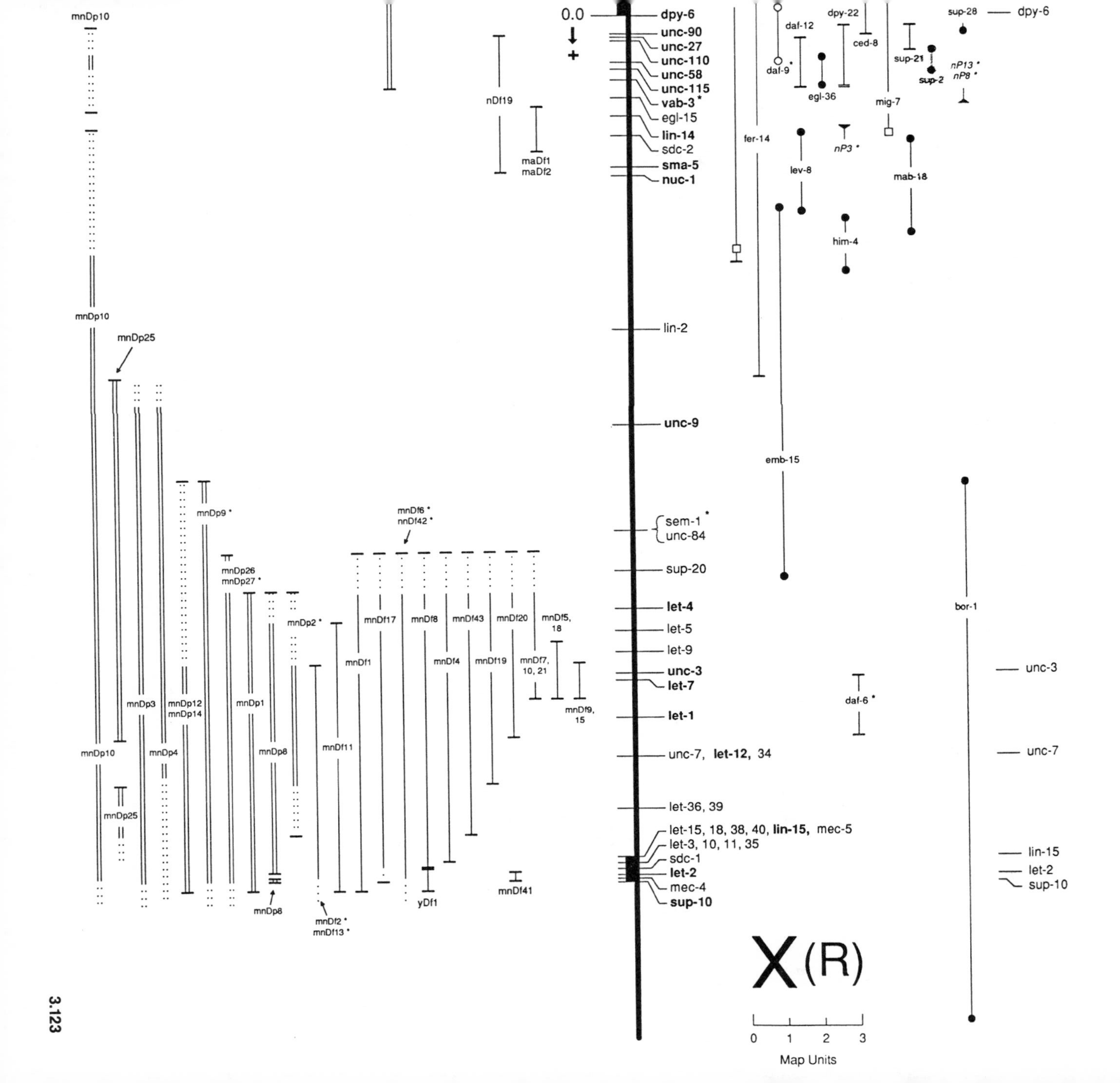

X (R)
0.0
dpy-6
unc-90
unc-27
unc-110
unc-58
unc-115
vab-3*
egl-15
lin-14
sdc-2
sma-5
nuc-1
lin-2
unc-9
sem-1*
unc-84
sup-20
let-4
let-5
let-9
unc-3
let-7
let-1
unc-7, let-12, 34
let-36, 39
let-15, 18, 38, 40, lin-15, mec-5
let-3, 10, 11, 35
sdc-1
let-2
mec-4
sup-10
daf-12
daf-9*
dpy-22
ced-8
egl-36
sup-21
sup-2
sup-28
dpy-6
nP13*
nP8*
mig-7
fer-14
lev-8
nP3*
mab-18
him-4
emb-15
bor-1
daf-6*
unc-3
unc-7
lin-15
let-2
sup-10
0
1
2
3
Map Units
mnDp10
nDf19
maDf1
maDf2
mnDp10
mnDp25
mnDp9*
mnDf6*
nnDf42*
mnDp26
mnDp27*
mnDp2*
mnDf17
mnDf8
mnDf43
mnDf20
mnDf5, 18
mnDf1
mnDf4
mnDf19
mnDf7, 10, 21
mnDp3
mnDp12
mnDp14
mnDp1
mnDf9, 15
mnDp10
mnDp4
mnDp8
mnDf11
mnDp25
mnDf41
mnDp8
yDf1
mnDf2*
mnDf13*

Gene	LG	Ref Allele	References
ace	abnormal ACEtylcholinesterase		
ace-1	X	p1000	338 464 513 514 608 622 642 675 761 795 796 1004 1039
ace-2	I	g72	464 513 514 527 608 622 642 675 761 795 796 1004 1039
ace-3	II	dc2	1039
act	ACTin		
act-1	V	st15	608 611 741 742 801 969
act-2	V	st15	608 611 741 742 801 969
act-3	V	st15	608 611 741 742 801 969 1092
aex	ABoc, EXpulsion defective: constipated		
*aex-1	I	sa9	
*aex-2	X	sa3	
*aex-4	X	sa22	
*aex-5	I	sa23	
age	AGEing alteration		
*age-1	II	hx546	1020 1054
ali	abnormal lateral ALAE		
ali-1	V	e1934	
ama	AMAnitin resistance		
ama-1	IV	m118	654 1019 1073 1074
ama-2	V	m323	1073
anc	abnormal nuclear ANChorage		
anc-1	I	e1753	562
ben	BENzimidazole resistance		
ben-1	III	e1880	
bli	BLIstered cuticle		
bli-1	II	e769	31 387 465 608 715
bli-2	II	e768	31 168 465 608
bli-3	I	e767	31 608 976 1057
bli-4	I	e937	31 387 608
bli-5	III	e518	31 554 608 1057
bli-6	IV	sc16	914
bor	BORdering behavior		
*bor-1	X	g320	
cad	abnormal CAthepsin D		
cad-1	II	j1	1041
caf	CAFfeine resistance		
*caf-1	IV	hf3	965
*caf-2	I	hf5	965
cal	CALmodulin related genes		
*cal-1	V		
cat	abnormal CATecholamine distribution		
cat-1	X	e1111	79 365 397 527 608 1105
cat-2	II	e1112	79 365 527 608
cat-4	V	e1141	365 527 608 1105
cat-6	V	e1861	793 932
ced	abnormal CEll Death		
ced-1	I	e1735	568 627 672 870
ced-2	IV	e1752	568 627 672
ced-3	IV	n717	568 627 635 671 672 731 756 870 1016
ced-4	III	n1162	731 870
*ced-5	IV	n1812	
*ced-6	III	n1813	
*ced-7	III	n1892	
*ced-8	X	n1891	

*Genes added since *Genetic Maps 1987*

Gene	LG	Ref Allele	References
ces	CEll death Specification		
*ces-1	I	n703	
*ces-2	I	n732	
cha	abnormal CHoline Acetyltransferase		
cha-1	IV	p1152	464 527 608 675 788 1004 1006
che	abnormal CHEmotaxis		
che-1	I	e1034	82 214 388 390 391 393 538 608 932
che-2	X	e1033	127 214 388 390 538 608 932
che-3	I	e1124	82 214 388 390 391 481 503 552 608 728 932
che-10	II	e1809	932
che-11	V	e1810	932
che-12	V	e1812	932
che-13	I	e1805	932
clb	CoLlagen Basement membrane		
*clb-2	III		
clr	CLeaR		
clr-1	II	e1745	1052
col	COLlagen		
col-1	V		569 722 746 748 806 1099 1101
col-2	IV		569 722 746 748 751 1099 1101
col-3	IV		569 748 751
col-4	IV		751
col-5	IV		751
col-6	II		748 751 1099
col-7	I		748 1099
col-8	III		748 751 1099
col-12	V		722 748
*col-33	IV		1101
*col-34	IV		1101
daf	abnormal DAuer Formation		
daf-1	IV	m40	316 503 504 505 635 680 932
daf-2	III	e1370	316 486 503 504 505 680 722 932 1127
daf-3	X	e1376	316 504 608 635 932
daf-4	III	m63	316 504 505 635 680 932
daf-5	II	e1386	504 608 635 932
daf-6	X	e1377	316 503 504 608 712 728 761 932 977
daf-7	III	e1372	316 486 504 505 635 680 706 932 1127
daf-8	I	e1393	316 504 505 635 680 932
daf-9	X	e1406	316 932 1035
daf-10	IV	e1387	82 316 481 503 504 505 552 608 932
daf-11	V	m47	504 680 932
daf-12	X	m20	504 608 932
daf-13	X	m66	504 608 932
daf-14	IV	m77	504 505 635 680 932
daf-15	IV	m81	932 1035
daf-16	I	m26	504 608 932
daf-17	I	m27	504 608 932
daf-18	IV	e1375	504 608 932
daf-19	II	m86	932
daf-22	II	m130	766
*daf-23	II	m333	
*daf-24	II	m336	
deb	DEnse Body		
*deb-1	IV	st555	
deg	DEGeneration of certain neurons		
deg-1	X	u38	
dig	DIsplaced Gonad		
*dig-1	III	n1321	
dpy	DumPY		
dpy-1	III	e1	31 168 486 554 608 609
dpy-2	II	e8	31 168 465 608 715
dpy-3	X	e27	31 465 608
dpy-4	IV	e1166	465 484 499 608
dpy-5	I	e61	31 168 321 554 608 976
dpy-6	X	e14	31 162 164 608
dpy-7	X	e88	31 164 165 465 608
dpy-8	X	e130	31 465 608
dpy-9	IV	e12	31 168 608 632
dpy-10	II	e128	31 55 127 163 164 168 465 468 715 769

Gene	LG	Ref Allele	References
dpy-11	V	e224	31 168 320 321 465 538 554 1073
dpy-13	IV	e184	31 55 164 168 468 484 608 1019 1101
dpy-14	I	e188	31 321 427 608 976
dpy-17	III	e164	31 486 554 608 769
dpy-18	III	e364	31 397 496 515 554 608 769 797
dpy-19	III	e1259	608
dpy-20	IV	e1282	608 609 797 813
dpy-21	V	e428	178 179 498 608 665 666 724 941 987 1003 1011 1042 1077 1115
dpy-22	X	e652	178 608 666 987 1003 1011 1042 1115
dpy-23	X	e840	608 666 987 1042
dpy-24	I	s71	427
dpy-25	II	e817	608
dpy-26	IV	n199	608 666 724 987 1011 1077 1115
dpy-27	III	rh18	941 1011 1077 1115
dpy-28	III	y1	941 1011 1077 1115
*dpy-29	V	y144	1115
*eP6	III		950
*eP7	III		950
*eP45	IV		
*eP56	IV		
*eP57	IV		

eat EATing: abnormal pharyngeal pumping

Gene	LG	Ref Allele	References
*eat-1	IV	ad427	

egl EGg Laying defective

Gene	LG	Ref Allele	References
egl-1	V	n487	635 1105 1133
egl-2	V	n693	635 1018
egl-3	V	n150	635
egl-4	IV	n478	635 680
egl-5	III	n486	635 1105 1125 1133
egl-6	X	n592	635
egl-7	III	n575	635
egl-8	V	n488	635
egl-9	V	n586	635
egl-10	V	n692	635 1105 1133
egl-11	V	n587	635
egl-12	V	n602	635
egl-13	X	n483	635
egl-14	X	n549	635
egl-15	X	n484	635
egl-17	X	e1313	635
egl-18	IV	n162	635 1105
egl-19	IV	n582	635
egl-20	IV	n585	635 1105
egl-21	IV	n611	635
egl-23	IV	n601	635
egl-24	III	n572	635
egl-25	III	n573	635
egl-26	II	n481	635
egl-27	II	n170	635 1105
egl-28	II	n570	635
egl-29	II	n482	635
egl-30	I	n686	635 914
egl-31	I	n472	635
egl-32	I	n155	635 680
egl-33	I	n151	635
egl-34	I	n171	635
egl-35	III	n694	635
egl-36	X	n728	635 914
egl-37	II	n742	635
egl-40	IV	n606	635 680
egl-41	V	n1077	944 1105 1133
egl-42	II	n995	1105 1133
egl-43	II	n997	1105 1133
egl-44	II	n1080	1105 1133
egl-45	III	n999	1105 1133
egl-46	V	n1127	1105 1133
egl-47	V	n1081	1105 1133
*egl-48	II	e1952	922
*egl-50	II	n1086	1105 1133

eha Egglaying Hormone (Aplysia homolog)

Gene	LG	Ref Allele	References
*eha-1	I		

emb abnormal EMBryogenesis

Gene	LG	Ref Allele	References
emb-1	III	hc57	440 448 449 501 543
emb-2	III	hc58	440 448 449 501 543
emb-3	IV	hc59	440 448 449 501 543
emb-4	V	hc60	440 448 449 501 543
emb-5	III	hc61	440 448 449 501 543 580 661 689
emb-6	I	hc65	440 448 449 501 543 661 689
emb-7	III	hc66	426 433 440 448 449 501 543 574
emb-8	III	hc69	440 448 449 501 543
emb-9	III	hc70	426 433 440 448 449 501 543 574 580 661 689
emb-10	I	hc63	543
emb-11	IV	g4	501 543 661
emb-12	I	g5	501 543 661
emb-13	III	g6	501 543 661 689
emb-14	I	g43	501 543 661
emb-15	X	g15	501 543 661 689
emb-16	III	g19	501 543 661 689
emb-17	I	g20	501 543 661 689
emb-18	V	g21	501 543 661 689
emb-19	I	g22	501 543 661
emb-20	I	g27	501 543 661
emb-21	II	g31	501 543 661 689
emb-22	V	g32	501 543 661 689
emb-23	II	g39	501 543 661 689
emb-24	III	g40	501 543 661 689
emb-25	III	g45	501 543 661 689
emb-26	IV	g47	501 543 661 666 689
emb-27	II	g48	501 543 649 661 689 715
emb-28	V	g49	501 543 661 689
emb-29	V	g52	501 543 661 689 959 1018 1108
emb-30	III	g53	501 543 661 689
emb-31	IV	g55	501 543 661 689
emb-32	III	g58	501 543 661 689
emb-33	III	g60	501 543 661 689
emb-34	III	g62	501 543 661 689
emb-35	IV	g64	501 543 661 689

exp EXPulsion defective: constipated

Gene	LG	Ref Allele	References
*exp-1	II	sa6	
*exp-2	V	sa26	

fem FEMinization

Gene	LG	Ref Allele	References
fem-1	IV	e1965	252 392 495 498 615 630 634 656 666 714 724 792 797 922 943 944 1037 1077
fem-2	III	e2105	615 649 714 792 922 943 944 1037
fem-3	IV	e1996 q20	922 943 944 1037 1053 1077

fer FERtilization defective

Gene	LG	Ref Allele	References
fer-1	I	hc1	392 393 495 529 537 545 558 559 649 1075
fer-2	III	hc2	447 495 529 536 537 545 559
fer-4	V	hc4	495 529 559
fer-6	I	hc6	495 529 559 1075
fer-7	I	hc34	495 1075
fer-14	X	hc14	
fer-15	II	hc15	558 658 659 662 781 926 1020 1054

fog Feminization Of Germline

Gene	LG	Ref Allele	References
fog-1	I	e2121	1037
fog-2	V	q71	1037

ges abnormal Gut ESterase

Gene	LG	Ref Allele	References
*ges-1	V	ca1	859 1018

glp Germ Line Proliferation

Gene	LG	Ref Allele	References
glp-1	III	q46	1007 1008 1053

gus abnormal GlucUronidaSe

Gene	LG	Ref Allele	References
*gus-1	I	b405	862
*hP2	IV		
*hP3	IV		
*hP4	I		
*hP5	I		
*hP6	I		
*hP7	I		
*hP9	I		

her HERmaphroditization

Gene	LG	Ref Allele	References
her-1	V	e1518 n659	498 597 630 635 665 666 731 922 943 944 1037 1057 1067 1077 1105 1115

Gene	LG	Ref Allele	References
him		High Incidence of Males	
him-1	I	e879	162 178 179 565 608 615 975
him-2	I	e1065	179 565 608
him-3	IV	e1147	179 565 608
him-4	X	e1267	179 676 739 1087
him-5	V	e1467	179 447 498 565 608 634 655 656 665 676 739 902 913 1005 1057
him-6	IV	e1423	179 565 608
him-7	V	e1480	179 565 608 902
him-8	IV	e1489	179 534 565 566 600 608 615 666 676 714 724 739 1134
him-9	II	e1487	179 565 608
him-10	III	e1511	565
*him-14	II	it23	1109
hsp		Heat Shock Protein	
*hsp-1	IV		
kin		protein KINase	
*kin-1	I		
*kin-4	IV		
kra		Ketamine Response Abnormal	
*kra-1	V	kh30	
lan		LANnate resistance	
*lan-5	III	e239	
let		LEThal	
let-1	X	mn119	241 497
let-2	X	mn153	241 497 501 543 661 689
let-3	X	mn104	241 497
let-4	X	mn105	241 497
let-5	X	mn106	241 497
let-6	X	mn130	241 497
let-7	X	mn112	241 497
let-9	X	mn107	241 497
let-10	X	mn113	241 497
let-11	X	mn116	241 497
let-12	X	mn121	241 497
let-14	X	mn120	241 497
let-15	X	mn127	497
let-16	X	mn117	241 497
let-18	X	mn122	497
let-19	II	mn19	163 715
let-21	II	e1778	
let-22	II	mn22	163 715
let-23	II	mn23	163 715 762 957
let-24	II	mn24	163 715
let-25	II	mn25	163 715
let-26	II	mn26	163 715
let-29	II	mn29	715 1109
let-30	II	mn30	163 715
let-31	II	mn31	163 715
let-33	X	mn128	497
let-34	X	mn134	497
let-35	X	mn135	497
let-36	X	mn140	497
let-37	X	mn138	497
let-38	X	mn141	497
let-39	X	mn144	497
let-40	X	mn150	497
let-41	X	mn146	497
let-49	I	st44	579 710
let-50	I	st33	579 710
let-51	IV	s41	499 592
let-52	IV	s42	592 813 1040
let-53	IV	s43	499 592
let-54	IV	s44	592
let-55	IV	s45	592
let-56	IV	s46	499 592 654 813 1040
let-59	IV	s49	499 592 813 1040
let-60	IV	s59	499 592 813 1040
let-61	IV	s65	592 813 1040
let-63	IV	s170	592 813 1040
let-64	IV	s216	592 813 1040
let-65	IV	s254	592 813 1040
let-66	IV	s176	592 813 1040
let-67	IV	s214	592 813 1040
let-68	IV	s680	813 1040
let-69	IV	s684	813 1040
let-70	IV	s689	813 1040
let-71	IV	s692	813 1040
let-72	IV	s52	813 1040
let-73	IV	s685	813 1040
let-75	I	s101	427
let-80	I	s96	427
let-84	I	s91	427
let-86	I	s141	427
let-87	I	s106	427
let-88	I	s132	427
let-89	I	s133	427

Gene	LG	Ref Allele	References
let-91	IV	s678	813 1040
let-92	IV	s504	813 1040
*let-93	IV	s734	1040
*let-96	IV	s1112	1040
*let-97	IV	s1121	1040
*let-98	IV	s1117	1040
*let-99	IV	s1201	1040
*let-100	IV	s1160	1040
let-201	I	e1716	710
let-202	I	e1720	710
let-203	I	e1717	710
let-204	I	e1719	710
let-205	I	e1722	710
let-206	I	e1721	710
let-207	I	e1723	710
let-208	I	e1718	710
let-236	II	mn88	715
let-237	II	mn208	715 1109
let-238	II	mn229	715
let-239	II	mn217	715
let-240	II	mn209	715
let-241	II	mn228	715
let-242	II	mn90	715
let-243	II	mn226	715
let-244	II	mn97	715
let-245	II	mn185	715
let-246	II	mn99	715
let-247	II	mn211	715
let-248	II	mn237	715
let-249	II	mn238	715
let-250	II	mn207	715
let-251	II	mn95	715
let-252	II	mn100	715
let-253	II	mn181	715
let-259	II	mn210	715 1041
let-263	II	mn240	715
let-264	II	mn227	715
let-265	II	mn188	715
let-266	II	mn194	715
let-268	II	mn189	715
let-272	IV	m243	
let-273	IV	m263	1019
let-274	IV	m256	1019
let-275	IV	m245	1019
let-276	IV	m240	1019
let-277	IV	m262	1019
let-278	IV	m265	1019
let-279	IV	m261	1019
let-280	IV	m259	1019
let-281	IV	m247	1019
let-282	IV	m258	1019
let-284	IV	m267	1019
*let-288	IV	m306	1019
*let-307	IV	s1171	1040
*let-308	IV	s1705	1040
*let-309	IV	s1115	1040
*let-311	IV	s1195	1040
*let-312	IV	s1234	1040
let-326	V	s238	750 1018 1108
let-327	V	s247	750 1108
let-329	V	s575	750 1108
let-330	V	s573	1108
let-331	V	s427	750 1108
let-332	V	s234	1108
*let-334	V	s383	1108
*let-335	V	s232	1108
*let-336	V	s1413	1018 1071 1108
*let-337	V	s825	1018 1108
*let-338	V	s1020	1071 1108
*let-339	V	s1019	1108
*let-340	V	s1022	1108
*let-341	V	n1613	1071 1108
*let-342	V	s1029	1071 1108
*let-343	V	s816	1018 1108
*let-344	V	s376	1071 1108
*let-345	V	s578	1071 1108
*let-346	V	s373	1071 1108
*let-347	V	s1035	1071 1108
*let-348	V	s998	1071 1108
*let-349	V	s217	1071 1108
*let-350	V	s250	1071 1108
*let-351	I	h43	975
*let-352	I	h45	975
*let-353	I	h46	975
*let-354	I	h79	975
*let-355	I	h81	975
*let-356	I	h83	975
*let-357	I	h89	975
*let-359	I	h94	975
*let-360	I	h96	975
*let-361	I	h97	975
*let-362	I	h86	975
*let-363	I	h111	975
*let-364	I	h104	975
*let-365	I	h108	975
*let-366	I	h112	975

Gene	LG	Ref Allele	References
let-367	I	h119	975
let-368	I	h121	975
let-369	I	h125	975
let-370	I	h128	975
let-371	I	h123	975
*let-372	I	h126	975
*let-373	I	h234	
*let-374	I	h251	
*let-375	I	h259	
*let-376	I	h130	975
*let-377	I	h110	975
*let-378	I	h124	975
*let-379	I	h127	975
*let-380	I	h80	975
*let-381	I	h107	975
*let-382	I	h82	975
*let-383	I	h115	975
*let-384	I	h84	975
*let-385	I	h85	975
*let-386	I	h117	975
*let-387	I	h87	975
*let-388	I	h88	975
*let-389	I	h106	975
*let-390	I	h44	975
*let-391	I	h91	975
*let-392	I	h120	975
*let-397	I	h228	
*let-398	I	h257	
*let-401	V	s193	1108
*let-402	V	s127	1108
*let-403	V	s120	1108
*let-404	V	s119	1108
*let-405	V	s116	1018 1108
*let-407	V	s118	1018 1108
*let-408	V	s195	1018 1108
*let-409	V	s206	1018 1108
*let-410	V	s815	1018 1108
*let-411	V	s223	1108
*let-412	V	s579	1108
*let-413	V	s128	1108
*let-414	V	s114	1108
*let-415	V	s129	1108
*let-416	V	s113	1108
*let-417	V	s204	1071 1108
*let-418	V	s1045	1071 1108
*let-419	V	s219	1108
*let-420	V	s1046	1071 1108
*let-421	V	s288	1071 1108
*let-422	V	s194	1071 1108
*let-423	V	s818	1018 1108
*let-424	V	s248	1108
*let-425	V	s385	1108
*let-426	V	s826	1018 1108
*let-427	V	s1057	1108
*let-428	V	s1070	1108
*let-429	V	s584	1108
*let-430	V	s1042	1071 1108
*let-431	V	s1044	1108
*let-436	V	s1403	
*let-437	V	s1405	
*let-438	V	s2114	1071
*let-440	V	s1411	
*let-441	V	s1414	
*let-442	V	s1416	
*let-443	V	s1417	
*let-445	V	s1419	
*let-447	V	s1654	1071
*let-448	V	s1363	
*let-450	V	s2160	1071
*let-452	V	s1434	
*let-453	V	s2167	
*let-454	V	s1423	
*let-455	V	s1447	
*let-456	V	s1479	
*let-458	V	s1443	
*let-459	V	s1432	
*let-461	V	s1486	
*let-601	I	h281	
*let-602	I	h283	
*let-651	IV	s1165	
*let-652	IV	s1086	
*let-653	IV	s1733	
*let-654	IV	s1734	
*let-655	IV	s1748	
*let-656	IV	s1753	

lev LEVamisole resistance

Gene	LG	Ref Allele	References
lev-1	IV	e211	31 464 484 538 608 622 950 998
lev-7	I	x13	464 484 608 622 998
lev-8	X	x15	464 484 538 622 998
lev-9	X	x16	464 484 622 998
lev-10	I	x17	464 484 622 710 998
lev-11	I	x12	464 484 579 608 622 710 998

lin abnormal cell LINeage

Gene	LG	Ref Allele	References
lin-1	IV	e1777	444 496 500 608 672 762 797 957
lin-2	X	e1309	444 496 500 608 762 957
lin-3	IV	e1417	496 500 608 762 957 1040
lin-4	II	e912	178 496 500 507 544 608 620 672 762 957 1002 1057
lin-5	II	e1348	8 354 496 500 578 672 715
lin-6	I	e1466	354 401 496 500 672
lin-7	II	e1413	496 500 608 635 762 957
lin-8	II	n111	486 496 500 672 762 957
lin-9	III	n112	496 500 672 762 957
lin-10	I	e1439	608 635 762 957
lin-11	I	n389	635 762 957
lin-12	III	n941 n137	635 646 672 708 731 762 812 950 957 1026 1080 1092
lin-13	III	n387	762 957
lin-14	X	n536n540 n536	544 620 672 731 737 749 944 966 1011 1057 1128
lin-15	X	n309	608 635 672 762 944 957 1081
lin-16	III	e1743	
lin-17	I	n671	635 731 762 957 1057 1096
lin-18	X	e620	608 762 957
lin-19	III	e1756	
lin-20	X	e1796	
lin-21	III	e1751	
lin-22	IV	n372	672 731 913 1057 1076
lin-23	II	e1883	
lin-24	IV	n432	635 762 957
lin-25	V	e1446	635 762 957
lin-26	II	n156	635 672 762 957
lin-28	I	n719	620 635 672 737 749 1057
lin-29	II	n333	620 635 672 715 737 749 1057
lin-31	II	n301	635 762 957
lin-32	X	e1926	672 1125
lin-33	IV	n1043	762 957
lin-34	IV	n1046	762 957
lin-35	I	n745	
lin-36	III	n766	
lin-37	III	n758	
lin-38	II	n751	
*lin-40	V	e2173	1071 1108
*lin-41	I	ma104	
*lin-42	II	n1089	

lon LONg

Gene	LG	Ref Allele	References
lon-1	III	e185	31 486 496 513 608 769 797
lon-2	X	e678	31 120 168 538 608
*lon-3	V	e2175	

mab Male ABnormal

Gene	LG	Ref Allele	References
mab-1	I	e1228	608 1057
mab-2	I	e1241	608 1057
mab-3	II	e1240	608 1057 1076
mab-4	III	e1252	608 1057
mab-5	III	e1239	444 608 629 708 913 1025 1057 1106
mab-6	II	e1249	608 1057
mab-7	X	e1599	608 1057
mab-8	II	e1250	608 1057
mab-9	II	e1245	608 1057
mab-10	II	e1248	608 1057
*mab-11	I	e2008	922
*mab-12	IV	e2166	
*mab-13	IV	ma117	
*mab-14	V	ma116	

mah reversible ts paralysis

Gene	LG	Ref Allele	References
*mah-2	X	cn110	788

mec MEChanosensory abnormality

Gene	LG	Ref Allele	References
mec-1	V	e1066	214 388 390 502 608 688 708 932
mec-2	X	u8	214 502 608
mec-3	IV	e1338	502 608 635 688 708 1052 1107
mec-4	X	u52	502 513 608 627 688 708 977 1052 1125
mec-5	X	e1340	502 608 688 708 1125
mec-6	I	e1342	502 608 1125
mec-7	X	e1343	502 550 560 608 688 708
mec-8	I	e398	427 502 608 932 1125
mec-9	V	u27	502 608
mec-10	X	e1515	502 608
mec-12	III	e1605	502 608 708 1125
mec-14	III	u55	1125
mec-15	II	u215	1125
*mec-16	II	u271	1125
*mec-17	IV	u265	1125

mei defective MEIosis

Gene	LG	Ref Allele	References
*mei-1	I	b284	

Gene	LG	Ref Allele	References
mel	Maternal Effect Lethal		
*mel-1	II	it19	1109
*mel-3	II	b281	1109
*mel-8	II	b309	1109
*mel-9	II	b293	1109
*mel-10	II	it10	1109
*mel-11	II	it26	1109
*mel-12	II	it42	1109
*mel-13	II	b306	1109
*mel-15	II	it38	1109
*mel-18	II	b300	1109
*mel-19	II	b310	1109
*mel-21	II	it9	1109
*mel-22	II	it30	1109
mig	abnormal cell MIGration		
mig-1	I	e1787	1105
*mig-7	X	rh84	
*mig-8	X	rh50	
*mig-10	III	ct41	1105
*mig-11	III	ct78	
mor	MORphological: rounded nose		
mor-1	III	e1071	214 608
mor-2	IV	e1125	608 724
msp	Major Sperm Protein		
*msp-3	II		1023
*msp-45	II		1023
*msp-56	IV		1023 1024
*msp-71	II		1023 1024
*msp-76	IV		1023
*msp-77	IV		1023
*msp-78	IV		1023
*msp-113	IV		1023 1024
*msp-142	II		1023 1024
*msp-152	II		1023 1024
myo	MYOsin heavy chain		
*myo-1	I		876 1124
*myo-3	V	st378	801 876 1124
*nP3	X		1128
*nP8	X		1128
*nP13	X		1128
ncl	abnormal NuCLeoli		
ncl-1	III	e1865	
ncl-2	IV	e1896	
nuc	abnormal NUClease		
nuc-1	X	e1392	8 363 397 565 608 627 672 1086
ooc	abnormal OOCyte formation		
ooc-1	II	mn250	715 1109
ooc-2	II	mn249	715
ooc-3	II	mn241	715
*ooc-4	III	e2078	
osm	defective OSMotic avoidance		
osm-1	X	p808	82 481 513 608 712 932 977
osm-3	IV	p802	82 491 552 608 932
osm-5	X	p813	82 481 608 932
osm-6	V	p811	82 481 608 932
*osm-7	III	n1515	
*osm-8	II	n1518	
*osm-9	IV	n1601	
*osm-10	III	n1602	
*osm-11	X	n1604	
*osm-12	III	n1606	
pal	Posterior ALae in males		
*pal-1	III	e2091	
*pal-2	IV	e2260	
par	abnormal cytoplasmic PARtitioning		
*par-1	V	b274	888 1032
*par-2	III	it46	1032
*par-3	III	e2074	1032
*par-4	V	it47	1032
*par-5	IV	it55	
pbo	PBOc defective: constipated		
*pbo-1	III	sa7	
*pbo-2	V	sa28	

Gene	LG	Ref Allele	References
pha	defective PHArynx development		
*pha-1	III	e2123	
plg	copulatory PLuG formation		
plg-1	III	e2001	
pvp	abnormal PVP neuron staining		
*pvp-1	X	rh114	
rad	abnormal RADiation sensitivity		
rad-1	I	mn155	565 663 726 733 1068 1093
rad-2	V	mn156	565 663 726 733 1068
rad-3	I	mn157	565 663 726 733 1000 1068 1093
rad-4	V	mn158	565 676 739 1068 1087
rad-6	III	mn160	565 1068
rad-7	IV	mn161	565 663 726 733 1068 1093
rad-8	I	mn163	565
rol	ROLler		
rol-1	II	e91	31 168 465 608 715
rol-3	V	e754	31 168 465 608 750 1018 1108
rol-4	V	sc8	465 608
rol-6	II	e187	465 608 715 914 1100
rol-8	II	sc15	906
*rol-9	V	sc148	
rpl	RibosomaL Protein		
*rpl-37	III		
rpo	RNA POlymerase		
*rpo-1	IV		
rrn	Ribosomal RNA		
*rrn-1	I		
rrs	Ribosomal RNA, Small		
*rrs-1	V		916
rtm	tRNA Met		
*rtm-3	X		
*sP1	I		
*sP3	IV		
*sP4	IV		
*sP5	IV		
*sP7	IV		
*sP8	IV		
sdc	Sex Determination & dosage Compensation		
*sdc-1	X	n485	635 944 1077
*sdc-2	X	y15	1077
sel	Suppressor/Enhancer of Lin-12		
*sel-1	V	e1948	
*sel-2	III	n655	
sem	SEx Muscle		
*sem-1	X	n1382	
*sem-2	I	n1343	
*sem-3	IV	n1655	
*sem-4	I	n1378	1105
*sem-5	X	n1779	
sma	SMAll		
sma-1	V	e30	31 388 608 609 1073
sma-2	III	e502	31 554 608 1057
sma-3	III	e491	31 554 608
sma-4	III	e729	31 608 1057
sma-5	X	n678	
sma-6	II	e1482	
*sma-8	V	e2111	914
spe	defective SPErmatogenesis		
spe-1	II	mn147	715
spe-2	II	mn63	715
spe-3	II	mn230	715
*spe-4	I	hc78	1075
*spe-5	I	hc93	1075
*spe-6	III	hc49	
*spe-8	I	hc40	1075
*spe-9	I	hc52	1075
*spe-10	V	hc104	
*spe-11	I	hc77	1075
*spe-12	I	hc76	1075
*spe-13	I	hc137	1075
*spe-15	I	hc75	1075

Gene	LG	Ref Allele	References
sqt	SQuaT		
sqt-1	II	sc1	465 608 715 769 906 1100
sqt-2	II	sc3	465 608 906
sqt-3	V	sc63	31 465 608 906 927
srf	SuRFace antigen		
* srf-1	II	yj1	1002
* stP1	IV		
sup	SUPpressor		
sup-1	III	e995	
sup-2	X	e997	
sup-5	III	e1464	397 484 496 502 515 563 564 579 666 729 768 797 1064
sup-6	II	st19	
sup-7	X	st5	496 502 515 563 564 579 630 646 680 724 729 768 933 1064
sup-8	V	e1563	
sup-9	II	n180	486 564 898
sup-10	X	n183	486 564 712 761 898
sup-11	I	n403	564 898
sup-12	X	st89	807 1018
sup-13	III	st210	
sup-17	I	n316	635
sup-18	III	n463	898
sup-19	V	m210	769
sup-20	X	n821	915
sup-21	X	e1957	797
sup-22	IV	e2057	797
sup-23	IV	e2059	797
* sup-24	IV	st354	1064
* sup-26	III	n1091	
* sup-27	V	n1092	
* sup-28	X	e2058	1064
* sup-29	IV	e1986	1064
sus	SUppressor of Suppressor		
sus-1	III	m156	769
tpa	TPA resistance		
tpa-1	IV	k501	632 1139
tra	TRAnsformer: XX and XO are males		
tra-1	III	e1099 e1575	178 498 597 615 630 665 666 714 724 731 797 922 944 999 1037 1057 1076 1077
tra-2	II	e1095 e2020	27 178 205 206 498 615 630 635 665 715 724 797 922 923 943 944 1037 1057 1076 1077 1105 1133
tra-3	IV	e1107	178 498 563 630 665 724 797 922 943 944 1037 1077
unc	UNCoordinated		
unc-1	X	e719 e94	31 162 318 513 608 914
unc-2	X	e55	31 164 608
unc-3	X	e151	31 241 486 497 513 538 608 712 761 977
unc-4	II	e120	31 127 163 461 608 715
unc-5	IV	e53	30 31 79 499 608
unc-6	X	e78	31 163 164 608 635 793 1105
unc-7	X	e5	31 162 214 513 608 724
unc-8	IV	e49	31 608 914
unc-9	X	e101	31 608 958
unc-10	X	e102	31 608
unc-11	I	e47	31 461 514 554 608 976
unc-13	I	e51	31 163 297 321 427 461 515 554 608 622 797 976 1064
unc-14	I	e57	31 608
unc-15	I	e73	31 318 399 427 461 485 515 536 608 644 645 715 755 769 1026 1051
unc-16	III	e109	31 608
unc-17	IV	e245	31 464 518 608 675 1004 1006
unc-18	X	e81	31 164 365 461 464 608
unc-20	X	e112	31 608
unc-22	IV	e66	245 461 464 484 499 518 536 571 592 608 622 727 784 813 877 915 980 998 1026 1029 1031
unc-23	V	e25	31 387 388 390 391 461 536 608
unc-24	IV	e138	31 163 318 513 608 797
unc-25	III	e156	31 608
unc-26	IV	e205	31 461 484 608
unc-27	X	e155	31 488 608
unc-29	I	e193	31 321 427 461 464 482 484 527 608 622 998
unc-30	IV	e191	30 31 79 464 592 608
unc-31	IV	e169	31 461 592 608 635 784 1020 1054
unc-32	III	e189	31 163 461 554 608 706
unc-33	IV	e204	31 461 608 675 688 708 793 1105
unc-34	V	e315	31 608 1018 1105

Gene	LG	Ref Allele	References
unc-35	I	e259	31 514 608 976
unc-36	III	e251	31 554 608
unc-37	I	e262	31 427 608
unc-38	I	e264	31 464 484 514 527 608 622 998
unc-39	V	e257	31 608
unc-40	I	e271	31 317 527 608 635 1105
unc-41	V	e268	31 318 461 608
unc-42	V	e270	31 163 318 321 461 554 608
unc-43	IV	e408	31 321 592 608 784 914
unc-44	IV	e362	31 461 608 793
unc-45	III	e286	31 128 461 488 536 608 635
unc-46	V	e177	31 608 750
unc-47	III	e307	31 608
unc-49	III	e382	31 608
unc-50	III	e306	31 464 484 608 622 998
unc-51	V	e369	31 365 461 464 484 608 635 793 1105
unc-52	II	e444	230 318 388 442 461 488 515 536 608 797
unc-53	II	e404	31 608 635
unc-54	I	e190	345 436 442 461 509 570 571 579 608 644 647 710 732 738 755 768 779 1026 1051 1065 1091 1124
unc-55	I	e402	31 608
unc-57	I	e406	31 464 514 608
unc-58	X	e665	31 565 608 914
unc-59	I	e261	31 496 500 578 608 635 672 710
unc-60	V	e723	31 461 488 536 608 1018
unc-61	V	e228	31 608
unc-62	V	e644	31 608 750 1108
unc-63	I	e384	31 464 484 514 527 608 622 998
unc-64	III	e246	31 513 554 608
unc-65	V	e351	31 608
unc-67	I	e713	31 554 608
unc-68	V	e540	31 320 464 484 608 622 998
unc-69	III	e587	31 461 608
unc-70	V	e524	31 608 914 1108
unc-71	III	e541	31 608 1105
unc-73	I	e936	31 514 608 1105
unc-74	I	e883	31 464 484 514 527 608 622 950 998
unc-75	I	e950	31 608 710
unc-76	V	e911	31 608 793 1105
unc-77	IV	e625	31 608
unc-78	X	e1217	164 461 488 536 608
unc-79	III	e1068	486 608 958 1063
unc-80	V	e1272	608 958 1065
unc-81	III	e1122	608
unc-82	IV	e1220	461 488 536 608 769
unc-83	V	e1408	496 500 608 672
unc-84	X	e1410	496 500 608 672
unc-85	II	e1414	486 496 500 578 608 635 672
unc-86	III	e1416	179 496 500 502 507 527 552 560 608 635 671 672 688 731 1025 1105 1107 1133
unc-87	I	e1216	318 427 461 488 536 608
unc-89	I	e1460	461 488 514 536 608
unc-90	X	e1463	461 488 536
unc-93	III	e1500 n392	486 564 608 712 761 797 898 977
unc-94	I	su177	488 608
unc-95	I	su106	488 608
unc-96	X	su151	488 608
unc-97	X	su110	488 608
unc-98	X	su130	488 608 635
unc-101	I	m1	
unc-103	III	e1597	608 914
unc-104	II	e1265	608 715
unc-105	II	n490	715 732 755 914 915
unc-108	I	n501	914
unc-109	I	n499	914
unc-110	X	e1913	
* unc-115	X	e2225	
* unc-116	III	e2310	
vab	Variable ABnormal morphology		
vab-1	II	e2	31 214 608
vab-2	IV	e96	31 214 608
vab-3	X	e648	214 608 1057
vab-6	III	e697	608
vab-7	III	e1562	498 608
vab-8	V	e1017	608 1057
vab-9	II	e1744	608 715
vab-10	I	e698	608
vit	VITellogenin		
* vit-1	X		
* vit-6	IV		1097
xol	XO Lethal		
* xol-1	X	y9	1077
zyg	defective ZYGote		
zyg-1	II	b1	426 433 501 543 574 713 1109
zyg-2	I	b10	426 433 501 543 574 661 689 713
zyg-3	II	b18	426 433 501 543 713
zyg-7	III	b187	426 433 501 543
zyg-8	III	b235	426 433 501 543 574
zyg-9	II	b244	426 433 501 543 574 649 667 867 1109
zyg-11	II	b2	426 715 867 1109
zyg-12	II		426 433 501 574
* zyg-13	IV	b126	

8 Albertson *et al.* Developmental Biology 63: 165 - 178 1978
27 Beguet & Gibert C.R. des Seances de l'Acad. Sci. Ser D 286: 989 - 992 1978
31 Brenner Genetics 77: 71 - 94 1974
79 Croll Annual Review of Phytopathology 15: 75 - 89 1977
82 Culotti & Russell Genetics 90: 243 - 256 1978
120 Dusenbery Journal of Nematology 8: 352 - 355 1976
127 Epstein *et al.* Journal of Comparative Physiology 110: 317 - 322 1976
128 Epstein & Thomson Nature 250: 579 - 580 1974
162 Herman *et al.* Genetics 83: 91 - 105 1976
163 Herman Genetics 88: 49 - 65 1978
164 Herman *et al.* Genetics 92: 419 - 435 1979
165 Hieb & Rothstein Science 160: 778 - 780 1968
168 Higgins & Hirsh Molecular & General Genetics 150: 63 - 72 1977
178 Hodgkin & Brenner Genetics 86: 275 - 287 1977
179 Hodgkin *et al.* Genetics 91: 67 - 94 1979
205 Klass *et al.* Developmental Biology 52: 1 - 18 1976
206 Klass *et al.* Developmental Biology 69: 329 - 335 1979
214 Lewis & Hodgkin Journal of Comparative Neurology 172: 489 - 510 1977
230 Mackenzie *et al.* Cell 15: 751 - 762 1978
241 Meneely & Herman Genetics 92: 99 - 115 1979
245 Moerman & Baillie Genetics 91: 95 - 104 1979
252 Nelson *et al.* Developmental Biology 66: 386 - 409 1978
297 Patel & McFadden Experimental Parasitology 44: 72 - 81 1978
316 Riddle Stadler Genetics Symposium 9: 101 - 120 1977
317 Riddle Journal of Nematology 10: 1 - 16 1978
318 Riddle & Brenner Genetics 89: 299 - 314 1978
320 Rose & Baillie Genetics 92: 409 - 418 1979
321 Rose & Baillie Nature 281: 599 - 600 1979
338 Russell *et al.* CGC Book 4050 Wilcox *et al.* (eds) 8: 359 - 371 1977
345 Schachat *et al.* Cell 15: 405 - 411 1978
354 Singh & Sulston Nematologica 24: 63 - 71 1978
363 Sulston Philosophical Trans. of the Royal Society of London 275B: 287 - 298 1976
365 Sulston *et al.* Journal of Comparative Neurology 163: 215 - 226 1975
387 Ward Proceedings of the National Academy of Sciences USA 70: 817 - 821 1973
388 Ward CGC Book 3030 Croll (ed) : 365 - 382 1976
390 Ward CGC Book 3025 Cowan, Ferrendelli (eds) : 1 - 26 1977
391 Ward CGC Book 3060 Hazelbauer (ed) : 141 - 168 1978
392 Ward & Carrel Developmental Biology 73: 304 - 321 1979
393 Ward & Miwa Genetics 88: 285 - 303 1978
397 Waterston & Brenner Nature 275: 715 - 719 1978
399 Waterston *et al.* Journal of Molecular Biology 117: 679 - 697 1977
401 White *et al.* Nature 271: 764 - 766 1978
426 Wood *et al.* Developmental Biology 74: 446 - 469 1980
427 Rose & Baillie Genetics 96: 639 - 648 1980
433 Hirsh CGC Book 4040 Subtelny, Konigsberg (eds) : 149 - 166 1979
436 MacLeod *et al.* CGC Book 3020 Celis, Smith (eds) : 301 - 312 1979
440 von Ehrenstein *et al.* CGC Book 3070 LeDouarin (ed) : 49 - 58 1979
442 Zengel & Epstein Proceedings of the National Academy of Sciences USA 77: 852 - 856 1980
444 Kimble *et al.* CGC Book 3070 LeDouarin (ed) : 59 - 68 1979
447 Nelson & Ward Cell 19: 457 - 464 1980
448 Schierenberg *et al.* Developmental Biology 76: 141 - 159 1980
449 Miwa *et al.* Developmental Biology 76: 160 - 174 1980
461 Waterston *et al.* Developmental Biology 77: 271 - 302 1980
464 Lewis *et al.* Neuroscience 5: 967 - 989 1980
465 Cox *et al.* Genetics 95: 317 - 339 1980
468 Hosono Journal of Experimental Zoology 213: 61 - 67 1980
481 Dusenbery Journal of Comparative Physiology 137: 93 - 96 1980
482 Kass *et al.* Proceedings of the National Academy of Sciences USA 77: 6211 - 6215 1980
484 Lewis *et al.* Genetics 95: 905 - 928 1980
485 Mackenzie & Epstein Cell 22: 747 - 755 1980
486 Greenwald & Horvitz Genetics 96: 147 - 164 1980
488 Zengel & Epstein Cell Motility 1: 73 - 97 1980
491 Edgar CGC Book 3065 Leighton, Loomis (eds) : 213 - 235 1980
495 Argon & Ward Genetics 96: 413 - 433 1980
496 Horvitz & Sulston Genetics 96: 435 - 454 1980
497 Meneely & Herman Genetics 97: 65 - 84 1981
498 Hodgkin Genetics 96: 649 - 664 1980
499 Moerman & Baillie Mutation Research 80: 273 - 279 1981
500 Sulston & Horvitz Developmental Biology 82: 41 - 55 1981
501 Cassada *et al.* Developmental Biology 84: 193 - 205 1981
502 Chalfie & Sulston Developmental Biology 82: 358 - 370 1981
503 Albert *et al.* Journal of Comparative Neurology 198: 435 - 451 1981
504 Riddle *et al.* Nature 290: 668 - 671 1981
505 Swanson & Riddle Developmental Biology 84: 27 - 40 1981
507 Chalfie *et al.* Cell 24: 59 - 69 1981
509 MacLeod *et al.* Nature 291: 386 - 390 1981
513 Johnson *et al.* Genetics 97: 261 - 279 1981
514 Culotti *et al.* Genetics 97: 281 - 305 1981
515 Waterston Genetics 97: 307 - 325 1981
518 Babu & Brenner Mutation Research 82: 269 - 273 1981
527 Russell CGC Book 3045 Gershon *et al.* (eds) : 113 - 128 1981
529 Ward *et al.* Journal of Cell Biology 91: 26 - 44 1981
534 Goldstein & Slaton Chromosoma 84: 585 - 597 1982
536 Nelson *et al.* Journal of Cell Biology 92: 121 - 131 1982
537 Roberts & Ward Journal of Cell Biology 92: 132 - 138 1982
538 Fodor *et al.* General & Comparative Endocrinology 46: 99 - 109 1982
543 Cassada *et al.* CGC Book 3015 Brown (ed) : 209 - 227 1981
544 Edgar *et al.* Journal of Nematology 14: 248 - 258 1982
545 Ward *et al.* Journal of Nematology 14: 259 - 266 1982
550 Chalfie & Thomson Journal of Cell Biology 93: 15 - 23 1982

552 Horvitz *et al.* Science 216: 1012 - 1014 1982
554 Rosenbluth *et al.* Genetics 99: 415 - 428 1981
558 Roberts & Ward Journal of Cell Biology 92: 113 - 120 1982
559 Ward & Klass Developmental Biology 92: 203 - 208 1982
560 Chalfie Cold Spring Harbor Symposia on Quantitative Biology 46: 255 - 261 1982
562 Hedgecock & Thomson Cell 30: 321 - 330 1982
563 Kimble *et al.* Nature 299: 456 - 458 1982
564 Greenwald & Horvitz Genetics 101: 211 - 225 1982
565 Hartman & Herman Genetics 102: 159 - 178 1982
566 Klass *et al.* Developmental Biology 93: 152 - 164 1982
568 Horvitz *et al.* Neuroscience Commentaries 1: 56 - 65 1982
569 Kramer *et al.* Cell 30: 599 - 606 1982
570 McLachlan & Karn Nature 299: 226 - 231 1982
571 Moerman *et al.* Cell 29: 773 - 781 1982
574 Hecht *et al.* Developmental Biology 94: 183 - 191 1982
578 White *et al.* Nature 297: 584 - 587 1982
579 Waterston *et al.* Journal of Molecular Biology 158: 1 - 15 1982
580 Hosono *et al.* Experimental Gerontology 17: 163 - 172 1982
592 Rogalski *et al.* Genetics 102: 725 - 736 1982
597 Hartman & Herman Molecular & General Genetics 187: 116 - 119 1982
600 Goldstein Chromosoma 86: 577 - 593 1982
608 Hodgkin Genetics 103: 43 - 64 1983
609 Hosono *et al.* Journal of Experimental Zoology 224: 135 - 144 1982
611 Files *et al.* Journal of Molecular Biology 164: 355 - 375 1983
615 Kimble & Sharrock Developmental Biology 96: 189 - 196 1983
620 Ambros & Horvitz Science 226: 409 - 416 1984
622 Culotti & Klein Journal of Neuroscience 3: 359 - 368 1983
627 Hedgecock *et al.* Science 220: 1277 - 1279 1983
629 Chalfie *et al.* Science 221: 61 - 63 1983
630 Hodgkin Nature 304: 267 - 268 1983
632 Tabuse & Miwa Carcinogenesis 4: 783 - 786 1983
634 Ward *et al.* Developmental Biology 98: 70 - 79 1983
635 Trent *et al.* Genetics 104: 619 - 647 1983
642 Johnson & Russell Journal of Neurochemistry 41: 30 - 46 1983
644 Karn *et al.* CGC Book 3095 Pearson, Epstein (eds) : 129 - 142 1982
645 Epstein *et al.* CGC Book 3095 Pearson, Epstein (eds) : 419 - 427 1982
646 Greenwald *et al.* Cell 34: 435 - 444 1983
647 Miller *et al.* Cell 34: 477 - 490 1983
649 Strome & Wood Cell 35: 15 - 25 1983
654 Sanford *et al.* Journal of Biological Chemistry 258: 12804 - 12809 1983
655 Roberts Cell Motility 3: 333 - 347 1983
656 Burke & Ward Journal of Molecular Biology 171: 1 - 29 1983
658 Klass *et al.* Mechanisms of Ageing & Development 22: 253 - 263 1983
659 Klass Mechanisms of Ageing & Development 22: 279 - 286 1983
661 Isnenghi *et al.* Developmental Biology 98: 465 - 480 1983
662 Bolanowski *et al.* Mechanisms of Ageing & Development 21: 295 - 319 1983
663 Hartman Photochemistry & Photobiology 39: 169 - 175 1984
665 Meneely & Wood Genetics 106: 29 - 44 1984
666 Hodgkin Molecular & General Genetics 192: 452 - 458 1983
667 Albertson Developmental Biology 101: 61 - 72 1984
671 Sulston Cold Spring Harbor Symposia on Quantitative Biology 48: 443 - 452 1983
672 Horvitz *et al.* Cold Spring Harbor Symposia on Quantitative Biology 48: 453 - 463 1983
675 Rand & Russell Genetics 106: 227 - 248 1984
676 Goldstein Canadian Journal of Genetics & Cytology 26: 13 - 17 1984
680 Golden & Riddle Proceedings of the National Academy of Sciences USA 81: 819 - 823 1984
688 Chalfie Bioscience 34: 295 - 299 1984
689 Denich *et al.* Roux's Archives of Developmental Biology 193: 164 - 179 1984
706 Nelson & Riddle Journal of Experimental Zoology 231: 45 - 56 1984
708 Chalfie Trends in Neurosciences 7: 197 - 202 1984
710 Anderson & Brenner Proceedings of the National Acad. of Sciences USA 81: 4470 - 4474 1984
712 Herman Genetics 108: 165 - 180 1984
713 Starck International Journal of Invert. Repro. & Develop. 7: 149 - 160 1984
714 Kimble *et al.* Developmental Biology 105: 234 - 239 1984
715 Sigurdson *et al.* Genetics 108: 331 - 345 1984
722 Cox *et al.* Molecular & Cellular Biology 4: 2389 - 2395 1984
724 Doniach & Hodgkin Developmental Biology 106: 223 - 235 1984
726 Hartman Mutation Research 132: 95 - 99 1984
727 Moerman & Waterston Genetics 108: 859 - 877 1984
728 Jansson *et al.* Experimental Parasitology 58: 270 - 277 1984
729 Bolten *et al.* Proceedings of the National Academy of Sciences USA 81: 6784 - 6788 1984
731 Sternberg & Horvitz Annual Review of Genetics 18: 489 - 524 1984
732 Eide & Anderson Genetics 109: 67 - 79 1985
733 Hartman Genetics 109: 81 - 93 1985
737 Marx Science 226: 425 - 426 1984
738 Eide & Anderson Molecular & Cellular Biology 5: 1 - 6 1985
739 Goldstein Mutation Research 129: 337 - 343 1984
741 Waterston *et al.* Journal of Molecular Biology 180: 473 - 496 1984
742 Landel *et al.* Journal of Molecular Biology 180: 497 - 513 1984
746 Kramer *et al.* Journal of Biological Chemistry 260: 1945 - 1951 1985
748 Cox & Hirsh Molecular & Cellular Biology 5: 363 - 372 1985
749 Kenyon Trends in Genetics 1: 2 - 2 1985
750 Rosenbluth *et al.* Genetics 109: 493 - 511 1985
751 Cox *et al.* Genetics 109: 513 - 528 1985
755 Eide & Anderson Proceedings of the National Academy of Sciences USA 82: 1756 - 1760 1985
756 Okamoto & Thomson Journal of Neuroscience 5: 643 - 653 1985
761 Herman & Kari Cell 40: 509 - 514 1985
762 Ferguson & Horvitz Genetics 110: 17 - 72 1985
766 Golden & Riddle Molecular & General Genetics 198: 534 - 536 1985
768 Wills *et al.* Cell 33: 575 - 583 1983
769 Brown & Riddle Genetics 110: 421 - 440 1985
779 Dibb *et al.* Journal of Molecular Biology 183: 543 - 551 1985
781 Johnson & McCaffrey Mechanisms of Ageing & Development 30: 285 - 297 1985

784 Baillie *et al.* Canadian Journal of Genetics & Cytology 27: 457 - 466 1985
788 Hosono *et al.* Journal of Experimental Zoology 235: 409 - 421 1985
792 Edgar & Hirsh Developmental Biology 111: 108 - 118 1985
793 Hedgecock *et al.* Developmental Biology 111: 158 - 170 1985
795 Kolson & Russell Journal of Neurogenetics 2: 69 - 91 1985
796 Kolson & Russell Journal of Neurogenetics 2: 93 - 110 1985
797 Hodgkin Genetics 111: 287 - 310 1985
801 Albertson EMBO Journal 4: 2493 - 2498 1985
806 Stinchcomb *et al.* Banbury Report 20: 251 - 263 1985
807 Francis & Waterston Journal of Cell Biology 101: 1532 - 1549 1985
812 Greenwald Cell 43: 583 - 590 1985
813 Rogalski & Baillie Molecular & General Genetics 201: 409 - 414 1985
859 McGhee & Cottrell Molecular & General Genetics 202: 30 - 34 1986
862 Sebastiano *et al.* Genetics 112: 459 - 468 1986
867 Kemphues *et al.* Developmental Biology 113: 449 - 460 1986
870 Ellis & Horvitz Cell 44: 817 - 829 1986
876 Miller *et al.* Proceedings of the National Academy of Sciences USA 83: 2305 - 2309 1986
877 Moerman *et al.* Proceedings of the National Academy of Sciences USA 83: 2579 - 2583 1986
888 Hirsh *et al.* Cold Spring Harbor Symposia on Quantitative Biology 50: 69 - 78 1985
898 Greenwald & Horvitz Genetics 113: 63 - 72 1986
902 Goldstein Journal of Cell Science 82: 119 - 127 1986
906 Kusch & Edgar Genetics 113: 621 - 639 1986
913 Kenyon Cell 46: 477 - 487 1986
914 Park & Horvitz Genetics 113: 821 - 852 1986
915 Park & Horvitz Genetics 113: 853 - 867 1986
916 Nelson & Honda Canadian Journal of Genetics & Cytology 28: 545 - 553 1986
922 Hodgkin Genetics 114: 15 - 52 1986
923 Doniach Genetics 114: 53 - 76 1986
926 Simpson *et al.* Nucleic Acids Research 14: 6711 - 6717 1986
927 Priess & Hirsh Developmental Biology 117: 156 - 173 1986
932 Perkins *et al.* Developmental Biology 117: 456 - 487 1986
933 Fire EMBO Journal 5: 2673 - 2680 1986
941 Meyer & Casson Cell 47: 871 - 881 1986
943 Barton *et al.* Genetics 115: 107 - 119 1987
944 Villeneuve & Meyer Cell 48: 25 - 37 1987
950 Lewis *et al.* Molecular Pharmacology 31: 185 - 193 1987
957 Ferguson *et al.* Nature 326: 259 - 267 1987
958 Sedensky & Meneely Science 236: 952 - 954 1987
959 Hecht *et al.* Journal of Cell Science 87: 305 - 314 1987
965 Hartman Genetical Research 49: 105 - 110 1987
966 Ambros & Horvitz Genes & Development 1: 398 - 414 1987
969 Krause & Hirsh Cell 49: 753 - 761 1987
975 Howell *et al.* Genetical Research 49: 207 - 213 1987
976 Kim & Rose Genome 29: 457 - 462 1987
977 Herman Genetics 116: 377 - 388 1987
980 Link *et al.* Proceedings of the National Academy of Sciences USA 84: 5325 - 5329 1987
987 Meneely & Wood Genetics 117: 25 - 41 1987
998 Lewis *et al.* Journal of Neuroscience 7: 3059 - 3071 1987
999 Hodgkin Genes & Development 1: 731 - 745 1987
1000 Keller *et al.* Photochemistry & Photobiology 46: 483 - 488 1987
1002 Politz *et al.* Genetics 117: 467 - 476 1987
1003 Donahue *et al.* Proceedings of the National Academy of Sciences USA 84: 7600 - 7604 1987
1004 Hosono *et al.* Journal of Neurochemistry 49: 1820 - 1823 1987
1005 Goldstein & Curis Mechanisms of Ageing & Development 40: 115 - 130 1987
1006 Sassa *et al.* Neurochemistry International 11: 323 - 329 1987
1007 Austin & Kimble Cell 51: 589 - 599 1987
1008 Priess *et al.* Cell 51: 601 - 611 1987
1011 DeLong *et al.* Genetics 117: 657 - 670 1987
1016 Avery & Horvitz Cell 51: 1071 - 1078 1987
1018 McKim *et al.* Genetics 118: 49 - 59 1988
1019 Rogalski & Riddle Genetics 118: 61 - 74 1988
1020 Friedman & Johnson Genetics 118: 75 - 86 1988
1023 Ward *et al.* Journal of Molecular Biology 199: 1 - 13 1988
1024 Klass *et al.* Journal of Molecular Biology 199: 15 - 22 1988
1025 Walthall & Chalfie Science 239: 643 - 645 1988
1026 Mori *et al.* Proceedings of the National Academy of Sciences USA 85: 861 - 864 1988
1029 Kiff *et al.* Nature 331: 631 - 633 1988
1031 Moerman *et al.* Genes & Development 2: 93 - 105 1988
1032 Kemphues *et al.* Cell 52: 311 - 320 1988
1035 Albert & Riddle Developmental Biology 126: 270 - 293 1988
1037 Schedl & Kimble Genetics 119: 43 - 61 1988
1039 Johnson *et al.* Neuron 1: 165 - 173 1988
1040 Clark *et al.* Genetics 119: 345 - 353 1988
1041 Jacobson *et al.* Genetics 119: 355 - 363 1988
1042 Meneely & Nordstrom Genetics 119: 365 - 375 1988
1051 Epstein *et al.* Journal of Cell Biology 106: 1985 - 1995 1988
1052 Way & Chalfie Cell 54: 5 - 16 1988
1053 Rosenquist & Kimble Genes & Development 2: 606 - 616 1988
1054 Friedman & Johnson Journal of Gerontology 43: B102 - B109 1988
1057 Link *et al.* Development 103: 485 - 1988
1063 Morgan *et al.* Anesthesiology 69: 246 - 251 1988
1064 Kondo *et al.* Molecular & Cellular Biology 8: 3627 - 3635 1988
1065 Pulak & Anderson Molecular & Cellular Biology 8: 3748 - 3754 1988
1067 Trent *et al.* Genetics 120: 145 - 157 1988
1068 Johnson & Hartman Journal of Gerontology 43: B137 - B141 1988
1071 Johnsen & Baillie Mutation Research 201: 137 - 147 1988
1073 Rogalski *et al.* Genetics 120: 409 - 422 1988
1074 Bullerjahn & Riddle Genetics 120: 423 - 434 1988
1075 L'Hernault *et al.* Genetics 120: 435 - 452 1988
1076 Shen & Hodgkin Cell 54: 1019 - 1031 1988
1077 Miller *et al.* Cell 55: 167 - 183 1988
1080 Yochem *et al.* Nature 335: 547 - 550 1988

1081 Sternberg Nature 335: 551 - 554 1988
1086 Hevelone & Hartman Biochemical Genetics 26: 447 - 461 1988
1087 Lerner & Goldstein Cytobios 55: 51 - 61 1988
1091 Bejsovec & Anderson Genes & Development 2: 1307 - 1317 1988
1092 McCoubrey *et al.* Science 242: 1146 - 1151 1988
1093 Coohill *et al.* Mutation Research 209: 99 - 106 1988
1096 Sternberg & Horvitz Developmental Biology 130: 67 - 73 1988
1097 Spieth *et al.* Developmental Biology 130: 285 - 293 1988
1099 Fields Journal of Molecular Evolution 28: 55 - 63 1989
1100 Kramer *et al.* Cell 55: 555 - 565 1988
1101 von Mende *et al.* Cell 55: 567 - 576 1988
1105 Desai *et al.* Nature 336: 638 - 646 1988
1106 Costa *et al.* Cell 55: 747 - 756 1988
1107 Finney *et al.* Cell 55: 757 - 769 1988
1108 Rosenbluth *et al.* Genetical Research 52: 105 - 118 1988
1109 Kemphues *et al.* Genetics 120: 977 - 986 1988
1115 Plenefisch *et al.* Genetics 121: 57 - 76 1989
1124 Dibb *et al.* Journal of Molecular Biology 205: 603 - 613 1989
1125 Chalfie & Au Science 243: 1027 - 1033 1989
1127 Wadsworth & Riddle Developmental Biology 132: 167 - 173 1989
1128 Ruvkun *et al.* Genetics 121: 501 - 516 1989
1133 Desai & Horvitz Genetics 121: 703 - 721 1989
1134 Herman & Kari Genetics 121: 723 - 737 1989
1139 Tabuse *et al.* Science 243: 1713 - 1716 1989
3015 Brown (ed) Academic Press, NY : 1 - 702 1981
3020 Celis & Smith (eds) Academic Press, London : 1 - 349 1979
3025 Cowan & Ferrendelli (eds) Society for Neuroscience : 1 - 461 1977
3030 Croll (ed) Academic Press, NY : 1 - 439 1976
3045 Gershon *et al.* (eds) Boxwood Press (Pacific Grove, CA) : 1 - 469 1981
3060 Hazelbauer (ed) Halsted Press, Wiley NY 5: 1 - 341 1978
3065 Leighton & Loomis (eds) Academic Press, NY : 1 - 478 1980
3070 Le Douarin (ed) Elsevier, NY : - 1979
3095 Pearson & Epstein (eds) Cold Spring Harbor Laboratory : - 1982
4040 Subtelny & Konigsberg (eds) Academic Press, NY : 1 - 333 1979
4050 Wilcox *et al.* (eds) Academic Press, NY : 1 - 449 1977

BIOCHEMICAL LOCI OF THE "FRUIT FLY" (*Drosophila melanogaster*)

Glen E. Collier

Department of Biological Sciences, Illinois State University
Normal, Illinois 61761

2N = 8

GENETIC MAP OF BIOCHEMICAL LOCI OF *Drosophila melanogaster*: Gene loci are listed according to chromosome or chromosome arm. The top entry in each column represents the leftmost locus and the bottom entry the rightmost locus.

X-CHROMOSOME

Locus	Gene
0.0	ac
0.0	arm
0.0	cin
0.0	elav
0.0	ewg
0.0	sc
0.0	su(b)
0.0	su(s)
0.0	l(1)nprl
0.1	su(w^a)
0.63	Pgd
0.8	pn
1.0	tko
?	<--Draf-1
1.0	z
1.2	per
?	<--fs(1)Ya
1.4	gustB1
1.5	w
3.0	N
3.6	Sgs4
6	norpA
12-16	sww
(14)	Act5C
(18)	Fum
19.2	Sxl
21±	fs(1)ll63
?	<--fs(1)410
?	<--rps141
23	l(1)myo
?	<--fsh
23.1	oc
23.1	s36, s38
23.2	otu
(27.7)	Mex
27.7	su(r),lz
28	gmp
29.2	Hex-A
29±	Ypl, Yp2
31	flp
32.8	ras
?	<--purl
?	<--mp20
33.0	v
33.2	sev
?	<--Glus2
34.8	l(1)dlg-1
35.4	Met
35.7	RpII215
?	<--Hspl0F
37.0	Flu
38	mfd
38.2	Sh
39.5	Lspl
42.0	l(1)ts403
42.6	Gpt
43.5	int, up
?	<--Pyk
44±	Yp3, Yp3R
44.5	fs(1)29
45±	rut
45.3	Eag
?	<--c-myb
?	Gapdh2
?	<--Gpro
54.4	Had
?	fu
?	Gap-43
59.5	hdp
62.9	Zw
64.8	ma-l
?	<--run
(65)	Cgl9-20, sdby
65.9	su(f)
66.0	bb

CHROMOSOME 2L

Locus	Gene
<0.1	Glusl
?	<--ninaA
1.3	Pkg-1
1.9	Lspl
3.0	Got2
4.0	dpp
5.0	fs(2)B
5.9	Pgk
9.0	msl2
?	<--Pkg-2
11.0	lcp5-8
13.0	dp
13.9	Sgsl
(15)	Cg25C
18.4	Fgarat
20.5	Gpdh, Gdt3
20±	Gal
?	<--Kf2
20.0	Gart
?	<--Dint-1
?	<--wg
22	ninaC
?	<--urox
?	<--lamB1
?	<--src4
?	<--numb
?	<--Pka
37±	Mdhl
(37)	Sucr
41.5	da
?	<--sal
44.0	abo
48.5	b
50.1	Adh
51.3	Bsh
52±	Ifm(2)l, Ifm(2)2, Ifm(2)3
(52)	Mhc36B
52.4	tyrl
53.3	msll
53.4±	Dox3
53.9	l(2)amd
53.9	Ddc
53.9	Dox2
53.9	l(2)37Bf
?	<--Top2
?	<--cad
(54)	Tpl
54	Hisl,His3, His4,His2a His2b(tandem repeats)
54.5	pr
(55)	Ifm(2)ll

CHROMOSOME 2R

Locus	Gene
55.2	Dip-A
55.2	disco
(55.4)	Act42A
55.7	bur
55.8	mle
?	<--Draf-2
57.5	cn
?	<--Gapdhl
58.6	Pgi
?	<--Pfk, eve, neup
62.0	en
62±	Lcpl, Lcp2, Lcp3, Lcp4
(62)	HDL family
64	Mar
?	<--Deb-A, Deb-B
?	<--Cmd
72±	l(2)me
73.5	Hex-C
?	<--slit
75±	Gpo
75.0	Gotl
?	<--kin
?	<--Pkc53E
77.7	Amy-p, Amy-d
?	<--Hsp54E
80	map
80.6	Phox=?Dox-1
?	<--Eip40
82.9	Aldox2
89±	sdh
?	<-- Tub56C
?	<-- Tub56D
92.9	Treh
(93)	Act57A
95±	5S RNA
97±	Pu
99	Der = flb = Elp
?	<--Dets
100	Adk-C
100	twist
?	<--CycB
104.5	bw
107±	Dat
?	<--Nac
107.0	sp
108	Kr

CHROMOSOME 3L

Locus	Gene
-1.41	Lspl
1	Sp
3.03	Aprt
?	<--Hsp83
?	<--Hsp63F
?	<--Ubi
?	<--Dmras64B
?	<--Achr
19	Amal
?	<--lamA
25.4	Idh2
26	Msl3
26.5	sl5, sl6, sl8, sl9
26.8	Argk
?	<--Hsp22,Hsp23, Hsp26,Hsp28
27.1	Idhl
?	<--rps17
?	<--lamB2
?	<-- Tub67C
34.4	hay
34.5	lxd
34.6	Sod
35.0	rs
?	<--CycA
35.5	Fuc
(36)	Sgs3, Sgs7, Sgs8
37±	Lsp2
36.8	Est6
?	<--Hscl
?	<--Eip28, Eip29
41.7	fz
?	<--gdl
42.0	Sgs6
43.4	Pgm
(44)	Dash
45	tra
?	<--Hsp73D
?	<--Rh4
?	<--Ars
45.9	DNase2
46.3	Aphl
47	Cat
?	<--kni, knrl
?	<--Pka
(47.5)	Act79B

CHROMOSOME 3R

Locus	Gene
?	<--Dsk
47.8	Antp
(47.8)	Tub84B
48	Gld
(48)	Tub84D
?	<--rn
48	Est-C
48	Dsx
?	<--nac
?	<--hb
48	Dhod
48.3	Ali
48.5	B2t
?	<--Dras85D
(49)	Mtn, Tub85E
49.2	Odh
(50)	Hsp70
(51)	Dip-C, Hsp70
51.7	Men
?	<--Hsc2
52.0	ry
?	<--sim
52.2	Ace
(52.3)	Act87E, Mhc87E
53.6	Dip-B
53.6	red
54.2	nc4
?	<--Hsp88B
?	<--su(Hw)
(55)	Ifm(3)l, Ifm(3)2, Ifm(3)4, Ifm(3)6, Tm
55.4	R3-55.4
56.7	m-Est
57.1	Act88F, Mhc88F, rsd
?	<--ea
57.2	lpo
57.2	Aldoxl
58.8	bx
58.8	Cbx
58.8	Ubx
58.8	bxd
58.8	pbx
61.8	DNasel
?	<--Sgs5
62	sr
62.6	Mdh2
64.5	Sodh
64.6	Cha
?	<--Rh2
66.2	Dl
66	ninaE = Rh1
?	<--Rh3
?	<--Kfl
70±	r-l
?	<--Atp
?	<--Hsr93D
?	<--Dirh, Dmt
?	<--Hsp68
81.7	Gdh
90	boss
91.1	ro
91.5	Ald
?	<--E(spl)
94.5	mod(?)
?	<--Toll
?	<--Tub97F
98.3	Lap-A, Lap-D
?	<--rpl1
?	<--Mlc98B
?	<--fkh, yema
?	<--Pkc,stg
100	Ama2
?	<--Rbp49, Glu?
?	<--sry, janus
101.1	Acphl
?	<--Mlc99E
101.3	Tpi
?	<--chp, awd
102.9	Kpn

BIOCHEMICAL LOCI OF THE "FRUIT FLY" (*Drosophila melanogaster*)

Glen E. Collier

Department of Biological Sciences, Illinois State University
Normal, Illinois 61761

2N = 8

This map is an updated version of earlier, similar maps (166, 174, 460, 485, 568, 882). New material surveyed through **July 1989** is included. Abstracts cited in earlier versions have either been superseded by published papers or deleted. Only selected references are listed for some well studied loci. Genetics symbols follow the rules (460, 461) for naming mutants as closely as feasible. Inferred genetic loci for cytological locations are shown parenthetically. No information is given for most transposable elements, tRNA genes, or mitochondrial genes.

ALPHABETICAL LISTING:

Gene Symbol	Genetic Map Locus	Cytological Location	Product or Regulatory Function	Reference
A (A^{53g}) Abnormal abdomen	1.45	3A5	Interferes with amino-acylation and protein synthesis	1108
abl (See *Dash*)				
abo abnormal oocyte	2-44.0	31F-32E	rDNA redundancy	406 416 495 614 660 815 261 461
ac achaete	1-0	1B1-5	1.1kb transcript; T4 & T5 encode myc related polypeptides; required for neural development	94 835 868 1087
Ace acetylcholinestrase(= *l(3)26*)	3-52.2	87E1-4	Acetylcholinesterase (EC 3.1.1.7)	38 44 111 231 267 290 292 311 519 547 700 825
---	(3-52.2)	87E4	Control of *Ace* expression in CNS	76
Achr acetylcholine receptor	(3L)	64B	3kb RNA; acetylcholine receptor homolog; present in CNS	1091
Acp accessory gland proteins (**Also see male-specific transcripts,** ***mst***);				
AcpGl	2-13.5		Accessory gland protein	771
AcpB	2-42.8			
AcpC	2-53.0	36D1-36E3,4		
AcpK	2-54.1			
Acp70A		70A	Accessory gland protein; 45aa, represses female receptivity and stimulates oviposition	877
Acphl acid phosphatase(= *Acph-l*)	3-101.1	99D-99E	Acid phosphatase-1 (EC 3.1.3.2) with *cis* control	209 356 357 483 542 543
Act actin genes;				
Act5C (= *act5C*)	(1-14)	5C3-4	Actins II & III, cytoplasmic	220 221 737 141 222 461 519 695
Act42A (= *act42A*)	(2-55.4)	42A	Actins II & III (?), cytoplasmic	221 737 222 519 461
Act57A (= *act57A*)	(2-93)	57A	Actin I, larval, pupal, adult muscle	220 221 737 222 519 461
Act79B (= *act79B*)	(3-47.5)	79B	Actin I (?), mostly in adult muscle	220 221 737 828 222 461 519 561 659
Act87E (= *act87E*)	(3-52.3)	87E9-12	Actin I (?), larval, pupal, adult muscle	220 221 737 222 519 461 976
Act88F (= *act88F*)	(3-57)	88F	Actin I, mostly adult muscle,	220 221 222 737 375 461 519 561 659
			5' reg seq	911
ade2 (See *Fgarat*)				
ade3 (See *Gart*)				
Adh	2-50.1	35B2-3	Alcohol dehydrogenase (EC 1.1.1.1), with *cis* control	11 14 51 52 53 54 55 115 117 167 244 252 255 301 414 419 428 497 498 509 519 570 616 661 750 754 798 799 1012
----	2-not near *Adh*		ADH *trans* control loci	315 731
Adk-C (= *Ak-C*)	2-100, ap-proximate		Adenylate kinase C (EC 2.7.4.3)	461 757
Ald	3-91.5	97AB	Aldolase (EC 4.1.2.13)	578 757 759
Aldox1 (= *Aldox= Ao= aldox*)	3-57.2	89A	Aldehyde oxidase (EC 1.2.1.3), with *cis* control	46 134 144 152 164 165 517 544 701 757 777 789
Aldox2	2-82.9	52B1-E8	Aldehyde oxidase-2; product has Mb+ binding site. Mutant has lowered AO, PO, XDH activity	47 518 852
Ali (= *ali= ali=est*)	3-48.3		Ali-esterase	576 577
ama amalgam (in ANT-C)			333aa sequence, member of Ig superfamily	1050
amd: see *l(2)amd*				
Amr1	3-19		a-amanitin resistance 1	605
Amr2	3-100		a-amanitin resistance 2	605
Amy-p *Amy-d*	2-77.7	54A1-B1	a-amylase (EC 3.2.1.1) proximal & distal (duplicate genes)	17 18 19 170 171 175 176 238 240 396 460 926 1048
---		53CD	Amylase pseudogene	240

Gene Symbol	Genetic Map Locus	Cytological Location	Product or Regulatory Function	Reference
Antennapedia Complex *(Ant-C)*:				
Antp Antennapedia (Includes *ftz*, *zen*, *bcd*,and *Dfd*)	3-47.8	84B1.2	3.5 and 5.0 kb *trans* cripts, tissue localizations, a family of transcription factors	96 104 162 184 226 288 299 378 445 456 457 460 672 720 963
Aph1	3-46.3		Alkaline phosphatase 1 (EC 3.1.3.1)	36 461 773
Aph2	2-not mapped		Alkaline phosphatase in adult hindgut	461 669
Aprt (= *aprt*)	3-3.03	62B9	Adenine phosphoribosyl transferase (EC 2.4.2.7) 20kD protein	358 359 460 461 944
Argk (= *Ak*)	3-26.5	66F3-6	Arginine kinase (EC 2.7.3.3)	217 218 942 991
arm armadillo	(1)	2B15	Two 3.2kb RNAs; 91kD with novel internal repeats	1028
Achaete-scute Complex (AS-C); (See *ac* and *sc*)				
Ars Arylsulfatase	(3L)	74A-79D	Arylsulfatase (EC 3.1.6.1)	461 484
Atp ATPase	(3R)	93B	Na^+, K^+ ATPase; 1038aa polypeptide	964
awd abnormal wing discs	(3R)	100C-D	0.8kb RNA abundant in 2nd & 3rd instar larvae; 153aa, 16kD protein similar to subunit V of yeast cyt oxidase	890 854
b black	2-48.5	34E5-35D1	ß-ureidopropionase(?), ß-alanine deficient	318 460 461
B2t (= $ms(3)KK^D$)	3-48.5	85D7-11	ß-tubulin, testis specific	383 384 385 461 519
bb bobbed	1.66.0, Y-proximal to *ks1* and *ks2*	right of 20F	rRNA: 2S 5.8S 18S 28S	120 130 157 169 242 249 250 251 362 363 386 390 391 401 402 403 422 472 -474 494 522 595 617 633 634 643 -645 666 678 684 787 809 821
bcd bicoid	(3R)	84A	Polarity gene; 55kD protein translated after egg deposition; May regulate *hb*	895 904 1065
Blastoderm-specific genes: (see *sry*)				
---	(2L)	25D3	2.7kb transcript	647
---	(3L)	75C	1.6kb "	
---	(3R)	99D4-8	3.6, 2.0kb transcripts	
---	(3L)	71A	gastrula-differential sequence	
---	(3R)	95C	1.5,1.0,0.7kb transcripts	
boss bride of sevenless	3-90	96F5-14	Required in R8 for R7 development	1027
bur burgundy (= *qua2?*)	2-55.7		Inosinate dehydrogenase inactivity in bur^{qua2-1}	361 460 555
bw brown	2-104.5	59D4-E1	2.8 & 3.0kb RNA; both encode 675aa protein with strong homology with *w* protein	894
Bithorax Complex *(BX-C)*: units in complex listed left to right				
bx bithorax			*cis* control unit	
Cbx Contrabithorax			*cis* control due to transposition of *pbx* DNA into middle of *Ubx* unit	
Ubx Ultrabithorax	3-58.8	89E1,2 (in or close to)	structural unit producing 4.3, 3.2, 1.6, & 1.4 kb poly(A)+ RNA transcripts and 4.7 kb poly(A)- RNA transcript	460 45 434 453 461 519 687
bxd bithoraxoid			structural unit producing 1.28 & 1.15 kb transcripts	
pbx postbithorax			*cis* control unit	
Bx Beadex	1-59.4	17C2-3	*cis* acting negative regulator of *hdp-a*	505
cad caudal	(2L)	38E	Maternal (2.4kb) & paternal (2.6kb) RNAs; 472aa ORF containing homeobox domain	984
Cat	3-47	75D-76A	Catalase (EC 1.11.1.6)	461 461 973
c-myb	(1)	13EF	3.0 & 3.8kb transcripts homologous to c-myb oncogene	376
Cha (= *Cat*)	3-64.6	91B-D	choline acetyltransferase (EC 2.3.1.6)	266 289 291 343 461
Cg9	(1)	9E	Collagen	858
Cg19-20	(1-65)	19E-20B	Collagen	461 519 536 558
Cg25C	(2-15)	25C	Collagen; 6kb RNA; 1774aa polypeptide	858 859 875 962
Chorion protein: see shell protein *(S)* genes				
chp chaoptin	(3R)	100B6-9	Chaoptin, 160kD glycoprotein on surface of developing rhabdomeres	1026 1084
cin cinnamon	1-0 (.0017cM left of *y*)		modifier of Mo hydroxylases (AO, PO, XDH) & sulfite oxidase	21 38 84 587 776 74 777
Cmd Calmodulin	(2R)	49A	Calmodulin; 1.6kb RNA; 148 aa protein	810 1054
cn cinnabar	2-57.5	43E3-14	Kynurenine hydroxylase (EC 1.99.1.5)	243 460 594 723 824 210

Gene Symbol	Genetic Map Locus	Cytological Location	Product or Regulatory Function	Reference
Collagen-like proteins:		(See *Cg*)		
cr compensatory response	(1)	20C1-2	rDNA redundancy	614 190
crn crooked neck		2E2-2F1	2.1kb transcript	282
Cuticle proteins: see *Lcp* and *Pcp*				
CycA Cyclin A	(3L)	68D/E	Cell cycle control; homologous to urchin & clam cyclin A & B; 2.3kb RNA; 491aa. product missing in *neo114* and *l(3)183*	968 1096
CycB Cyclin B	(2R)	59A	Cell cycle control; 2.3kb RNA; 491aa product	1096
cyt cytochrome c	(2L)	36A10,11	cytochrome c-like sequences, 2 copies	459
da daughterless	2-41.5	31D-F	Interacts with *Sxl*; maternal product 694aa, paternal product 710aa, both with myc homology	873 886 887
Dash (=*S16*) (*abl* oncogene homologue)	(3-44)	73B1-2	Tyrosine protein kinase; may interact with *dab*	519 685 910 924 925
Dat	2-107	60B1-10	Dopamine acetyl transferase (EC 2.3.1.5)	329 501
Ddc	2-53.9 (.025cM right of *hk*)	37C1,2	Dopa decarboxylase (EC 4.1.1.26) with *cis* control	191 245 314 317 500 502 519 670 801 -805 929
Der	(2R)	57F	EGF receptor homologue	471 668
Dets	(2R)	58A/B	c-ets oncogene homolog	1016
Dfd Deformed	(3R)	84A	Homeotic selector gene of ANT-C; 586aa polypeptide with homeobox domain	1025
Dhod	3-48	85A-C	Dihydroorotate dehydrogenase (EC 1.3.3.1), mitochondrial	620 624 946
Dint-1	(2L)	28A1-2	*Int-1* oncogene homolog; 2.9kb RNA; 469aa	1083
Dip-A ---	2-55.2	41A	Dipeptidase-A; modifier within 0.029mu	579 757 432 918 927
Dip-B	3-53.6	87F12-88C3	Dipeptidase-B	579 432
Dip-C	(3-51)	87B5-10	Dipeptidase-C	579 432
Dirh	(3R)	93E	Insulin receptor homolog; 11 & 8.6kb RNA	905 996 1010
disco	1-53		Plays role in neuronal cell recognition in optic ganglia	1060
dl dorsal	2.52.9	36C2-E1	*c-rel* oncogene homolog; 2.5kb RNA, 677aa nuclear protein found in syncytial & cellular blastoderm	1063 1064
Dl Delta	(3R)	92A1-2	Product homologous to mammalian EGF	836 956 1085 1086
Dll Distal-less	(3L)	60E5-6	2.5, .4kb RNA; Encodes homeodomain protein	881
Dmras64B (*ras* homologue)	(3L)	64B	22.7Kd protein	549
Dmt Drs. mouse t complex		94B1-2	577aa homolog of mouse t complex polypeptide I	1082
DNase1	3-61.8	90C2-E	deoxyribonuclease-1 (EC 3.1.4.5)	163 272 715
DNase2	3-45.9		Deoxyribonuclease-2	272
dnc dunce	1-4.6	3D4	cAMP-phosphodiesterase, form II (EC 3.1.4.17) with *cis* control, *cf*, .09 cM right of *dnc*	88 154 155 181 393 394 379 395 652 656 681 696
Dox-1	(2R)	55A-B	A1 phenoloxidase; allelism with *Bc* not yet established	891
Dox-2	2-53.9	37B10-13	Phenoloxidase (A2 proenzyme)	601 602
Dox-3	between *rdo* (53.1) and *Tft* (53.6)	36C-F	Phenoloxidase (A3 proenzyme)	636 637
dp dumpy	2-13.0	25A1-2	Orotate phosphoribosyl transferase(EC 2.4.2.10)	73 259 460 752
dpp decapentaplegic	2-4.0	22F1-3	3.5-5kb RNA; 588 or 455aa protein with homology to TGF beta protein family	1003
Draf-1	(1)	2F5-6	*c-raf* oncogene homolog; 3.2kb RNA; 666aa polypeptide	979 997
Draf-2	(2R)	43A2-5	*c-raf* oncogene homolog; 3.2kb RNA	979
Dras85D	(3R)	85D	*ras* oncogene homolog	1099
Dras64B	(3L)	64B	*ras* oncogene homolog	1099
Dsk	(3R)	81F	Drosulfakinin; homolog of vertebrate neuropeptide cholecystokinine; 128aa polyprotein	995
Dsrc (=*S24*) (*v-src* homologue)	(2L)	29A,C		519 685
Dsrc (=*S24*) (*v-src* homologue)	(3L)	64B	Potentially encodes 62Kd polypeptide	519 685 686
dsx double-sex	3-48	84E1-2	YP regulation; 3.7 & 1.65 kb larval transcripts; common primary transcript with sex-specific splicing; controls female pheromones	460 80 456 461 519 795 842 865 941
E(spl) Enhancer of split	(3R)	96F11-14	Neurogenic gene; 719aa product with homologies to b subunit of G-protein	921 966 1018 1107

Gene Symbol	Genetic Map Locus	Cytological Location	Product or Regulatory Function	Reference
ea caster	(3R)	88F2	Dorso-ventral gene; 1.5kb RNA; 392aa (43kD) protein with serine protease homology	876
Eag Eagle	1-45.3		Control of K currents in muscle and nerve membranes	807
Ecdysteroid-inducible genes for polypeptide (EIPs), proteins & RNAs: (see *Acs, Act, Ddc, l(2)37Bf, Lcp, Lsp, Sgs, Yp*)				
Ecdysone-dependent proteins for pupal cuticle development:				
Edg	(2R)	42A	Ecd dep. genes necessary for formation of pupal cuticle	898
	(3L)	64CD		
	(3R)	78E		
	(3R)	84A1		
Ecdysone-inducible proteins:				
Eip28	(3L)	71C3-D2	EIP 28 (I, II, III)	109 110 112 113
Eip29			EIP 29 (I, II)	519 662 1045 1046
Eip40	(2R)	55B-D	EIP 40 (I & II)	
Ecdysone-inducible membrane proteins from imaginal discs:				
Imp-E1	(3L)	66C		560
Imp-E2	(3L)	63E		560
Imp-E3	(3R)	84E		560
Imp-L1	(3L)	70A		560
Imp-L2	(3L)	64B		560
Imp-L3	(3L)	65B		560
Ecdysone-inducible RNAs:				
---	(1)	2B5	ecdysone-inducible mRNA (4.5 kb)	106
---	(1)	5C	ecdysone-inducible RNA	142
---	(1)	5C2-5	ecdysone-inducible RNA	142
---	(1)	7	ecdysone-inducible RNA	142
---	(2L)	25A-C	ecdysone-inducible RNA	142
---	(2L)	37AB	ecdysone-inducible RNA	142
---	(3L)	63F2-4	ecdysone-inducible mRNA	345
---	(3L)	71E	6 "late" genes and 1 "early" gene within 10kb	628
---	(3L)	74EF	ecdysone-inducible RNA (2.7 kb)	86 519 541
---	(3L)	95B-C	ecdysone-inducible RNA	142
---	(3L)	97D1-3	ecdysone-inducible RNA	142
Epidermal Growth Factor homologs (See *slit*):				
	(2L)	27DE		1032
	(2R)	51F-52A		
	(3R)	95F		
	(3R)	97F		
elav embryonic lethal, abnormal visual system	(1)	1B5-9	Req. for proper dev of all neurons; 4.7kb RNA, 483 aa (50.8kD) protein; contains 3 RNP concensus motifs	870 871 1020 1021
Elp (=*flb*) Ellipse				

Gene Symbol	Genetic Map Locus	Cytological Location	Product or Regulatory Function	Reference
en engrailed (see **Homeo box**)	2-62.0	48A1-4	3.5 kb & 7.5 kb RNA transcripts; 2.7kb embryonic transcript with 1.7kb ORF with homeo box	187 227 409 410 420 435 460 540 609
en-r engrailed-related (see **Homeo box**)		17kb downstream from *en*	Homeo box containing transcript	420
Est-C	3-47.7	84D3.4-11.12	Esterase-C	37 106 578 580 742
Est6 esterase-6	3-36.8	69A1	Carboxylesterase (EC 3.1.1.1); 1.7 and 1.8kb RNA; 548aa protein	2 37 128 209 493 630 641 642 800 998
Est9	2-not mapped		Esterase-9	477
ets (See *Dets*)				
eve even-skipped	2-59	46c3-11	Homeobox gene encoding 376aa protein which binds adjacent to *eve* & *en*	930
F	---	---	F factor, dispersed retroposon with ORF encoding 859aa reverse transcriptase homolog	893
fas fasciclin	(2L)	36E1	Fasciclin III; surface glycoprotein of axon bundles of some neurons	1006 1109
FB Foldback	---	---	Dispersed element; 3 ORFs, longest encodes 71aa protein localized to egg chambers	1071
Follicle cell protein genes:				
Fc	1-21.0	7C1-9	Follicle cell protein	462
Fgarat (=*ade2*)	2-18.4		Phosphoribosylglycinamidine synthase (FGARAT)	361
fkh fork head	(3R)	98D2,3	Region specific homeotic gene; 4.2kb RNA; 510aa nuclear protein	1093
flb faint little ball	2-99	57F	=*to*=*Der;* EGF receptor, dominant allele called Ellipse (Elp)	843 994 1017 1041
fs(1)29 (= *fs29*)	1-44.5	12E1-F1	Sequestration of YPs	775
fs(1)410		7C1-3	Encodes s-130 that is cleaved to s76 & s85	847
fs(1)1163 female sterile (1) 1163	1-21		=Yolk protein-1; 1 aa *from* wild-type	78 79 224 612
fs(1)K10	(1)	2E2-F1	3.2 & 6kb transcripts involved in establishing dorsoventral axis	282
fs(1)K313	1-48	12E1-13A5	=Yolk protein 2; 2 aa diff from wild type	1097

Gene Symbol	Genetic Map Locus	Cytological Location	Product or Regulatory Function	Reference
fs(1)Ya	(1)	3B4-6	Required for 1st embr. mitotic division; tightly linked to *fs(1)Yb*; 2.4kb RNA in nurse cells and oocytes	970 971
fs(2)B female sterile (2) Bridges	2-5.0		Thymidylate synthetase (EC 2.1.1.6)	99
fs(3)272-9	3-between *st*(44) and *cu*(50)		*Trans* regulation of major chorion proteins, amplification defective	694
fs(3)293-19	3-between *st*(44) and *ss*(58.5)		*Trans* regulation of major chorion proteins, amplification defective	694
fsh female sterile hom.	(1)	7D5-6	205kD ovarian & 110kD embryonic transmembrane proteins	923
Female-specific transcripts (fst);				
fst241	(3L)	69C	3.3kb RNA	892
ftz fushi tarazu	(3R)	84B	1.8kb transcript, tissue localization, developmental profile, *cis* control	421 780 313 928 959
fu fused	(1)	17C	Segment polarity gene; 3.5, 2.5, 1.6, 1.3 kb RNA	978
aFuc	3-35.5		a-fucosidase	484
Fum (= *Fuh*)	1-(left of *cm*)	5F1-6D2	Fumarase (fumaratehydratase)	489 606 439
fz frizzled	3-41.7		Required to coordinate cytoskeleton of epidermal cells; 581aa; seven trans-membrane domains	915 1089
ßGal	2-20+	26A7-9	ß-galactosidase (EC 3.2.1.23)	400 412
GAP-43	(1)	17E3-9	441aa homolog of mammalian neural growth factor GAP-43	993
Gapdh1	(2R)	43EF	Glyceraldehyde-3-phosphate dehydrogenase	217 218 724 1066
Gapdh2	(1)	13F	Glyceraldehyde-3-phosphate dehydrogenase	822 1066 1067
Gart (=*ade3* = *GART*= *ade8*)	(2L)	27C	Glycinamide ribotide transformylase (GART); phosphoribosylamineglycine ligase (GARS); Phosphoribosylformylglycinamidine cyclo-ligase (AIRS)	302 304 305 307 308 361 461 519 555
Gdh	3-81.7		Glutamate dehydrogenase, NAD-dependent (EC 1.4.1.2-3)	90
gdl gonadal	(3L)	71CD	1.3 & 1kb RNA in ovaries 1.5 & 1.2kb RNA in testes	1047

Gene Symbol	Genetic Map Locus	Cytological Location	Product or Regulatory Function	Reference
Gdt3	2-20.5		Temporal & tissue-specific *cis*-control for *Gpdh*	59 60 677
Gl Glued	3-41.4	70C2	5.5kb transcript; 1319aa (148kD) homolog of filamentous proteins	725 1068
Gld (= *Hex-1=Go)*	3-48	84C8-D1	FADglucose dehydrogenase (EC 1.1.99.10); 2.8kb RNA	100 103 104 105 1095
Glue protein (See *Sgs*):				
Glu	3-distal end	98F-100F	ß-glucuronidase, I & II activity	429
Glus1	2-<0.1	21B	Glutamine synthetase I (EC 6.3.1.2); multimer of 64 & 43kD subunits	91 870
Glus2	(1)	10B8-10	Glutamine synthetase II (EC 6.3.1.2); 42kD subunit	92
Go : see *Gld*				
Got1	2-75.0		Glutamate oxaloacetate transaminase-1	273 101
Got2	2-3.0 (or 4.87)	22B1-4	Glutamate oxaloacetate transaminase-2	271 273 102
Glucose-6-phosphate dehydrogenase: see *Zw*				
Gpdh (= *Gpdh)*	2-20.5	26A2-3	sn-glycerol-3-phosphate dehydrogenase (EC 1.1.1.8), with *cis* systemic regulator	60 62-64 131 268 412 413 565 -567 677 788 883 900 1069 1090
Gpo	2-75.5	52D2-5	a-glycerophosphate oxidase (EC 1.1.99.5), mitochondrial	298 300 653 889
Gpro G-protein	(1)	13F	Guanine nucleotide binding protein homolog; 5.2-1.9kb RNAs; 340aa	1102
Gpro alpha G-protein	(3L)	65C	G-protein alpha subunit; 2.3, 1.7kb RNA; 355aa	1019
Gpt	1-42.6	11F1-12A2	Glutamate pyruvate transaminase (EC 2.6.1.2)	440 757
gsb gooseberry	(3L)	60E9-F1	Segment polarity gene; transcripts in specific embryonic domains	884
gustB gustatory B	1-1.38		Alters chemosensory specificity	838
h hairy	3-25.5	66D10.11	Transcriptional repressor of *ftz*; nuclear protein	324 874 936 937
Had	1-54.4		Hydroxy acid dehydrogenase (EC 1.1.1.45)	75 738
hb hunchback		85A5-B3	Gap gene; control of homeotic gene expression domains; 5 overlapping transcripts 2.6-3.5kb	786 850 967

Gene Symbol	Genetic Map Locus	Cytological Location	Product or Regulatory Function	Reference
hay	3-34.3	67F	Fails to complement	1024
haywire			tubulin mutations;	
			product interacts	
			with tub. subunits	
HDL gene family:				
H44D			Gene-specific mRNAs in	
D44D	(2R)	44D	lst, 2nd, early 3rd	
L44D			larval instars & adult	693
			(Ca -activated secre-	
			tory proteins?)	
Histone gene cluster:				
His1			Histone 1	
His2a			Histone 2A	31 70 114 115
His2b	2-54	39D2-F2	Histone 2B	330 380 458
His3			Histone 3	539 589 519 537
His4			Histone 4	657 682 717
Hex-A	1-29.2	8D4-E	Hexokinase-A	547 757
(= *Hex-A,B*)				
Hex-C	2-73.5		Hexokinase-C	355 490 547
(= *Hex-3* = *Fk*)			(fructokinase)	550
Homeo box containing genes:				
ANT-C;	(3R)			
Antp		84B		226 513
ftz		84B		513
Scr		84B		296
(Sex combs reduced)				
Dfd		84A		513 830
(Deformed)				
BX-C:	(3R)			
Ubx		89E		45 513
iab-2		89E		513 830
(infra-abdominal-2)				
iab-7		89E		513 830
Others;				
en	(2R)	48A		831 609
(engrailed)				
cad	(2L)	38E		530
(caudal)				
---		84A1,2		319
---		38EF	(=*cad*?)	
Heat shock (*HS*) genes and related elements:				13
Heat shock cognates (*hsc*) - not HS induced, homology with Hsp70:				
Hsc1	(3L)	70C	500 b RNA transcript	
Hsc2	(3R)	87D	375 b RNA transcript	336 149 519
Hsc3	(3R)	88E	400 b RNA transcript;	
			70K hsc4 protein	

Gene Symbol	Genetic Map Locus	Cytological Location	Product or Regulatory Function	Reference
Heat shock protein (*hsp*) genes:				
Hsp22			hsp22 protein	137 147 323
Hsp23			hsp23 protein;	337 338 381
			cis hormonal reg.	464 519 520
Hsp26	(3L)	67B	hsp26 protein	604 632 687
Hsp28			hsp28 protein;	698 707 762
(= *hsp27*)			*cis* hormonal reg.	763
---			Gene 2 transcript	1007
			overlaps *hsp22*; 0.78	
			& 0.56kb RNAs; 111aa	
			protein unrelated to hsps	
Hsp68	(3R)	95D	hsp68 protein	13 322 323
				519 663 707
Hsp70	(3-51)	87A7	hsp70 protein, with	11 13 66 89
		87C1	*cis* control	138 148 229 230
				256 285 303 322
				323 335 339 -342
				347 371 441 455
				463 -465 467 468
				482 504 514 519
				525 526 561 604
				667 706 707 736
				740 806
Hsp83	(3L)	63BC1	hsp83 protein	13 286 322 323
(= *hsp82*)				464 519 569 707
				806
Other HS genes and related elements:				
Hs	(2R)	42B,	HS RNAs:	
Hs	(3R)	87Cl, &	RNA	285 463 464
Hs		chromocenter	RNA	465 466 504 519
Hsr1,2, and *3*	(3L)	67B	small hsp proteins	15
(= *genes 1, 4,* and *5* of Sirotkin & Davidson (687))				
Hsr93D	(3R)	93D4-9	HS RNA mostly nuclear &	225 423 424
			and not likely	442 535 770
			translated with *cis* control	
Hsp10F	(1)	10F1		
Hsp54E	(2R)	54E1		
Hsp63F	(3L)	63F1	minor hsp's	464
Hsp73D	(3L)	73D1		
Hsp88B	(3R)	88B1		
Idh1	3-27.1	66B-67C	NADP-isocitrate	208 622 714 50
(= *Idh-NADP*)	(bet.*h*	(66D1-67C)	dehydrogenase-1	
	& *th*)		(EC 1.1.1.42), with	
			cis control (?)	
Idh2	3-25.4		NADP-isocitrate	578 580
(= *Idh*)	(bet. *jv*		dehydrogenase-2	
	& *se*)		(EC 1.1.1.42)	
Indirect flight muscle (IFM) myofibrillar protein genes:				289 533
(see actin, muscle protein, TM genes)				

Gene Symbol	Genetic Map Locus	Cytological Location	Product or Regulatory Function	Reference
Aberrant IFM Morphology Loci			**IFM Protein Affected**	
Bsh Bashed (= *Ifm(2)*- locus?)	2-51.3		Several IFM proteins	
ewg erect wing	1-0.0	1A	80K & 90K proteins	
Flu Flutter	1-37.0	10F7-11D1	80D & 90K proteins	
flp flapwing	1-31	9B1-10A1	Actin III (80K?), para-Myosin (90K?), myosin light chains, 55K (54K?) proteins	289 460 531 540
gmp gumper	1-28	8E-9D	80K & 90K proteins	
gnd grounded	1-58		80K & 90K proteins	
hdp heldup	1-59.5		Similar to *flp*	
Ifm(2)1				
Ifm(2)2	2-near 52		Mhc allele	879
Ifm(2)3				
Ifm(2)11	2-near 55			
Ifm(3)1				
Ifm(3)2			Gene-specific array of IFM proteins, actin I & III, tropomyosin	289 374 375 531 533
Ifm(3)3 (= *Ifm(3)5*, see *Tm*)	3-near 55	88F2-3		
Ifm(3)4				
Ifm(3)5 (= *Ifm(3)3*, see *Tm*)				
Ifm(3)6				
Ifm(3)7 (= *Act88F*)				
int indented thorax (= *up*?)	1-43.5	12A1-7	Similar to flp	
l(1)93p	1-near 21	7D1-6	80K & 90K proteins	289 159
mfd myofibrillar-defective	1-38	6-7	4 proteins absent (abn. phosphorylation?)	
rsd raised (see *rsd*)	3-57.1		similar to *flp*	289 426 460 531 159
sdby standby	1-right of *car*	19F-20F	80K & 90K proteins	289 159
sr stripe	3-62.0		80K & 90K proteins	
up upheld (= *wupB* ?)	1-43.5	12A1-7	similar to *flp*	289 532 158 325 159
vtw vertical wings	1-near 21	7D1-6	80K & 90K proteins	289 540
----	(3-50)	87B	22.5 K protein	56 618
jan A&B janus		99D	Overlapping genes encoding homologous 115 & 140aa proteins	1100

Gene Symbol	Genetic Map Locus	Cytological Location	Product or Regulatory Function	Reference
Jonah (dispersed, clustered multigene family)		25B(3 copies) 44E(1) 65A(3) 66C(2) 99C (2) 99C (3)	28Kd polypeptide (=peritrophic membrane proteins?)	97
Kf1	(3R)	91B-93F	Kynurenine formamidase-I	538
Kf2	(2L)	25A-27E	Kynurenine formamidase-II (EC 3.5.1.9)	538
kin		53A	Kinesin heavy chain; 115kD	1039 1101
kl-2 *kl-3* *kl-5*		YL	High MW sperm proteins (Dynein arm components?)	258 298
kni knirps		77E1-2	Gap gene; 2.5, 2.2kb RNA; 429aa; steroid/thyroid receptor superfamily	992
knrl knirps-related		77E1-2	648aa; steroid/thyroid receptor superfamily	1001
Kpn Killer of prune	3-102.9	100CD	=*awd*;153aa, 16kD protein similar to subunit V of yeast cyt oxidase	854
Kr Kruppel	2-108		Gap gene product; protein contains 4 Zn fingers	906 907 1030
kz kurz		2E2-2F1	4.5, 3.2, & 6kb overlapping transcripts	282
l(1)dlg-1	1-34.8	10B7-8	Recessive oncogene; 5 transcripts, 1.9-6kb	1009 1098
l(1)myo lethal(1) myospheroid	1-23	7D1-5	Membrane protein homologous to b subunit of vertebrate integrins; b subunit of PS1 & PS2	969 974
l(1)npr1 no puff regulator 1	1-0.0	2B5	*Trans*control for *Sgs3, Sgs7, Sgs8*	662
l(1)ts403 lethal(1)temperature sensitive-403	1-42.0		Control of heat shock proteins	193
l(2)37Bf lethal(2) 37Bf right of *hk*	2-53.9,	37D	Diphenoloxidase	601
l(2)amd lethal(2) -methyl dopa	2-53.9+	37C1,2	a-methyl dopa hyper-sensitive; 2kb distal to Ddc	500 699 800 -803 805 856
l(2)gl lethal(2)giant larvae		21A	130Kd cell surface protein homologous to cadherins;alt splicing produces two RNAs encoding 161 & 708 aa polypeptides	516 940 955
l(2)me lethal(2)meander	2-72		Protease activity, iso-acceptor tRNA4Glu reduced	107 179 418 460 769 829

Gene Symbol	Genetic Map Locus	Cytological Location	Product or Regulatory Function	Reference
l(2)thin	2-85.6		Larval muscle structure disrupted	23
l(3)c21R	3-67.8	92A11-B3	Protein modification function responsible for acidification of three proteins abundant in ovaries and salivary gland.	108 600 878
lamA		65A10-11	Laminin homolog, 400kD	987
lamB1		28D	" " , 220kD	
lamB2		67C	" " , 180kD	
Lap-A	3-98.3		Leucine aminopeptidase-A	37 196
Lap-D	3-98.3		Leucine aminopeptidase-D	37 196 653
Lap-P(pupal)	3-98.3		Lap-D of ref 37	918
Lap-G(gut)	3-98.3		Lap-D of Kalker& Williamson	918
Larval cuticle protein (*LCP*) genes :				
Lcp1			L3CP 1 (3rd instar)	
Lcp2	2-62	44D	L3CP 2 (3rd instar)	
Lcp3			L3CP 3 (3rd instar)	119 188 189
Lcp4			L3CP 4 (3rd instar)	214 215 463 519
Lcp5	3-11		L3CP 5 (3rd instar)	690-3
Lcp6			L3CP 6 (3rd instar)	
Lcp8			L3CP 8 (3rd instar)	
lpo	3-57.2, .009cM left of *Aldox1*	89A	Pyridoxal oxidase (low)	132 168 776
Larval serum proteins (*LSP*) genes:				
Lsp1	1-39.5	11A7-B9	LSP-1	81 639 688 507 82 427
Lsp1	2-1.9	21D3-4	LSP-1	81 444 639 640 688
Lsp1	3-(-1.41)	61A1-6	LSP-1	82 427 507 554 816
Lsp2	3-37(?)	68E3-4	LSP-2	2 3 444 933
lxd low xanthine dehydrogenase	3-34.5 (0.1 distal to *Sod*)	68A4-9	Modifier of Mo hydroxylases (AO, PO & XDX) & sulfite oxidase	74 105 134 182 246 460 534 671 776 777
lz lozenge	1-27.7	8D4-E1	Monophenoloxidase diphenoloxidase	34 519 529 597 598 775
ma-l maroon-like (= *mal*)	1-64.8	19D1-3	Controls incorporation of cyanolyzable sulfur into Mo hydroxylases (AO, PO, XDH)	74 121 144 180 202 -207 213 246 247 588 765 -767 776 777
mam mastermind		50CD	Neurogenic locus	1093 1102
map midgut activity pattern	2-80		*Trans* control for *Amy-p* & *Amy-d*	1 172 173 176
Mar	2-64		Major gene for malathion resistance & increased cytochrome P450	935
---	3-58		Minor gene for same	935

Gene Symbol	Genetic Map Locus	Cytological Location	Product or Regulatory Function	Reference
Maternal differential transcripts:				
		3BC	2.5, 2kb RNA	1081
		4C	7, 3, 2.2kb RNA	
		7D	4.1, 2.9, 1.7, 1.33	
		31A	2.7, 2.2kb RNA	
		33B	2.5, 2kb RNA	
		44EF	3.5kb RNA	
		52D9-15	2.9, 2.6kb RNA	
		67C	1.8kb RNA	
		69D4	2.7, 2.2kb RNA	
		85E	2.5kb RNA	
		93B	4.9kb RNA	
		98F	3.3kb RNA	
		99	3.5, 3.2kb RNA	
Maternal specific transcripts:				
		14D1,2	4.2, 3.3, 2.3, 2kb RNA	1081
		45B	2.2, 1.7kb RNA	
		52A1,2	4.3, 4kb RNA	
		52A3,5	3.2kb RNA	
		98F	3.8kb RNA	
		23BC	2kb cDNA of maternal RNA	
		55F	1.8 & 1.5 kb cDNA of maternal RNA	1061
Mdh1 (= *aMdh*)	2-37	31B-E	NAD-malate dehydrogenase (cytoplasmic) (EC 1.1.1.37)	5 269 564 565 757 758
Mdh2 (= *mMdh*)	3-62.6	90C-91A3	NAD-malate dehydrogenase (mitochondrial) (EC 1.1.1.37)	757 758
Men (= *Mdh-NADP*)	3-51.7	87D1-2	NADP-malate dehydrogenase (malic enzyme: EC 1.1.1.40), with *cis* control)	211 234 578 757 760 50 790
Mesoderm specific genes:				
		17A		981
		60A		981
m-Est (= *m-est*)	3-56.7		Esterase-6 and leucine aminopeptidase modifier	126 127
Met methroprene tolerant	1-35.4	10C2-10D4	JHIII resistant, complements l(1)L5	792
Mex X-linked modifier of *ME*	(1)	8D10 to 9A2	*Trans* regulation of $NADP^+$ malic enzyme	48
Mhc			Myosin heavy chain; isoforms-alt splicing	908 1092
mod	3 between e^s and *ca*		Post-translational modification of Ifm proteins 74 and 100	492

Gene Symbol	Genetic Map Locus	Cytological Location	Product or Regulatory Function	Reference
Male-specific lethal loci:				22 479 -480
mle maleless	2-55.8	41A-43A	Control of X-linked enzymes (G6PD, 6PGD, FUM, -HAD)	40 219 726
msl1 male-specific lethal-1	2-53.3	36F7-37B8 (= msl-1)	Control of X-linked enzymes (G6PD, 6PGD, FUM)	40 41
msl2 male-specific lethal-2 (= msl-2)	2-9.0	23E1-F6	Control of X-linked enzymes (G6PD, 6PGD, FUM)	40 41 508
msl3 male-specific lethal-3 (= msl-3)	3-26		(Control of X-linked enzymes?)	479 481
Male-specific transcripts(Also see accessory gland proteins, *Acp*);				
mst(2)ag-1	(2R)	57D	0.45kb transcript	665
mst(2)ag-35	(2R)	51F	0.66kb transcript	
mst(3)ag-2	(3L)	75C	0.45kb transcript	
mst(3)ag-3	(3R)	95F	0.45kb transcript	
mst(3)gl-9	(3R)	87F	0.74kb transcript; produced in mid-testis; 56aa protein	960
mst(2)355a&b	(2L)	26A	Two genes 20bp apart; Both proteins transferred to female genital tract and altered; One is homologous to oviposition hormone of Aplysia	985
mst(2)323	(2L)	25F	0.35kb RNA	892
mst(2)325	(2R)	47A	1.0kb RNA	
mst(3)349	(3L)	66D	2.6 kb RNA	
mst(3)316	(3R)	95E	0.35kb RNA	
mst(3)345	(3R)	95EF	1.6, 1.1kb RNA	
mst(3)336	(3R)	98C-E	1.6, 1.3kb RNA	
Mtn	(3-49)	85E10-15	Metallothionein	431 499 587 1002
Muscle protein genes: (see actin, IFM, TM genes)				
Mhc36B	(2-52)	36A8-B12	Myosin heavy chain	57 519 534 646
Mhc87E	(3-52.3)	87E	Myosin heavy chain (?)	56 618
Mhc88F	(3-57)	88F	Myosin heavy chain (?)	56 58 618
Mlc98B	(3R)	98B	Myosin alkali light chain	197
Mlc99E =*Mlc-2*	(3-101)	99E1-2	Myosin light chain	519 591 1077
mp20 muscle protein 20		9F9-13	Muscle protein (20kD) in most synchronous muscle	840
-----	(2L)	32CD		
-----	(2R)	46D-F & chromocenter	Gene-specific, unidentified muscle proteins translated from abundant mRNAs	716
-----	(2R)	47F-48D		
-----	(3L)	62CD		
-----	(3L)	63A-C		
-----	(3L)	69F		
-----	(3R)	88F		
Myosin see **muscle protein genes**				

Gene Symbol	Genetic Map Locus	Cytological Location	Product or Regulatory Function	Reference
N Notch (complex)	1-3.0	3C7	Lowered activities of; choline DH(EC 1.1.99.1), dihydro-orotic acid DH (EC 1.3.99.9), -glycerophosphate DH (EC 1.1.99.5), NADH DH & NADH oxidase (EC 1.6.99.3), succinate DH (EC 1.3.99.1), xanthine DH (EC 1.2.99.1); *opa* repetitive transcribed region; EDGF-like sequences tandemly repeated 36 times	12 277 388 389 392 460 519 732 733 734 735 781 782 783 952 1075
--		3C6-7	Interband contains reg. seq. for *N*	1034
Nac Na^+ channel		60DE	Na^+ channel gene; 1622aa protein composed of 4 homologous membrane-spanning domains	1035
nac	3R)	84F	Mutant eliminates neuronal carbohydrate epitope	950
nc4	3-54.2	88B-D	Fails to complement *B2t* mutation	913
neup neuropeptide		46C	1.7, 0.7kb RNA encoding 39kD neuropeptide precursor; homologous to molluscan FMRF-amide	1043
NHC protein genes:				
---	(2L)	29A	19Kd nuclear protein	351
D1	(3R)	85D1-2	Chromosomal protein D1; binds to AT-rich sat DNA; 355aa	839
nina A neither inactivation nor after potential-A		21D1-E2	0.95kb RNA; 237aa homolog cyclophilin	1044 1053
nina C neither inactivation nor after potential-C	2-22	28A1-3	Two overlapping transcripts, potential polypeptide with protein kinase-like domain and MHC-like domain	986
ninaE (=*Rh1*) neither inactivation nor after potential-E	3-66	92B6-7	41.4Kd opsin (rhodopsin) *cis* regulatory elements	664 575 827 982 1110
norpA no receptor potential		4B6-C1	Putative 1095aa product homologous to bovine phospholipase C	857
numb		30B	Involved in sensory neuron determination; Zygotic product 556aa; Maternal product 514aa (limited to germ line)	1080

Gene Symbol	Genetic Map Locus	Cytological Location	Product or Regulatory Function	Reference
oc ocelliless (associated with In(1)7F1,2;8A1,2)	1-23.1	7F1 to 8A2	Stable position effect on shell protein-36 & shell protein-38	438 460 704 705 709
Ocd^{ts-1} Out-cold	(1)		Affects mt succinate cyt. c reductase activity	1056 1057 1058
Odh	3-49.2	86D1-4	Octanol dehydrogenase (EC 1.1.1.73)	124 139 145 580
Oncogene homologues: see *c-myb, Dash, Dets, Dint, Dmras64B, Draf, Dsrc, dl, l(1)dlg-1, twist,* and *src4*				
otu	1-23.2	7F11	Ovarian tumor gene 3.2 kb RNA	953 990
Proteins-miscellaneous genes:				
P1	(3L)	70CD	110K larval fat body protein P1; *cis* hormonal control	444 503 554 687
P6	(3L)	70CD	29K larval fat body protein P6	444
-----	(3L)	80C	26K embryonic, cytoplasmic protein	67
Pupal cuticle protein genes:				
Pcp	(2L)	27C	0.9kb transcript; included within *Gart* intron	306
pcx pecenex	(1)	2E2-2F1	12kb transcript(?)	282
per period	(1)	3B1-2	571 a.a. proteoglycan associated with biological rhythms; 4.5kb RNA; 3 alternative products sharing 862aa of N-term; most common RNA yeilds 1218aa protein	626 627 689 846 880 939 972 1104
Pfk	(2R)	45F-47E	Phosphofructokinase (EC 2.7.1.11)	551
Pgd (= *6Pgd*)	1-0.63	2D3-4	6-phosphogluconate dehydrogenase (EC 1.1.1.44)	61 129 241 280 -282 284 819 916
Pgi	2-58.6		Phosphoglucose isomerase	757 760
Pgk	2-5.9	22D-23E3	3-phosphoglycerate kinase (EC 2.7.2.3)	116 759
Pgm	3-43.4	72D1-5	Phosphoglucomutase (EC 2.7.5.1)	316 616 744 759
Phox (= *Bc*?)	2-80.6		Phenol oxidase	29 270 635
Pka30C Protein kinase	(2L)	30C1	cAMP protein kinase, 352aa catalytic subunit	899 947
Pka77F Protein kinase	(3L)	77F	cAMP protein kinase, regulatory subunit	947
Pkc98F Protein kinase	(3R)	98F	Protein kinase C homolog expressed throughout development	1040

Gene Symbol	Genetic Map Locus	Cytological Location	Product or Regulatory Function	Reference
Pkc53E Protein kinase	(2R)	53E4-7	Protein kinase C homolog specific to photoreceptors	1031 1040
Pkg-1 Protein kinase	(2L)	21D	cGMP dependent protein kinase; 2.8kb RNA; 653aa	899 948
Pkg-2 Protein kinase	(2L)	24A	cGMP dependent protein kinase; 4.6, 4.4, 3.6kb RNA;1088aa	948
pn prune	1-0.8	2D5-6	GTP-cyclohydrolase control	192 460 486
pr purple	2-54.5	37B2-40B2	Ramiopterin synthase sepiapterin synthase	177 178 417 460 739 818
PS2 [see *l(1)myo*] Position specific antigen			Cell surface antigen related to fibronectin; alpha & beta subunits	860
Pu Punch	2-97	57C5-6	GTP cyclohydrolase (EC 3.5.4.16); 1.7 kb RNA in eye	460 486 999 1000
pur1	1-left of v	9E1-3	Purine 1	360 556
Pyk	(1)	12A-C	Pyruvate kinase (EC 2.7.1.40)	651
r rudimentary	1-55.3	15A1	Carbamyl phosphate synthetase (EC 2.7.2.9) aspartate transcarbamylane (EC 2.1.3.2) dihydroorotase (EC 3.5.2.3); 7.2kb RNA; 2236aa protein with all four activities	83 98 194 195 200 352 353 354 437 460 519 563 585 621 623 625 674 675 747 902 943
R3-55.4 (= *R3-55.4, r3-55.4)*	3-55.4		sn-GPDH *trans* control (a-GPDH modifier)	398 399
raf (See *Draf*)				
ras raspberry	1-32.8	9E1-3	GTP-cyclohydrolase control	192 460 486 556 826
red red Malpighian tubules	3-53.6	88A-C	GTP-cyclohydrolase control	460 486
Ribosomal protein genes:				
r7/8	(1)	5D		85
Rbp49	(3R)	99D	Ribosomal protein 49	408 519 796
S14A&B	(1)	7C5-9	0.65kb RNA; tandemly duplicated copies of 151aa homolog of human ribosomal protein s14	863
S17	(3L)	67B1-5	Ribosomal small subunit protein 17; 131aa	975
S18	(1)	15B		85
L1	(3R)	98AB	Homolog of rpl1 of X. laevis; 1400bp cDNA; 407aa	1022 1023
L12	(3L)	62E		85
21		80	26kD ribosomal protein	951

Gene Symbol	Genetic Map Locus	Cytological Location	Product or Regulatory Function	Reference
r-l rudimentary-like (= *ral*)	3-70	93B4-13	Orotate phosphoribosyl-transferase (EC 2.4.2.10) orotidylate decarboxy-lase (EC 4.1.1.23)	136 430 619 620
Rh1 (See *ninaE*)				
Rh2 Rhodopsin 2	(3R)	91D1-2	381 aa opsin found only in photoreceptor cell 8; ocellus-specific; 1.7kb RNA	146 981 1013
Rh3 Rhodopsin 3	(3R)	92D1	Expressed in subset of R7; 1.5kb RNA; 383aa	1106
Rh4 Rhodopsin 4	(3R)	73D3-5	Expressed in subset of R7; 1.5kb RNA; 378aa	988
rn rotund	(3R)	84D3-4	Deletes specific distal parts of adult extremities; 1.7kb RNA accumulates steadily 5.3kb RNA accumulates abruptly in early pupae	833
RpII215 (= *AmaC4*= *l(1)L5*= *PolII*= *Ubl*)	1-35.7	10C1-2	RNA polymerase II, 215K subunit (EC 2.7.7.6); 7kb RNA	143 263 264 265 334 519 544 545 673 761 945
RpII140		88AB	RNA polymerase II, 140K subunit (EC 2.7.7.6)	199
Rp-related		9C-11A 70C-74A 76A-78A 91B-94F 96A-98E	Additional dosage sensi-tive regions for RNA polymerase activity	791
RNAs - unidentified gene products, developmentally expressed:				
Embryonic poly(A)+ RNAs:			**Stage Expressed**	
----	(2L)	25D	blastoderm-differential	
(02 gene)	(2L)	31BC	maternal-differential	443 730
Deb-A	(2R)	48EF	developmental-embryonic	
Deb-B	(2R)	48EF	developmental-embryonic	756
----	(3L)	71A	gastrula-differential	
----	(3R)	94F-95A	blastoderm-differential	443
----	(3R)	99D	blastoderm-differential	443
----	(3R)	99E1-3	blastoderm-differential	443
Larval instar poly(A)+ RNAs:				
----	(1)	4F-5A	late instar IV	795
----	(3L)	67B	late instar III + pupa	687
----	(3L)	71DE	late instar I	
----	(3L)	71DE	late instar II + III	
Intermolt poly(A)+ RNAs:				795
----	(1)	3C7-D1	intermolt I	
----	(3L)	68C.	intermolt II	
----	(3L)	68C	intermolt III	
----	(3L)	68C	intermolt IV	
----	(3R)	90BC	intermolt V	

Gene Symbol	Genetic Map Locus	Cytological Location	Product or Regulatory Function	Reference
Adult poly(A)+ RNAs:	(1)	15AB	Head-specific RNA for each locus (some with multiple genes per site)	451 452 519
	(2L)	28A, 28C, 32AB, 34F		
	(2R)	43AB,44CD, 46E,47E,48F 51B,57C		
	(3L)	66D,72BC, 73DE	head-specific RNA, continued	
	(3R)	82F,92CD,99C,100B		451 452 519
rRNA genes: (see *bb*)				10 31 43 309
---- 5S RNA gene	2-95	56F	5S-rRNA	365 476 607 613 615 650 728 729 745 746 748 793 794 808
ro rough	3-91.1	97D5-7	1.3kb RNA, 350aa, homeobox gene required for R2 & R5 inductive interactions	1079
rs rose	3-35.0		GTP-cyclohydrolase control	460 486
rsd raised	3-57.1		reduced levels of act88F	493
run runt		19E2	Pair rule gene; 2.6kb RNA	909
rut rutabaga	1-45±	12F5-7	Calcium/calmodulin activation of adenylate cyclase (EC 4.6.1.1) is missing	469 470
ry rosy	3-52.0	87D12	Xanthine dehydrogenase, with *cis* control (EC 1.2.1.37)	44 121 122 123 125 140 200 231 236 238 292 311 506 571 648 700 711 817
Shell protein genes (chorion protein genes):				367
s15 (= *c15* = *A1*)			Shell protein 15 (15K chorion protein) with *cis* control	160 275 276 461 670 702 797 811 812
s16 (= *c16* = *A2*)			Shell protein 16 (16K chorion protein)	160 275 276 461 670 702 797 811 812
s18 (= *c18* = *B1*)	3-26.5	66D11-15	Shell protein 18 (18K chorion protein) with *cis* control of amplification	160 275 276 461 670 702 797 811 812
s19 (= *c19* = *B2*)			Shell protein 19 (19K chorion protein) with *cis* control	160 275 276 461 670 702 797 811 812
s35 (=*c36*= *1*)	1-23.1	7E11-7F2	Shell protein 36 (36K chorion protein)	702 703 704

Gene Symbol	Genetic Map Locus	Cytological Location	Product or Regulatory Function	Reference
s38 (= *c38* = *C2)*			Shell protein 38 (38K chorion protein)	708 709 811 461
s70	1-near *y*	2B3-6	Shell protein 70 (70K chorion protein)	603 814
----	1-between *y* and *cv*		100K shell protein	813
(K451)	1-near	12A6-7	Control of amplification	404 584
(K575ts)		4F5-4A1,2	Control of amplification	404 405
(K1214)	1-18.7	5D5-6C12	Control of amplification	404 584
(K79)	1-bet. *ct-v*	8E-9B1		
(K254ts)	1-17	5D5-6E1		
(K499)	1-bet. *g-f*		Defective shell (chorion) protein mutants	404
(K1563TS)	1-44.7	12D3-E1 (?)		
(384)	1-20.5	7B8-C3		
(473)	1-20.5	7D10-8A5		
sal spalt		32F-33A	Homeotic gene; 142aa polypeptide	901
sc scute	1-0	1B1-5	1.2 & 1.6kb transcripts; T4 & T5 encode myc related polypeptides; required for neural development	94 868 1087
sdh	2-89		Succinate dehydrogenase (EC 1.3.99.1), mitochondrial	433
sev sevenless	1-33.2	10A1-2	8.2kb RNA expressed in eye disc; transmembrane protein in R7	844 845 862 917 1078
Salivary gland structural (SGS) protein genes for glue polypeptides:				
Sgs1	2-13.9	25A3-D2	SGS-1	752
Sgs3			SGS-3 with *cis* control	
Sgs4	1-3.6	3C11-12	SGS-4, with *cis* control, with *trans* control	35 232 406 407 510 511 512 515 519 552 553 411 931 932 957
Sgs5 (= *P4= group V protein)*	3-60.8	90B3-8	15.5Kd glue protein	278 279 934
Sgs6	3-42.0	71C1-F5	SGS-6	680 751 753
Sgs7	(3-36)	68C3-5	SGS-7	2 77 151 406 521
Sgs8			SGS-8	228 232 629
Sh Shaker	1-38.2	16A	Control of K currents in muscle and nerve membranes; encodes a putative K^+ channel; 643aa & 616aa alternative product;A second channel unaffected by *Sh*	807 938 949 1004 1014 1049 1070 1076 1055
sim single minded		87D,E	3.0-3.5kb RNAs; 655 aa protein similar to *per* product	885 1074

Gene Symbol	Genetic Map Locus	Cytological Location	Product or Regulatory Function	Reference
slit		52D	9.5kb RNA; protein EGF-like	1032
snRNP genes:	see U genes			
Sod (= *To)*	3-34.6 (or 32.5?)	68A2-C1	Superoxide dismutase (tetrazolium oxidase)	93 102 210 260 355 568 954 961 1051 1052
Sodh (= *SoDH)*	3-64.5	91B-93F	NAD-sorbitol dehyrogenase, cytoplasmic	71 72
sp speck	2-107.0	60B13-C5	Phenol oxidase	460
Spe spectrin		62B	Spectrin homolog; >9kb RNA	867 896
src4	(2L)	28C	Oncogene homologue, related to *Dsrc* & *Dash*	764 1088
sry serendipity (= *EH8)*	(3R)	99D	5 overlapping blastoderm specific transcripts; delta and beta encode related Cys2/His2 finger proteins. Transcribed in nurse cells, transferred to oocytes	751 755 1008
stg string	(3R)	98F-99A	Cell cycle control; 2.3kb RNA; 479aa product homologous to C-term of *cdc25* gene of S. pombe	897
su(b) suppressor of black	1-0.0	1B4-C4	-alanine level	460 679
Su(b)	1-55.5	(15A1)	Control of *r* gene products	596
su(f) suppressor of forked	1-65.9	20D or 20EF	Affects processing or translatability of SGS-1, SGS-3, SGS-4	294 295 460 666
su(Hw)		88BC	3.3kb RNA; 109kD DNA-binding protein; binds to gypsy seq.	1005 1059
su(r) suppressor of rudimentary	1-27.7		Dihydrouracil dehydrogenase	20 194 718 719
su(s) suppressor of sable	1-0.0	1B11-13	Suppresses *pr, s, sp* & *v* (tRNATyr?)	65 300 348 349 418 487 519 749 784
su(w^a) suppressor of white apricot	1-0.1	1D-F	3.5-5.2 kb family of ubiquitous transcripts	1105
Sucr	(2-37)	31CD-EF	Sucrase	582
sww swallow	1-12-16	5E	2.1cDNA of maternal specific RNA; Necessary for proper *bcd* transcript localization	1061 1062
Sxl Sex lethal	1-19.2	6F	354aa female product similar to RNP; male product truncated	849 1036 1112

Gene Symbol	Genetic Map Locus	Cytological Location	Product or Regulatory Function	Reference
tko	1-1.0	3A2	0.68kb RNA,140aa protein homologous to rpS12	1033
Tm-c			Tropomyosin, cytoplasmic	
Tm1	(3-57)	88F2-5	Tropomyosin I, muscle	27 28 32 33
Tm2			Tropomyosin II, muscle	374 519 716 919 920
Toll	(3R)	97D1,2	5.3kb maternal RNA; sequence suggests integral membrane protein	922
Top2	(2L)	37D2-6	DNA topoisomerase II	562
Tp1 (= *T1*)	(2-54)	39CD	Small acidic temporal protein (TI)	216
Tpi	3-101.3	99B-E	Triosephosphate isomerase	757 759 30 32 519 716
tra transformer	3-45	73A5-10	1kb female specific RNA; 1.2kb RNA in both sexes	861 980
tra-2 transformer-2	(2R)	51B	*trans* sex-specific regulation of YP1,2,&3; 179aa with homologies to RNA & ssDNA-binding proteins	42 837 912
***trans* control of spatial expression of homeotic genes:** (Also see *hb*)				
Asx Additional sex combs	2-72	51AB		66
Pcl Polycomb-like	2-83	55AF		366
Psc Posterior sex combs	2-67	49EF		366
Scm Sex comb on midleg	3-49	85EF		366
Pc Polycomb	3-47.1			453 779
sxs super sex combs	2-55.3	41		333
abd-A and *Abd-B*				722 832
Transfer RNA genes: see 547, 419 and E. Kubli in this volume				
Treh	2-92.9	55B-E	Trehalase	86 583 866
(Troponin gene)	(2R)	49F	Troponin C	527
tryp	(2R)	47D-F	Trypsin-like enzymes, cluster of 4 genes	155
trx trithorax		88B	Homeotic gene	989
Tubulin genes: (see *B2t*)				
Tub67C	(3L)	67C4-6	-tubulin subunit	368 559 658
Tub84B	(3-47.8)	84B3-6	-tubulin subunits (2 genes)	315 368 559
Tub84D	(3-48)	84D4-8	-tubulin subunit	658
Tub85E	(3-49)	85E6-10	-tubulin subunit	
Tub56C	(2R)	56C	-tubulin subunit	559 590
Tub56D	(2R)	56D4-12	-tubulin subunit	557
Tob60C	(3L)	60C	-tubulin subunit	559
Tub85D	(3R)	85D	-tubulin subunit	559
Tub97F	(3R)	97F	-tubulin subunit	557 559 590

Gene Symbol	Genetic Map Locus	Cytological Location	Product or Regulatory Function	Reference
twist	2-100	59C3-D2	myc homolog; 490aa	1073
tyr1 tyrosinase-1	2-52.4 (= *tyr-1*)		Monophenoloxidase (tyrosinase) diphenoloxidase (dopa oxidase)	454 529 676 775
U-genes:				
U1	(1)	11B?		
	(2L)	21E	U1 = snRNA2	7 382 508 549 654 655
	(2R)	61A		
	(3R)	82E, 95C		
U2	(2L)	34AB, 38AB	U2	6
	(2L)	34BC	Two copies	8
	(3R)	84C	Two copies	8
U4	(2L)	40AB	U4 = snRNA1	
	(1)	14B		
U5	(2L)	23D, 34AB 35EF, 39B	U5 = snRNA3	654 655
	(2R)	63A		
U6	(3R)	96A	U6 = snRNA4	655 888
Ubi Ubiquitin	(3L)	63F	Polyubiquitin; 4.4kb RNA; 18 repeats of 228bp ubiquitin (76aa) encoding monomer	965
Ubl : see *RpII*				
Ubx (see ANT-C)	3-58.8	89E1-2	44Kd protein binds to its own & *Antp* promotors	785 848
urox	(2L)	28C	Urate oxidase; dimer, 41kD subunit	903 958
v vermilion	1-33.0	10A1	Tryptophan oxygenase (EC 1.13.1.12)	11 348 377 800 727 772 826
Vitelline membrane proteins:				
---		26A		310 523
---		32EF		523
---		34		523
---		39DE	14K polypeptide?	198
---		42A	23-4K & 17.5 K polypeptide	
w white	1-1.5	C1-2	2.7kb RNA transcript, proximal *cis* control	68 69 133 254 255 262 364 373 447 448 572 573 574 588 592 608 649 823
wg wingless	(2L)	28A	2.9kb RNA, 1419 ORF, 468aa polypeptide homologous to int-1 oncogene	841 869 1029
Xh	1	20C1-2	rDNA redundancy	593 660

Gene Symbol	Genetic Map Locus	Cytological Location	Product or Regulatory Function	Reference
y yellow	1-0	1B1-5	1.9kb transcript	94
yema "mother" in Berber		98F3-10	Four maternal transcripts; Disappear at gastrulation	834
Yolk proteins (*YP*) genes (see *fs(1)29*, *fs(1)1163*)				
Yp1 (= *fs(1)1163*)	1-29	8F-9A	YP-1	25 26 78 80 326 327 328
Yp2	1-29	8F-9A	YP-2	524 610 611 631
Yp3	1-44	12BC	YP-3	778
			YP1, 2, & 3 related to triacylglycerol lipase family	1072
Yp3R	1-near Yp3		YP-3 control, *cis*-acting	612
z zeste	1-1.0	3A3-4	Represses expression of *w*, *bx*, *dpp*; protein binds to sites upstream from *Ubx* & *w*, also *bxd* & *Sgs-4*	237 346 364 460 851 853 977 1011
Zw Zwischenferment	1-62.9	18D	Glucose-6-phosphate dehydrogenase (EC 1.1.1.49) with *cis* control	129 223 233 283 496 519 524 711 712 714 752 819 820

REFERENCES

1 Abraham I & WW Doane (1978) Proc Natl Acad Sci USA 75:4446-4450
2 Akam ME et al. (1978a) Cell 13:215-225
3 Akam ME et al. (1978b) Biochem Genet 16:101-119
4 Akam M et al. (1983) Trends in Biochm Sci 8:173-177
5 Alahiotis S (1979) Comp Biochem Physiol 62B:375-380
6 Alonso A et al. (1983) J Mol Biol 169:691-705
7 Alonso A et al. (1984) J Mol Biol 180:825-836
8 Alonso A et al. (1984) Nuc Acids Res 12:9543
9 Anderson SM & JF McDonald (1983) Proc Natl Acad Sci USA 80:4798-4802
10 Artavanis-Tsakonas S et al. (1977) Cell 12:1057-1067
11 Artavanis-Tsakonas S et al. (1979) Cell 17:9-18
12 Artavanis-Tsakonas S et al. (1983) Proc Natl Acad Sci USA 80:1977-1981
13 Ashburner M & JJ Bonner (1979) Cell 17:241-254 (review)
14 Ashburner M et al. (1983) Genetics 104:405-431
15 Aymes A and A Tissieres (1985) EMBO J 5:2949-2954
16 Baglioni C (1960) Heredity 15:87-96
17 Bahn E (1967) Hereditas 58:1-12
18 Bahn E (1971a) Hereditas 67:75-78
19 Bahn E (1971b) Hereditas:79-82
20 Bahn E (1972) Dros Inform Serv 49:98
21 Baker BS (1973) Devel Biol 33:429-440
22 Baker BS & JM Belote (1983) Ann Rev Genet 17:345-393 (review)
23 Ball E et al. (1985) Dev Gen 6:77-92
24 Bargiello TA and MW Young (1984) Proc Natl Acad Sci USA 81:2142-2146
25 Barnet T et al. (1980) Cell 21:729-738
26 Barnett T (1983) (see ref 519)
27 Basi GS and RV Storti (1986) J Biol Chem 261:817-827
28 Basi GS et al. (1984) Mol Cell Biol 4:2828-2836
29 Batterham P & SW McKechnie (1980) Genetica 54:121-126
30 Baum HJ et al. (1983) Nuc Acids Res 11:5569-5587
31 Bauman JGJ et al. (1981) Chromosoma 84:1-18
32 Bautch VL & RV Storti (1983) Proc Natl Acad Sci USA 80:7123-7127
33 Bautch VL et al. (1982) J Mol Biol 162:231-250
34 Beadle GW & E Tatum (1941) Amer Nat 75:107-116

37 Beckman L & FM Johnson (1964) Hereditas 51:221-230
38 Bell JB et al. (1972) Biochem Genet 6:205-216
39 Belote JM (1983) Genetics 105:881-896
40 Belote JM & JC Lucchesi (1980) Nature 285:573-575
41 Belote JM & JC Lucchesi (1980) Genetics 96:165-186
42 Belote JM et al. (1985) Cell 40:339-348
43 Bencze JL et al. (1979) Exp Cell Res 120:365-372
44 Bender W et al. (1983) J Mol Biol 169:17-33
45 Bender W et al. (1983) Science 221:23-29
46 Bentley MM (1986) Biochem Gen 24:291-308
47 Bentley MM & JH Williamson (1979) Z f Naturforsch 34:304-305
48 Bentley MM & JH Williamson (1979) Can J Genet Cytol 21:457-471
49 Bentley MM and JH Williamson (1985) Can J Genet Cytol 27:322-333
50 Bentley MM et al. (1983) Biochem Genet 21:725-733
51 Benyajati C et al. (1980) Nuc Acids Res 8:5649-5667
52 Benyajati C et al. (1981) Proc Natl Acad Sci USA 78:2717-2721
53 Benyajati C et al. (1982) Nuc Acids Res 10:7261-7272
54 Benyajati C et al. (1983) Cell 33:125-133
55 Benyajati C et al. (1983) Mut Res 111:1-7
56 Bernstein SI et al. (1981) Genetics 97:s10
57 Bernstein S et al. (1983) Nature 302:393-397
58 Bernstein SP et al. (1986) Mol Cell Biol 6:2511-2519
59 Bewley GC (1981) Devel Genet 2:113-129
60 Bewley GC (1983) In: Isozymes MC Rattazzi JG Scandalios GS Whitt eds Vol 9 pp 33-62 Alan R Liss NY (review)
61 Bewley GC & JC Lucchesi (1975) Genetics 79:451-457
62 Bewley GC & S Miller (1979) In:Isozymes MC Rattazzi JG Scandalios & GS Whitt eds Vol 3 pp 23-52 Alan R Liss NY
63 Bewley GC et al. (1980) Mol Gen Genet 178:301-308
64 Bienz M & II Deak (1978) Insect Biochem 8:449-455
65 Bienz M & E Kubli (1981) Nature 284:188-191
66 Bienz M & HRB Pelham (1982) EMBO J 1:1583-1588
67 Biessmann H et al. (1981) Chromosoma 82:493-503
68 Bingham PM & BH Judd (1981) Cell 25:705-711
69 Bingham PM et al. (1982) Cell 29:995-1004
70 Birnsteil ML et al. (1973) In:Molecular Cytogenetics BS Hamkala & J Papaconstantinou eds pp 75-93 Plenum Press NY
71 Bischoff WL (1976) Biochem Genet 14:1019-1039
72 Bischoff WL (1978) Biochem Genet 16:485-507
73 Blass DH & DM Hunt (1980) Mol Gen Genet 178:437-442
74 Bogaart AM & LF Bernini (1981) Biochem Genet 19:929-946
75 Borack LI & W Sofer (1971) J Biol Chem 246:5345-5350
76 Bossy B et al. (1984) EMBO J 3:2537-2541
77 Bourouis M and G Richards (1985) Cell 40:349-357
78 Bownes M & BD Hames (1978) J Embryol Exp Morph 47:111-120
79 Bownes M & BA Hodson (1980) Mol Gen Genet 180:411-418
80 Bownes M et al. (1983) J Embryol Exp Morph 75:249-268
81 Brock HW & DB Roberts (1981) Chromosoma 83:159-168
82 Brock HW & DB Roberts (1983) Genetics 103:75-92
83 Brothers VM et al. (1978) Biochem Genet 16:321-331
84 Browder LW & JH Williamson (1976) Devel Biol 53:241-249
85 Burns DK et al. (1984) Mol Cell Biol 4:2643-2652
86 Burtis K (1986) Diss Abst 47:507-B
87 Burton RS and A LaSpada (1986) Biochem Gen 24:715-719
88 Byers D et al. (1981) Nature 289:79-81
89 Caggese C et al. (1979) Proc Natl Acad Sci USA 76:2385-2389
90 Caggese C et al. (1982) Biochem Genet 20:449-461
91 Cagesse C and Dell'Aquila (1983) Atti Ass Gen Ital 29:81-82
92 Cagesse C et al 1986 Mol Gen Genet 204:208-213
93 Campbell SD et al. (1986) Genetics 112:205-215
94 Campuzano S et al. (1985) Cell 40:327-338
95 Capdevila MP & A Garcia-Bellido (1981) Wilhelm Roux's Arch 190:339-350
96 Carlson J (1982) PhD thesis Dept Biochem Stanford Univ CA
97 Carlson JR and DS Hogness (1985) Dev Bio 108:341-354 and 355-368
98 Carlson PS (1971) Genet Res Camb 17:53-81
99 Carpenter NJ (1973) Genetics 75:113-122
100 Cavener D (1980) Biochem Genet 18:929-938

105 Cavener D et al. (1986) EMBO J 5:2939-2948
106 Chao A and GM Guild (1986) EMBO J 5:13-150
107 Chen PS (1978) In: Biochemistry of Insects M Rockstein ed pp 145-203 Acad Press New York (review)
108 Cheney CM et al. (1984) Proc Natl Acad Sci USA 81:6422-6426
109 Cherbas L & P Cherbas (1981) Adv Cell Culture 1:91-124
110 Cherbas L et al. (1986) J Mol Biol 189:617-631
111 Cherbas P et al. (1977) Science 197:275-277
112 Cherbas PL et al. (1981) Amer Zool 21:743-750
113 Cherbas P et al. (1983) In: Gene Structure and Regulation in Development S Subtelny & FC Kafatos eds 41st Symp SDB pp 95-112 Alan R Liss NY (review)
114 Chernyshev AI et al. (1980) Mol Gen Genet 178:663-668
115 Chernyshev AI et al. (1981) Mol Biologia (Moscow) 15:387-393
116 Chew GK & DW Cooper (1973) Biochem Genet 8:267-270
117 Chia W et al. (1985) J Mol Biol 186:689-706
118 Chihara CJ et al. (1982) Devel Biol 89:379-388
119 Chihara CJ and DA Kimbrell (1986) Genetics 114:393-404
120 Chooi WY & KR Leiby (1981) Mol Gen Genet 182:245-251
121 Chovnick A et al. (1969) Genetics 62:145-160
122 Chovnick A et al. (1976) Genetics 84:233-255
123 Chovnick A et al. (1977) Cell 11:1-10 (review)
124 Clark BA (1983) Biochem Genet 21:375-390
125 Clark SH et al. (1986) Genet Res 47:109-116
126 Cochrane BJ & RC Richmond (1978) Isoz Bull 11:54
127 Cochrane BJ & RC Richmond (1979a) Biochem Genet 17:167-183
128 Cochrane BJ & RC Richmond (1979b) Genetics 93:461-478
129 Cochrane BJ et al. (1983) Genetics 105:601-613
130 Coen ES & GA Dover (1982) Nuc Acids Res 10:7017-7026
131 Collier GE (1979) Genetics 91:s23
132 Collins JF & E Glassman (1969) Genetics 61:833-839
133 Collins M & GM Rubin (1982) Cell 30:71-79
134 Collins JF et al. (1971) Biochem Genet 5:1-14
135 Collins M & GN Rubin (1983) Nature 303:259-260
136 Conner TW & JM Rawls Jr (1982) Biochem Genet 20:607-619
137 Corces V et al. (1980) Proc Natl Acad Sci USA 77:5390-5393
138 Corces V et al. (1981) Proc Natl Acad Sci USA 78:7038-7042
139 Costa R et al. (1977) Dros Inform Serv 52:92
140 Coté B et al 1986 Genetics 112:249-265
141 Couderc J L et al 1983 J Mol Biol 164:419-430
142 Couderc JL et al 1984 Chromosoma 89:338-342
143 Coulter DE & AL Greenleaf 1982 J Biol Chem 257:1945-1952
144 Courtright JB 1967 Genetics 57:25-39
145 Courtright JB et al 1966 Genetics 54:1251-1260
146 Cowman AF et al 1986 Cell 44:705-710
147 Craig EA & BJ McCarthy 1980 Nuc Acids Res 8:4441-4457
148 Craig EA et al 1979 Cell 16:575-588
149 Craig EA et al 1983 Devel Biol 99:418-426
150 Crosby MA and EM Meyerowitz 1986 Genetics 112:785-802
151 Crowley TE et al 1983 Mol Cell Biol 3:623-634
152 David J et al 1978 Biochem Genet 16:203-211
153 Davis CA et al 1985 Nuc Acids Res 13:6605-6619
154 Davis RL & JA Kiger Jr 1981 J Cell Biol 90:101-107
155 Davis RL and N Davidson 1984 Mol Cell Biol 4:358-367
156 Dawid IB et al 1981 Cell 25:399-408
157 Dawid IB & MI Robbert 1981 Nuc Acids Res 9:5011-5020
158 Deak II 1977 J Embryol Exp Morphol 40:35-63
159 Deak II et al 1982 J Embryol Exp Morph 69:61-81
160 deCiccio DV and AC Spradling 1984 Cell 38:45-54
161 Denell RE & RD Frederick 1983 Devel Biol 97:34-47
162 Denell RE et al 1981 Devel Biol 81:43-50
163 Detwiler C & RJ MacIntyre 1978 Biochem Genet 16:1113-1132
164 Dickinson WJ 1970 Genetics 66:487-496
165 Dickinson WJ 1978 J Exp Zool 206:333-342
166 Dickinson WJ & DT Sullivan 1975 Gene Enzyme Systems in Drosophila Springer-Verlag New York 163 pp
167 Dickinson WJ 1983 In: Isozymes M C Rattazzi J G
169 DiNocera PP & IB Dawid 1983 Nuc Acids Res 11:5475-5482
170 Doane WW 1967 J Exp Zool 164:363-378
171 Doane WW 1969 J Exp Zool 171:31-41
172 Doane WW 1977 J Cell Biol 75(2 Pt 2):147a
173 Doane WW 1980 Dros Inform Serv 55:36-39
174 Doane WW & LG Treat-Clemons 1984 In: Genetic Maps SJ O'Brien ed Vol 3 pp 309-323 Cold Spring Harbor Publications; and in Dros Inform Serv 58:41-59
175 Doane WW et al 1975 In: Isozymes: Genetics and Evolution C L Markert ed Vol 4 pp 585-608 Acad Press New York
176 Doane WW et al 1983 In: Isozymes MC Rattazzi JG Scandalios GS Whitt eds Vol 9 pp 63-90 Alan R Liss NY (review)
177 Dorsett D et al 1979 Biochemistry 12:2596-2600
178 Dorsett D et al 1982 Biochemistry 16:3892-3899
179 Dubendorfer K et al 1974 Biochem Genet 12:203-211
180 Duck P & A Chovnick 1975 Genetics 79:459-466
181 Dudai Y et al 1976 Proc Natl Acad Sci USA 73:1684-1688
182 Duke EJ et al 1975 Biochem Genet 13:53-62
183 Duncan IM 1982 Genetics 102:49-70
184 Duncan IM & TC Kaufman 1975 Genetics 80:733-752
185 Dura JM & P Santamaria 1983 Mol Gen Genet 189:235-239
186 Eanes WF 1983 Biochem Genet 21:703-711
187 Eberlein S & MA Russell 1983 Devel Biol 100:227-237
188 Eissenberg JC & SCR Elgin 1983 Genetics 104:s23-24
189 Eissenberg JC & SCR Elgin 1983 Mol Cell Biol 3:1724-1729
190 Endow SA 1983 Proc Natl Acad Sci USA 80:4427-4431
191 Estelle M and RB Hodgetts 1984 Mol Gen Genet 195:434-446
192 Evans BA & AJ Howell 1978 Biochem Genet 16:13-26
193 Evgen'ev M et al 1979 Mol Gen Genet 176:275-280
194 Falk DR & EA deBoer III 1980 Mol Gen Genet 180:419-424
195 Falk DR et al 1977 Genetics 86:765-777
196 Falke E & RJ MacIntyre 1966 Dros Inform Serv 41:165-166
197 Falkenthal S et al 1984 Mol Cell Biol 4:956-965
198 Fargnoli J and GL Waring 1984 Dev Biol 105:41-47
199 Faust DM et al 1986 EMBO J 5:741-746
200 Fausto-Sterling A 1977 Biochem Genet 15:803-815
201 Ferre J 1983 Insect Biochem 13:289-294
202 Finnerty V 1976 In: The Genetics and Biology of Drosophila M Ashburner & E Novitski eds Vol Ib pp 721-765 Acad Press NY (review)
203 Finnerty V & GB Johnson 1979 Genetics 91:695-722
204 Finnerty V et al 1970 Proc Natl Acad Sci USA 65:939-946
205 Finnerty V et al 1979 Mol Gen Genet 172:37-43
206 Forrest HS et al 1961 Genetics 46:1455-1463
207 Forrest HS et al 1961 Biochem Biophys Acta 50:596-598
208 Fox D J 1971 Biochem Genet 5:69-80
209 Franklin IR 1971 Dros Inform Serv 47:113
210 Franklin IR & GK Chew 1971 Ibid:38
211 Franklin I R & W Rumball 1971 Ibid:37
212 Freund JN et al 1986 J Mol Biol 189:25-36
213 Friedman TB 1973 Biochem Genet 8:37-45
214 Fristrom JW et al 1978 Biochemistry 17:3917-3924
215 Fristrom J et al 1978 Biochemistry 19:3917-3924
216 Fruscoloni P et al 1983 Proc Natl Acad Sci USA 80:3359-3363
217 Fu L-J & GE Collier 1981 Genetics 97:s37-38
218 Fu L-J & GE Collier 1983 Bull Inst Zool Acad Sinica 22:25-35
219 Fukunaga A et al 1975 Genetics 81:135-141
220 Fyrberg EA et al 1980 Cell 19:365-378
221 Fyrberg EA et al 1981 Cell 24:107-116
222 Fyrberg EA et al 1983 Cell 33:115-123
223 Ganguly R et al 1985 Gene 5:91-101
224 Gans M et al 1975 Genetics Princeton 81:683-704
225 Garbe JC and ML Pardue 1986 Proc Natl Acad Sci USA 83:1393-1397
226 Garber RL et al 1983 EMBO J 2:2027-2036
227 Garcia-Bellido A et al 1979 Sci Amer 241:102-110 (review)
228 Garfinkel MD et al 1983 J Mol Biol 168:765-789
229 Gausz J et al 1979 Genetics 93:917-934
230 Gausz J et al 1981 Genetics 98:775-789
231 Gausz PJ et al 1986 Genetics 112:65-78
232 Gautam N 1983 Mol Gen Genet 189:495-500

233 Geer BW et al 1974 J Exp Zool 187:77-86
234 Geer BW et al 1979 Biochem Genet 17:867-879
235 Gehring WJ 1985 Cold Spring Harbor Symp Quant Biol 50:243-251
236 Gelbart W and A Chovnick 1979 Genetics 92:849-859
237 Gelbart WM & C Wu 1982 Genetics 102:179-189
238 Gelbart W et al 1976 Genetics 84:211-232
239 Gemmill RM et al 1983 J Cell Biol 97:132a
240 Gemmill RM et al 1985 Genetics 110:299-312
241 Gerasimova TI & EV Ananjev 1972 Dros Inform Serv 48:93
242 Gersh ES 1968 Science 162:1139
243 Ghosh D & HS Forrest 1967 Genetics 55:423-431
244 Gibson JB et al 1981 In: Genetic Studies of Drosophila Populations Proc Kioloa Conf JB Gibson & JG Oakeshott eds pp 251-267 Australian National Univ Canberra
245 Gilbert D et al 1984 Genetics 106:679-694
246 Glassman E 1965 Fed Proc 24:1243-1251
247 Glassman E & HK Mitchell 1959 Genetics 44:153-167
248 Glassman E & HK Mitchell 1959 Genetics 44:547-554
249 Glover DM 1981 Cell 26:297-298 (review)
250 Glover DM & DS Hogness 1977 Cell 10:167-176
251 Glover DM et al 1975 Cell 5:149-157
252 Goldberg DA 1980 Proc Natl Acad Sci USA 77:5794-5798
253 Goldberg DA et al 1983 Cell 34:59-73
254 Goldberg ML et al 1982 EMBO J 1:93-98
255 Goldberg ML et al 1983 Proc Natl Acad Sci USA 80:5017-5021
256 Goldschmidt-Clermont M 1980 Nuc Acids Res 8:235-252
257 Goldstein LSB et al 1982 Proc Natl Acad Sci USA 79:7405-7409
258 Goldstein LSB et al 1982 Proc Natl Acad Sci USA 79:7405-7409
259 Grace D 1980 Genetics 94:647-662
260 Graf J-D and FJ Ayala 1986 Biochem Gen 24:153-168
261 Graziani F et al 1981 Proc Natl Acad Sci USA 78:7662-7664
262 Green MM 1982 Proc Natl Acad Sci USA 79:5367-5369
263 Greenleaf AL et al 1979 Cell 18:613-622
264 Greenleaf AL et al 1980 Cell 21:785-792
265 Greenleaf AL 1983 J Biol Chem 258:13403-13406
266 Greenspan RJ 1980 J Comp Physiol 137:83-92
267 Greenspan RJ et al 1980 J Comp Neurol 189:741-774
268 Grell EH 1967 Science 158:1319-1320
269 Grell EH 1969 Dros Inform Serv 44:47
270 Grell EH 1969 Dros Inform Serv 44:46
271 Grell EH 1973 Genetics 74:s100
272 Grell EH 1976 Genetics 83:s23
273 Grell EH 1976 Genetics 83:753-764
274 Grell EH et al 1965 Science 149:80-82
275 Griffin-Shea R et al 1980 Cell 19:915-922
276 Griffin-Shea R et al 1982 Devel Biol 91:325-336
277 Grimwade BG et al 1985 Dev Bio 107:503-519
278 Guild GM 1984 Dev Bio 102:4642-470
279 Guild GM and EM Shore 1984 J Mol Biol 179:289-314
280 Gvozdev VA et al 1970 Mol Biologia (Moscow) 6:876-887
281 Gvozdev VA et al 1973 Dros Inform Serv 50:34
282 Gvozdev VA et al 1975 Genetika (Russ) 11:73-79
283 Gvozdev VA et al 1976 Fed Lettr 64:85-88
284 Gvozdev VA et al 1978 Dros Inform Serv 53:143-144
285 Hackett RW & JT Lis 1981 Proc Natl Acad Sci USA 78:6196-620
286 Hackett RW & JT Lis 1983 Nuc Acids Res 11:7011-7030
287 Haenlin M et al 1985 Cell 40:827-837
288 Hafen E et al 1983 EMBO J 2:617-623
289 Hall JC 1982 Quart Rev Biophysics 15(2):223-479 (review)
290 Hall JC & DR Kankel 1976 Genetics 83:517-535
291 Hall JC et al 1979 Soc Neurosci Symp 4:1-42
292 Hall LM C et al 1983 J Mol Biol 169:83-96
293 Hall LMC and P Spierer 1986 EMBO J 5:2949-2954
294 Hansson L et al 1981 Wilhelm Roux's Arch 190:308-312
295 Hansson L & A Lambertsson 1983 Mol Gen Genet 192:395-401
296 Harding K et al 1985 Science 229:1236-1242
297 Hardy RW et al 1981 Chromosoma 83:593-617
298 Hardy RW et al 1984 Genetics 107:591-610
299 Hazelrigg T et al 1982 Chromosoma 87:535-559
301 Henikoff S 1983 Nuc Acids Res 11:4735-4752
302 Henikoff S & CE Furlong 1983 Nuc Acids Res 11:789-800
303 Henikoff S & M Meselson 1977 Cell 12:441-451
304 Henikoff S et al 1981 Nature 289:33-37
305 Henikoff S et al 1983 Cell 34:405-414
306 Henikoff S et al 1986 Cell 44:33-42
307 Henikoff S et al 1986 Proc Natl Acad Sci USA 83:3919-3923
308 Henikoff et al 1986 Proc Natl Acad Sci USA 83:720-724
309 Hershey ND et al 1977 Cell 11:585-598
310 Higgins HJ et al 1984 Dev Biol 105:155-165
311 Hilliker AJ et al 1981 Genetics 95:95-110
312 Hilliker AJ et al 1981 Dros Inform Serv 56:65-72
313 Hiromi Y et al 1985 Cell 43:603-613
314 Hirsh J & N Davison 1981 Mol Cell Biol 6:475-485
315 Hisey BN et al 1979 Experientia 35:591-592
316 Hjorth JP 1970 Hereditas 64:146-148
317 Hodgetts RB 1975 Genetics 79:45-54
318 Hodgetts RB 1972 J Insect Physiol 18:937-947
319 Hoey T et al 1986 Proc Natl Acad Sci USA 83:4809-4813
320 Hoffman-Frank H et al 1983 Cell 32:589-598
321 Hoffman FM et al 1983 Cell 35:393-401
322 Holmgren R et al 1979 Cell 18:1359-1370
323 Holmgren R et al 1981 Proc Natl Acad Sci USA 78:3775-3778
324 Holmgren R 1984 EMBO J 3:569
325 Hotta Y & S Benzer 1972 Nature 240:527-535
326 Hovemann B & R Galler 1982 Nuc Acids Res 10:2261-2274
327 Hovemann B et al 1981 Nuc Acids Res 9:4721-4734
328 Hung M-C & PC Wensink 1981 Nuc Acids Res 9:6407-6419
329 Huntley MD 1978 PhD Thesis Univ Virginia Charlottesville
330 Ikenaga H & K Saigo 1982b Proc Natl Acad Sci USA 79:4143-4147
315 Indik ZK & KD Tartof 1980 Nature 284:477-479
332 Indik ZK & KD Tartof 1982 Nuc Acids Res 10:4159-4172
333 Ingham PW 1984 Cell 37:815-823
334 Ingles CJ et al 1983 Proc Natl Acad Sci USA 80:3396-3400
335 Ingolia TD & EA Craig 1981 Nuc Acids Res 9:1627-1642
336 Ingolia TD & EA Craig 1982 Proc Natl Acad Sci USA 79:525-529
337 Ingolia TD et al 1980 Cell 21:669-679
338 Ingolia TD & EA Craig 1982 Proc Natl Acad Sci USA 79:2360-2364
339 Ish-Horowicz D & SM Pinchin 1980 J Mol Biol 142:231-245
340 Ish-Horowicz D et al 1977 Cell 12:643-652
341 Ish-Horowicz D et al 1979 Cell 17:565-571
342 Ish-Horowicz D et al 1979 Cell 18:1351-1358
343 Itoh N et al 1986 Proc Natl Acad Sci USA 83:4081-4085
344 Iwabuchi M et al 1986 Biochem Gen 24:319-327
345 Izquierdo M et al 1981 Chromosoma 83:353-366
346 Jack JW & BH Judd 1979 Proc Natl Acad Sci USA 76:1368-1372
347 Jack R S et al 1981 Cell 24:321-331
348 Jacobson KB 1978 Nuc Acids Res 5:2391-2404
349 Jacobson KB 1982 Cell 30:817-823
350 Jacobson KB et al 1982 Cell 30:817-823
351 James TC and SCR Elgin 1986 Mol Cell Biol 6:3862-3872
352 Jarry B 1979 Mol Gen Genet 172:199-202
353 Jarry B & D Falk 1974 Mol Gen Genet 135:113-122
354 Jarry BP 1983 Insect Biochem 13:171-176
355 Jelnes JE 1971 Hereditas 67:291-293
356 Johns MA 1979 Genetics 91:s54
357 Johns MA and JH Postlethwaite 1985 Biochem Gen 23:465-482
358 Johnson DH & TB Friedman 1981 Science 212:1035-1036
359 Johnson DH & TB Friedman 1983 Proc Natl Acad Sci USA 80:2990-2994
360 Johnson MM et al 1979 Mol Gen Genet 174:287-292
361 Johnstone ME et al 1985 Biochem Gen 23:539-555
362 Jolly DJ & CA Thomas Jr 1980 Nuc Acids Res 8:67-84
363 Jordan BR et al 1980 FEBS Lettr 117:227-231
364 Judd BH 1976 In: The Genetics and Biology of Drosophila M Ashburner & E Novitski eds Vol 1b pp 767-799 Acad Press NY (review)
365 Junakovic N 1980 Nuc Acids Res 8:3611-3622
366 Jurgens G 1985 Nature 316:153-156
367 Kafatos FC 1983 In: Gene Structure and Regulation in Developoment S Subtelny & F C Kafatos eds pp 33-61 Alan R Liss NY (review)

369 Kalfayan L et al 1981 CSH Symp Quant Biol 46:185-191
370 Kalfayan L & PC Wensink 1982 Cell 29:91-98
371 Karch F et al 1981 J Mol Biol 148:219-230
372 Karch F et al 1985 Cell 43:81-96
373 Karess RE & GM Rubin 1982 Cell 30:63-69
374 Karlik CC et al 1984 Cell 37:469-481
375 Karlik CC et al 1984 Cell 38:711-719
376 Katzen AL et al 1985 Cell 41:449-456
377 Kaufman S 1962 Genetics 47:807-817
378 Kaufman TC et al 1980 Genetics 94:115-133
379 Kauver LM 1982 J Neurosci 2:1347-1358
380 Kedes LH 1979 Ann Rev Biochem 48:837-870
381 Keene MA et al 1981 Proc Natl Acad Sci USA 78:143-146
382 Kejzlarová-Lepesant J et al 1984 Nuc Acids Res 12:8835-8846
383 Kemphues KJ et al 1979 Proc Natl Acad Sci USA 76:3991-3995
384 Kemphues KJ et al 1980 Cell 21:445-451
385 Kemphues KJ et al 1983 Genetics 105:345-356
386 Kennison JA 1981 Genetics 98:529-548
387 Kennison JA 1983 Genetics 103:219-234
388 Keppy DO & WJ Welshons 1977 Genetics 85:495-506
389 Keppy DO & WJ Welshons 1980 Chromosoma 76:191-200
390 Kidd SJ & DM Glover 1980 Cell 19:103-119
391 Kidd SJ & DM Glover 1981 J Mol Biol 151:645-662
392 Kidd S et al 1983 Cell 34:421-433
393 Kiger Jr JA & E Golanty 1977 Genetics 85:609-622
394 Kiger Jr JA & E Golanty 1979 Genetics 91:521-535
395 Kiger JA et al 1981 Adv Cyclic Nucleotide Res 14:273-288
396 Kikkawa H 1964 Jap J Genet 39:401-411
397 King JJ & JF McDonald 1981 22nd Ann Dros Conf Chicago
398 King JJ & JF McDonald 1982 Genetics 100:s36
399 King JJ & JF McDonald 1983 Genetics 105:55-69
400 Knipple DC and RJ MacIntyre 1984 Mol Gen Genet 198:75-83
401 Kohorn BD & PMM Rae 1983 Proc Natl Acad Sci USA 80:3265-3268
402 Kohorn BD & PMM Rae 1983 Nature 304:179-181
403 Kolchinskii AM et al 1981 Molec Biol 14:868-877
404 Komitopoulou K et al 1983 Genetics 105:897-920
405 Komitopoulou K et al 1986 Dev Gen 7:75-80
406 Korge G 1975 Proc Natl Acad Sci USA 72:4550-4554
407 Korge G 1977 Chromosoma 62:155-174
408 Korge G 1981 Chromosoma 84:373-390
409 Kornberg T 1981 Devel Biol 86:363-372
410 Kornberg T 1981 Proc Natl Acad Sci USA 78:1095-1099
411 Kornher JS and D Brutlag 1986 Cell 44:879-883
412 Kotarski MA et al 1983 Genetics 105:371-386
413 Kotarski MA et al 1983 Genetics 105:387-407
414 Kreitman M 1983 Nature 304:412-417
415 Krider HM & BI Levine 1975 Genetics 81:501-513
416 Krider HM et al 1979 Genetics 92:879-889
417 Krivi GG & GM Brown 1979 Biochem Genet 17:371-390
418 Kubli E 1982 Adv Genet 21:123-172 (review)
419 Kubli E et al 1982 Nuc Acids Res 10:7145-7152
420 Kuner JM et al 1985 Cell 42:309-316
421 Kuroiura A et al 1984 Cell 37:825-831
422 Labella T et al 1983 Mol Gen Genet 190:487-493
423 Lakhotia SC & T Mukherjee 1980 Chromosoma 81:125-136
424 Lakhotia SC & T Mukherjee 1982 Chromosoma 85:369-374
425 Lambt C and O Schmidt 1986 EMBO J 5:2955-2961
426 Lang AB et al 1981 Nature 291:506-508
427 Langer-Safer PR et al 1982 Proc Natl Acad Sci USA 79:4381-4385
428 Langley CH et al 1982 Proc Natl Acad Sci USA 79:5631-5635
429 Langley S et al 1983 J Biol Chem 258:7416-7424
430 Lastowski DM & DR Falk 1980 Genetics 96:471-478
431 Lastowski-Perry D et al 1985 J Biol Chem 260:1527-1530
432 Laurie-Ahlberg CC 1982 Biochem Genet 20:407-424
433 Lawrence PA 1981 J Embryol Exp Morphol 64:321-332
434 Lawrence PA & G Morata 1983 Cell 35:595-601 (review)
435 Lawrence PA & G Struhl 1982 EMBO J 1:827-833
436 Lefevre G 1969 Genetics 63:589-600
437 Lefevre G 1976 In: The Genetics and Biology of Drosophila M Ashburner & E Novitski edsl Vol 1a pp 32-64 Acad Press New York (review)
438 Lefevre G in reference 377
439 Lefevre G & ML Sparling 1982 personal communication
440 Leigh Brown AJ & RA Voelker 1980 Biochem Genet 18:303-309
441 Leigh Brown AJ 1983 Proc Natl Acad Sci USA 80:5350-5354
442 Lengyel JA et al 1980 Chromosoma 80:237-252
443 Lengyel J 1983 (see ref 519)
444 Levine M et al 1981 Proc Natl Acad Sci USA 78:2417-2421
445 Levine M et al 1983 EMBO J 2:2037-2046
446 Levine M et al 1985 Cold Spring Harbor Symp Quant Biol 50:209-222
447 Levis R & GM Rubin 1982 Cell 30:543-550
448 Levis R et al 1982 Cell 30:551-565
449 Levis R et al 1982 Proc Natl Acad Sci USA 79:564-568
450 Levy JN et al 1985 Genetics 110:313-324
451 Levy LS & JE Manning 1982 Devel Biol 94:465-476
452 Levy LS et al 1982 Devel Biol 94:451-464
453 Lewis EB 1978 Nature 276:565-570
454 Lewis HW & HS Lewis 1961 Proc Natl Acad Sci USA 47:78-86
455 Lewis M et al 1975 Proc Natl Acad Sci USA 72:3604-3608
456 Lewis RA et al 1980 Genetics 95:367-381
457 Lewis RA et al 1980 Genetics:383-397
458 Lifton RP et al 1977 C S H Symp Quant Biol 42 (Pt 2:)1047-1057
459 Linbach KJ and R Wu 1985 Nuc Acids Res 13:631-644
460 Lindsley DL & EH Grell 1968 Genetic Variations of Drosophila melanogaster Carnegie Inst Wash Publ No 627
461 Lindsley DL & G Zimm 1984 and 1986 Dros Infor Ser 62 and 64
462 Lineruth K et al 1985 Mol Gen Genet 201:375-398
463 Lis JT et al 1978 Cell 14:901-919
464 Lis JT et al 1981 Gene 15:67-80
465 Lis JT et al 1981 Nuc Acids Res 9:5297-5310
466 Lis J et al 1981c Devel Biol 83:291-300
467 Lis JT et al 1983 Cell 35:403-410
468 Livak KF et al 1978 Proc Natl Acad Sci USA 75:5613-5617
469 Livingstone MS et al 1984 Cell 37:205-215
470 Livingstone MS 1985 Proc Natl Acad Sci USA 82:5992-5996
471 Livneh E et al 1985 Cell 40:599-607
472 Long EO & IB Dawid 1979 Cell 18:1185-1196
473 Long EO & IB Dawid 1979 Nuc Acids Res 7:205-215
474 Long EO et al 1981 Proc Natl Acad Sci USA 78:1513-1517
475 Long EO et al 1981 Mol Gen Genet 182:377-384
476 Louis C et al 1980 Cell 22:387-392

477 Loukas M 1981 Dros Inform Serv 56:85
478 Lubinsky S & GC Bewley 1979 Genetics 91:723-742
479 Lucchesi JC 1983 In: Isozymes MC Rattazzi JG Scandalios GS Whitt eds Vol 9 pp 179-188 Alan R Liss NY (review)
480 Lucchesi JC 1983 Devel Genet 2:275-282 (review)
481 Lucchesi JC et al 1982 Genetics 100:s42
482 Mace HAF et al 1983 Nature 304:555-557
483 MacIntyre RJ 1966 Genetics 53:461-474
484 MacIntyre RJ 1974 Isoz Bull 7:23-24
485 MacIntyre RJ & SJ O'Brien 1976 Ann Rev Genet 10:281-318 (review)
486 Mackay WJ & JM O'Donnell 1983 Genetics 105:35-53
487 Maddern RH 1977 Dros Inform Serv 52:110-111
488 Maddern RH 1981 Genet Res Camb 38:1-7
489 Madhavan K & H Ursprung 1973 Mol Gen Genet 120:379-380
490 Madhavan K et al 1972 J Insect Physiol 18:1523-1530
491 Mahaffey JW et al 1985 Cell 40:101-110
492 Mahaffey JW et al 1985 Cell 40:101-110
493 Mane SD Tepper CS & RC Richmond 1983 Biochem Genet 21:1019-1040
494 Mandal RK & IB Dawid 1981 Nuc Acids Res 9:1801-1811
495 Mange AP & L Sandler 1973 Genetics 73:73-86
496 Manning JE 1983 personal communication
497 Maroni G 1978 Biochem Genet 16:509-523
498 Maroni G & CC Laurie-Ahlberg 1983 Genetics 105:921-933
499 Maroni G et al 1986 Genetics 112:93-504
500 Marsh JL & TRF Wright 1979 Genetics 91:s74-75
501 Marsh JL & TRF Wright 1980 Devel Biol 80:379-387
502 Marsh JL and TRF Wright 1986 Genetics 112:249-265
503 Maschat F et al 1985 EMBO J 5:583-588
504 Mason PJ et al 1982 J Mol Biol 156:21-35
505 Mattox WW and N Davidson 1984 Mol Cell Biol 4:1343-1353
506 McCarron M et al 1979 Genetics 91:275-293
507 McClelland A et al 1981 J Mol Biol 153:257-272
508 McClelland A 1985 J Mol Biol 185:649
509 McDonald JF & FJ Ayala 1978 Genetics 89:371-388
510 McGinnis W et al 1980 Proc Natl Acad Sci USA 77:7367-7371
511 McGinnis W et al 1983 Cell 34:75-84
512 McGinnis W et al 1983 Proc Natl Acad Sci USA 80:1063-1067
513 McGinnis W et al 1984 Nature 308:428-433
514 McKenzie SL et al 1975 Proc Natl Acad Sci USA 72:1117-1121
515 McNabb SL and SK Beckendorf 1986 EMBO J 5:2331-2340
516 Mechler BM et al 1985 EMBO J 4:1551-1557
517 Meidinger EM & JH Williamson 1978 Can J Genet Cytol 20:489-497
518 Meidinger RG and MM Bentley 1986 Biochem Gen 24:683-699
519 Merriam J et al 1986 Dros Infor Ser 63:173-263 (Clone list)
520 Mestril R et al 1985 EMBO J 5:1667-1673
521 Meyerowitz EM & DS Hogness 1982 Cell 28:165-176
522 Miller JR et al 1983 Nuc Acids Res 11:11-19
523 Mindrinos MN et al 1985 EMBO J 4:147-153
524 Minoo Pand JH Postlethwaite 1985 Biochem Gen 23:913-932
525 Mirault et al 1979 Proc Natl Acad Sci USA 76:5254-5258
526 Mirault M-E et al 1982 EMBO J 1:1279-1285
527 Mischke D & ML Pardue 1982 J Mol Biol 156:449-446
528 Mischke D & ML Pardue 1983 J Submicrosc Cytol 15:367-370
529 Mitchell HK & UM Weber 1965 Science 148:964-966
530 Mlodzik M et al 1986 EMBO J 4:2961-2969
531 Mogami K & Y Hotta 1981 Mol Gen Genet 183:409-417
532 Mogami K et al 1981 Jpn J Genet 56:51-65
533 Mogami K et al 1982 J Biochem 91:643-650
534 Mogami K et al 1986 Proc Natl Acad Sci USA 83:1393-1397
535 Mohler J & M L Pardue 1982 Chromosoma 86:457-467
536 Monson J M et al 1982 Proc Natl Acad Sci USA 79:1761-1765
537 Moore G D et al 1983 Genetics 105:327-344
538 Moore G P & D T Sullivan 1978 Biochem Genet 16:619-634
539 Moore G D et al 1979 Nature 282:312-314
540 Morata G & P A Lawrence 1975 Nature 255:614-617
541 Moritz Th et al 1984 EMBO J 3:235-243
542 Morrison W J 1973 Isoz Bull 6:13
543 Morrison W J & R J MacIntyre 1978 Genetics 88:487-497
544 Mortin M A & G Lefevre Jr 1981 Chromosoma 82:237-247
545 Mortin MA and TC Kaufman 1984 Dev Bio 103:343-354
546 Morton R A & R S Singh 1982 Biochem Genet 20:179-198
547 Moser D et al 1980 J Biol Chem 255:4673-4679
548 Mount S M & J A Stietz 1981 Nuc Acids Res 9:6351-6368
549 Mozer B et al 1985 Mol Cell Biol 5:885-889
550 Mukai T & RA Voelker 1977 Genetics 86:175-185
551 Munneke L and GE Collier 1985 Biochem Gen 23:847-857
552 Muskavitch MAT & DS Hogness 1980 Proc Natl Acad Sci USA 77:7362-7366
553 Muskavitch MAT & DS Hogness 1982 Cell 29:1041-1051
554 Nakanishi Y & A Garen 1983 Proc Natl Acad Sci USA 80:2971-2975
555 Nash D & JF Henderson 1982 Adv Comp Physiol Biochem 8:1-51 (review)
556 Nash D et al 1981 Can J Genet Cyhtol 23:411-423
557 Natzle JE & JW Fristrom 1983 (see ref 519)
558 Natzle JE et al 1982 Nature 296:368-371
559 Natzle JE and BJ McCarthy 1984 Dev Bio 104:187-198
560 Natzle JE et al 1986 J Biol Chem 261:5575-5583
561 Nierman WC et al 1983 J Biol Chem 258:12618-12623
562 Nolan JM et al 1986 Proc Natl Acad Sci USA 83:3664-3668
563 Nørby S 1973 Hereditas 73:11-16
564 O'Brien SJ 1973 Biochem Genet 10:191-205
565 O'Brien SJ & RC Gethman 1973
566 O'Brien SJ & RJ MacIntyre 1972a Genetics 71:127-138
567 O'Brien SJ & RJ MacIntyre 1972b Biochem Genet 7:141-161
568 O'Brien SJ & RJ MacIntyre 1978 In: The Genetics and Biology of Drosophila M Ashburner & TRF Wright eds Vol 2a pp 395-551 Acad Press New York (review)
569 O'Connor D & JT Lis 1981 Nuc Acids Res 9:5075-5092
570 O'Donnell J et al 1977 Genetics 86:553-566
571 O'Donnell JM et al 1978 Genetics 88:s72-73
572 O'Hare K & GM Rubin 1983 Cell 34:25-35
573 O'Hare K et al 1983 Proc Natl Acad Sci USA 80:6917-6921
574 O'Hare K et al 1984 J Mol Biol 180:437-455
575 O'Tousa JE et al 1985 Cell 40:349-357
576 Ogita Z 1961 Botyu Kogoku 26:93-97 (Sci Insect Control)
577 Ogita Z 1962 Dros Inform Serv 36:103
578 Ohnishi S & RA Voelker 1979 Jap J Genet 54:203-209
579 Ohnishi S & RA Voelker 1981 Biochem Genet 19:75-85
580 Ohnishi S & RA Voelker 1982 Dros Inform Serv 58:121
581 Ohnishi S & RA Voelker 1982 Dros Inform Serv 58:121
582 Oliver MJ & JH Williamson 1979 Biochem Genet 17:897-907
583 Oliver MJ et al 1978 Biochem Genet 16:927-940

584 Orr W et al 1984 Proc Natl Acad Sci USA 81:3773-3777
585 Osgood CJ 1983 Genetics 104:s55
586 Otto E et al 1986 Proc Natl Acad Sci USA 83:6025-6029
587 Padilla HM & WG Nash 1977 Mol Gen Genet 155:171-177
588 Paradi E et al 1983 Mut Res 111:145-159
589 Pardue ML et al 1977 Chromosoma 63:135-151
590 Pardue ML 1983 (see ref 519)
591 Parker VP et al 1985 Mol Cell Biol 5:3058-3068
592 Paro R et al 1983 EMBO J 2:853-860
593 Parry DM & L Sandler 1974 Genetics 77:535-539
594 Paton DR & DT Sullivan 1978 Biochem Genet 16:855-865
595 Pavlakis GN et al 1979 Nuc Acids Res 7:2213-2238
596 Pedersen MB 1982 Carlsberg Res Commun 47:391-400
597 Peeples EE et al 1969a Biochem Genet 3:563-569
598 Peeples EE et al 1969b Genetics 62:161-170
599 Pelham HRB 1982 Cell 30:517-528
600 Pentz ES and A Shearn 1979 Dev Bio 70:149-170
601 Pentz ES et al 1986 Genetics 112:823-841
602 Pentz ES and TRF Wright 1986 Genetics 112:843-859
603 Peterson JS and WH Petri 1986 Dros Infor Ser 63:158
604 Peterson NS et al 1979 Genetics 92:891-902
605 Phillips JP et al 1982 Can J Genet Cytol 24:151-162
606 Pipkin SB et al 1977 J Hered 68:245-252
607 Pirrotta V & C Tschudi 1978 In: Genetic Engineering HW Boyer & S Nicosia eds pp 127-134 Elsevier/North Holland Biomed Press
608 Pirrotta V et al 1983 EMBO J 2:927-934
609 Poole SJ et al 1985 Cell 40:37-43
610 Postlethwait JH & T Jowett 1980 Cell 20:671-678
611 Postlethwait JH & R Kaschnitz 1978 FEBS Lettr 95:247-250
612 Postlethwait JH and PD Shirk 1981 Amer Zool 21:687-700
613 Procunier JD & RJ Dunn 1978 Cell 15:1087-1093
614 Procunier JD & KD Tartof 1978 Genetics 88:67-79
615 Quincey RV 1971 Biochem J 123:227-233
616 Rabinow L & WJ Dickinson 1981 Mol Gen Genet 183:264-269
617 Rae PMM 1981 Nuc Acids Res 9:4997-5010
618 Raghavan KV 1981 Wilhelm Roux's Arch 190:297-300
619 Rawls Jr JM 1980 Mol Gen Genet 178:43-39
620 Rawls Jr JM 1981 Mol Gen Genet 184:174-179
621 Rawls Jr JM & JW Fristrom 1975 Nature 255:738-740
622 Rawls Jr JM & JC Lucchesi 1974 Genet Res 24:59-72
623 Rawls Jr JM & L Porter 1979 Genetics 93:143-161
624 Rawls Jr JM et al 1981 Biochem Genet 19:115-128
625 Rawls JM et al 1986 Mol Gen Genet 202:493-499
626 Reddy P et al 1984 Cell 38:701-710
627 Reddy P et al 1986 Cell 46:53-61
628 Restifo LL and GM Guild 1986 J Mol Biol 188:517-528
629 Richards G et al 1983 EMBO J 2:2137-2142
630 Richmond RC & CS Tepper 1983 In: Isozymes MC Rattazzo JG Scandalios GS Whitt eds Vol 9 pp 91-106 Alan R Liss NY (review)
631 Riddell DC et al 1981 Nuc Acids Res 9:1323-1338
632 Riddihough G and HRB Pelham 1985 EMBO J 5:1653-1658
633 Ritossa F 1976 In: The Genetics and Biology of Drosophila M Ashburner & E Novitski eds Vol 1b pp 801-846 Acad Press New York (review)
634 Ritossa F et al 1966 Genetics 54:819-834
635 Rizki TM et al 1980 Wilhelm Roux's Arch 188:91-99
636 Rizki TM et al 1985 Mol Gen Genet 201:7-13
637 Rizki TM and R Rizki 1985 Genetics 110:s98
638 Robbins LG 1983 Genetics 103:633-648
639 Roberts DB & S Evans-Roberts 1979 Genetics 93:663-679
640 Roberts DB et al 1985 Genetics 109:145-156
641 Rodino E & GA Danielli 1972 Dros Inform Serv 48:77
642 Rodino E & A Martini 1971 Dros Inform Serv 46:139-140
643 Roiha H & DM Glover 1980 J Mol Biol 140:341-355
644 Roiha H & DM Glover 1981 Nuc Acids Res 9:5521-5532
645 Roiha H et al 1981 Nature 290:749-753
646 Rozek CE & N Davidson 1983 Cell 32:23-34
647 Rozek M et al 1985 Dev Bio 109:476-488
648 Rubin GM & AC Spradling 1982 Science 218:348-353
649 Rubin GM et al 1982 Cell 29:987-994
650 Rudkin GT & BD Stollar 1977 Nature 265:472-473
651 Rust K and GE Collier 1984 J Hered 76:39-44
652 Salz H and JA Kiger 1984 Genetics 108:377-392
653 Sakai RK et al 1969 Mol Gen Genet 105:24-29
654 Saluz HP 1984 personal communication
655 Saluz HP et al 1983 Nuc Acids Res 11:77-90
656 Salz HC et al 1982 Genetics 100:587-596
657 Samal B et al 1981 Cell 23:401-409
658 Sanchez F et al 1980 Cell 22:845-854
659 Sanchez F et al 1983 J Mol Biol 163:533-551
660 Sandler L 1977 Genetics 86:567-582
661 Sampsell B 1977 Biochem Genet 15:971-988
662 Savakis C et al 1984 EMBO J 3:235-243
663 Scalenghe F & F Ritossa 1976 Atti Acad Nat Lincei 13:439-528
664 Scavarda NJ et al 1983 Proc Natl Acad Sci USA 80:4441-4445
665 Schaffer U 1986 Mol Gen Genet 202:219-225
666 Schalet A & G Lefevre Jr 1976 In: The Genetics and Biology of Drosophila M Ashburner & E Novitski eds Vol 1b pp 847-902
667 Schedl TB et al 1978 Cell 14:921-929
668 Schejter ED et al 1986 Cell 46:1091-1101
669 Schneiderman HA et al 1966 Science 151:461-463
670 Scholnick SB et al 1983 Cell 34:37-45
671 Schott DR et al 1986 Biochem Gen 24:509-527
672 Scott MP et al 1983 Cell 35:763-776
673 Searles L et al 1983 Cell 31:585-592
674 Segraves WA et al 1983 Mol Gen Genet 189:34-40
675 Segraves WA et al 1984 J Mol Biol 175:1-17
676 Seybold W et al 1975 Biochem Genet 13:85-108
677 Shaffer JB & GC Bewley 1983 J Biol Chem 258:10027-10033
678 Sharp ZD et al 1983 Mol Gen Genet 190:438-443
679 Sherald AF 1981 Mol Gen Genet 183:102-106
680 Shermoen AW & SK Beckendorf 1982 Cell 29:601-607
681 Shotwell SL 1983 J Neurosci 3:739-747
682 Siegel JG 1981 Genetics 98:505-527
683 Silverman S et al 1979 Nuc Acids Res 6:421-433
684 Simeone A et al 1982 Nuc Acids Res 10:8263-8272
685 Simon MA 1983 Nature 302:837-839
686 Simon MA et al 1985 Cell 42:831-840
687 Sirotkin K & N Davidson 1982 Devel Biol 89:196-210
688 Smith DF et al 1981 Cell 23:441-449
689 Smith RF and R JKonopka 1981 Mol Gen Genet 183:243-251

690 Snyder M et al 1981 Cell 25:165-177
691 Snyder M et al 1982 Cell 29:1027-1040
692 Snyder MP et al 1982 Proc Natl Acad Sci USA 79:7430-7434
693 Snyder M & N Davidson 1983 J Mol Biol 166:101-118
694 Snyder PB et al 1986 Proc Natl Acad Sci USA 83:3341-3345
695 Sodja A et al 1982 Chromosoma 86:293-298
696 Solti M et al 1983 Biochem Biophys Res Commun 111:652-658
697 Sore EM and GM Guild 1986 J Mol Biol 190:149
698 Southgate R et al 1983 J Mol Biol 165:35-57
699 Sparrow JC & TRF Wright 1974 Mol Gen Genet 130:127-141
700 Spierer P et al 1983 J Mol Biol 168:35-50
701 Spillmann E & R Nothiger 1978 Dros Inform Serv 53:124
702 Spradling AC 1981 Cell 27:193-201
703 Spradling AC & AP Mahowald 1979b Cell 16:589-598
704 Spradling AC & AP Mahowald 1980 Proc Natl Acad Sci USA 77:1096-1100
705 Spradling AC & AP Mahowald 1981 Cell 27:203-209
706 Spradling AC et al 1975 Cell 4:395-404
707 Spradling AC et al 1977 J Mol Biol 109:559-578
708 Spradling AC et al 1979 Cell 16:609-616
709 Spradling AC et al 1980 Cell 19:905-914
710 Spradling AC & GM Rubin 1983 Cell 34:47-57
711 Steele MW et al 1968 Biochem Genet 2:159-175
712 Steele MW et al 1969 Biochem Genet 3:359-370
713 Stephenson RS et al 1983 In: Biology of Photoreceptors D Cosens & D Vince-Prue eds pp 471-495 Cambridge Univ Press
714 Stewart BR & JR Merriam 1974 Genetics 76:301-309
715 Stone JC et al 1983 Can J Genet Cytol 25:129-138
716 Storti RV & AE Szwast 1982 Devel Biol 90:272-283
717 Strausbaugh LD & ES Weinberg 1982 Chromosoma 85:489-505
718 Stroman P 1974 Hereditas 78:157-168
719 Stroman P et al 1973 Hereditas 78:239-246
720 Struhl G 1983 J Embryol Exp Morph 76:297-331
721 Struhl G 1981 Nature 293:36-41
722 Struhl G and RAH White 1985 Cell 43:507-519
723 Sullivan DT et al 1973 Genetics 75:651-661
724 Sullivan DT et al 1985 J Biol Chem 260:4345-4350
725 Swaroop A et al 1986 Mol Cell Biol 7:833-841
726 Tanaka A et al 1976 Genetics 84:257-266
727 Tartof KD 1969 Genetics 62:781-795
728 Tartof KD 1975 Ann Rev Genet 9:355-385 (review)
729 Tartof KD & RB Perry 1970 J Mol Biol 51:171-183
730 Thomas SR et al 1983 J Cell Biol 97:145a
731 Thompson JN Jr 1977 Nature 270:363
732 Thorig GEW & W Scharloo 1982 Genetica 57:219-225
733 Thorig GEW et al 1981 Mol Gen Genet 182:31-38
734 Thorig GEW et al 1981 Genetics 99:65-74
735 Thorig GEW 1983 PhD Diss Univ of Utrecht Utrecht The Netherlands
736 Tissieres A et al 1974 J Mol Biol 84:389-398
737 Tobin SL et al 1980 Cell 19:121-131
738 Tobler JE & EH Grell 1978 Biochem Genet 16:333-342
739 Tobler JE et al 1979 Biochem Genet 17:197-206
740 Torok I & F Karch 1980 Nuc Acids Res 8:3105-3123
741 Treat-Clemons LG & WW Doane 1982 Isoz Bull 15:7-24
742 Triantaphyllidis CD & C Christodoulou 1973 Biochem Genet 8:383-390
743 Trippa G et al 1970 Biochem Genet 4:665-667
744 Trippa G et al 1978 Biochem Genet 16:299-305
745 Tschudi C & V Pirrotta 1982 In: The Cell Nucleus H Busch & L Rothblum eds Vol 11 pp 29-44 Acad Press N Y (review)
746 Tschudi C et al 1982 EMBO J 1:977-985
747 Tsubota SI & JW Fristrom 1981 Mol Gen Genet 183:270-276
748 Tschudi C & V Pirrotta 1980 Nuc Acids Res 8:441-451
749 Twardzik DR et al 1971 J Mol Biol 57:231-246
750 Ursprung H & J Leone 1965 J Exp Zool 160:147-154
751 Vaslet CA et al 1980 Nature 285:674-676
752 Velissariou V & M Ashburner 1980 Chromosoma 77:13-27
753 Velissariou V & M Ashburner 1981 Chromosoma 84:173-185
754 Vigue C & W Sofer 1974 Biochem Genet 11:387-396
755 Vincent A et al 1985 J Mol Biol 186:149-166
756 Vincent III WJ & ES Goldstein 1983 personal communication
757 Voelker RA et al 1978 Dros Inform Serv 53:200
758 Voelker RA et al 1979 Biochem Genet 17:947-956
759 Voelker RA et al 1979 Ibid:769-783
760 Voelker RA et al 1981 Biochem Genet 19:525-534
761 Voelker RA et al 1985 Mol Gen Genet 201:437-445
762 Voellmy R et al 1981 Cell 23:261-270
763 Wadsworth SC et al 1980 Proc Natl Acad Sci USA 77:2134-2137
764 Wadsworth S et al 1985 Nuc Acids Res 13:2153-2170
765 Wahl RC et al 1982 J Biol Chem 257:3958-3962
766 Wahl RC & KV Rajagopalan 1982 J Biol Chem 257:1354-1359
767 Wahl RC et al 1983 J Biol Chem 257:3958-3962
768 Wakimoto BT & TC Kaufman 1981 Devel Biol 81:51-64
769 Waldmer-Stiefelmeier RD 1967 Z Vergl Physiol 56:268-274
770 Waldorf U et al 1984 EMBO J 3:2499-2504
771 Walen M and TG Wilson 1986 Genetics 114:77-92
772 Walker AR et al 1986 Mol Gen Genet 202:102-107
773 Wallis BB & AS Fox 1968 Biochem Genet 2:141-158
774 Waring GL et al 1983 Devel Biol 100:452-463
775 Warner CK et al 1974 Biochem Genet 11:359-365
776 Warner CK et al 1980 Mol Gen Genet 180:449-453
777 Warner CK & V Finnerty 1981 Mol Gen Genet 184:92-96
778 Warren TG & AP Mahowald 1979 Devel Biol 68:130-139
779 Wedeen C et al 1986 Cell 44:739-748
780 Weiner AJ et al 1984 Cell 37:843-851
781 Welshons W J & D O Keppy 1975 Genetics 80:143-155
782 Welshons W J & D O Keppy 1981 Mol Gen Genet 181:319-324
783 Wharton KA et al 1985 Cell 40:55-62
784 White BN et al 1973 J Mol Biol 74:635-651
785 White RA and M Wilcox 1984 Cell 39:163-171
786 White RAH and R Lehmann 1986 Cell 47:311-321
787 White RL & DS Hogness 1977 Cell 10:177-192
788 Wilkins JR et al 1982 Devel Genet 3:129-142
789 Williamson JH et al 1978 Can J Genet Cytol 20:545-553
790 Williamson JH 1982 Can J Genet Cytol 24:409-416
791 Willms J and J Phillips 1984 Can J Gen Cytol 26:682-691
792 Wilson TG and J Fristrom 1986 Dev Bio 118:190-201
793 Wimber DE & DM Steffenson 1970 Science 170:639-641
794 Wimber DE & DM Steffenson 1970 J Cell Biol 47:228A
795 Wolfner M et al 1983 (in ref 661)
796 Wong Y-C et al 1981 Nuc Acids Res 9:6749-6762
797 Wong Y-C et al 1985 Chromosoma 92:124-135

798 Woodruff RC & M Ashburner 1979 Genetica 92:117-132
799 Woodruff RC & M Ashburner 1979 Genetica 92:133-149
800 Wright TRF 1963 Genetics 48:787-801
801 Wright TRF 1977 Amer Zool 17:707-721
802 Wright TRF et al 1976 Genetics 84:267-286
803 Wright TRF et al 1976 Genetics 84:287-310
804 Wright TRF et al 1981 Devel Genet 2:223-235
805 Wright TRF et al 1982 Mol Gen Genet 188:18-26
806 Wu C 1980 Nature 286:854-860
807 Wu C-F et al 1983 Science 220:1076-1078
808 Wu M & N Davidson 1981 Proc Natl Acad Sci USA 78:7059-7063
809 Yagura T et al 1979 J Mol Biol 133:533-547
810 Yamanaka M et al 1983 J Cell Biol 97:145a
811 Yannoni CZ & WH Petri 1980 Wilhelm Roux's Arch 189:17-245
812 Yannoni CZ & WH Petri 1981 Wilhelm Roux's Arch 190:301-303
813 Yannoni CZ & WH Petri 1983 Genetics 104:s73
814 Yannoni CZ and WH Petri 1984 Dev Bio 102:504-508
815 Yedvobnick B et al 1980 Genetics 95:661-672
816 Yedvobnick B & M Levine 1982 Nature 297:239-241
817 Yen TT & E Glassman 1965 Genetics 52:977-981
818 Yim JJ et al 1977 Science 198:1168-1170
819 Young WJ 1966 J Heredity 57:58-60
820 Young WJ et al 1964 Science 143:140-141
821 Youvan DC & JE Hearst 1981 Nuc Acids Res 9:1723-1741
822 Yun J et al 1985 J Biol Chem 260:8220-8228
823 Zachar Z & PM Bingham 1982 Cell 30:529-541
824 Zacharopoulou A et al 1981 Dros Info Serv 56:166-167
825 Zadar E et al 1986 Mol Gen Genet 204:469-472
826 Zhimulev IF et al 1981 Chromosoma 82:25-40
827 Zuker CS et al 1985 Cell 40:851-858
828 Zulauf E et al 1981 Nature 292:556-558
829 Zust H et al 1972 Verh Schweiz Naturforsch Ges p
830 Regulski MK et al 1985 Cell 43:71-80
831 Fjose A et al 1985 Nature 313:284-289
832 Sanchez-Herrera E et al 1985 Nature 313:108-113
833 Agnel M et al.(1989) Genes & Develop 3:85-95
834 Ait-Ahmed O et al. (1987) Dev Biol 122:153-162
835 Alonso MC & CV Cabrera (1988) EMBO J 7:2585-2591
836 Alton AK et al. (1989) Dev Genet 10:261-272
837 Amrein H et al. (1988) Cell 55:1025-1035
838 Arora K et al. (1987) Nature 330:62-63
839 Ashley CT et al. (1989) J Biol Chem 264:8394-8401
840 Ayme-Southgate A et al. (1989) J Cell Biol 108:521-531
841 Babu P & SG Bhat (1986) Mol Gen Genet 205:483-486
842 Baker BS & MF Wolfner (1988) Genes & Develop 2:477-489
843 Baker NE & GM Rubin (1989) Nature 340:150-153
844 Banerjee U et al. (1987) Cell 49:281-291
845 Banerjee U et al. (1987) Cell 51:151-158
846 Bargiella et al. (1987) Nature 328:686-691
847 Bauer BJ & GL Waring (1987) Dev Biol 121:349-358
848 Beachy PA et al. (1988) Cell 55:1069-1081
849 Bell LR et al. (1988) Cell 55:1037-1046
850 Bender M et al. (1988) Dev Genet 9:715-732
851 Benson M & V Pirotta (1987) EMBO J 6:1387-1392
852 Bentley MM et al.(1989) Biochem. Genet. 27:99-118
853 Biggin MD et al. (1988) Cell 53:713-722
854 Biggs J et al.(1988) Genes & Develop 2:1333-1343
855 Biggs J et al.(1988) Genes & Develop 2:1333-1343
856 Black BB et al. (1987) Mol Gen Genet 209:306-312
857 Bloomquist BT et al. (1988) Cell 54:723-733
858 Blumberg B et al. (1987) J Biol Cehm 262:5947-5950
859 Blumberg B et al. (1988) J Biol Chem 263:18328-18337
860 Bogaert T et al. (1987) Cell 51:929-940
861 Boggs RT et al. (1987) Cell 50:739-747
862 Bowell DDL et al. (1988) Genes & Develop 2:620-634
863 Brown SJ et al. (1988) Mol Cell Biol 8:4314-4321
864 Burke T et al. (1987) Dev Biol 124:441-450
865 Burtis KC & BS Baker (1989) Cell 56:997-1010
866 Burton RS & A LaSpada (1986) Biochem. Genet 24:715-719
867 Byers TJ et al. (197) J Cell Biol 105:2105-2110
868 Cabrera CV et al. (1987) Cell 50:425-433
869 Cabrera CV et al. (1987) Cell 50:659-663
870 Caggese C et al. (1988) Biochem. Genet. 26:571-584.
871 Campos et al. (1985) J Neurosci 2:197-218
872 Campos et al. (1987) EMBO J 6:425-31
873 Candy M et al. (1988) Cell 55:1061-1067
874 Carroll SB et al. (1988) Genes & Develop 2:883-890
875 Ceccini JP et al. (1987) Eur J Biochem 165:587-597
876 Chasan R & KV Anderson (1989) Cell 56:391-400
877 Chen PS et al. (1988) Cell 54:291-298
878 Cheng CM & TJ Lang (1988) Dev Biol 130:551-557
879 Chun M & S Falkenthal (1988) Genome 30, suppl. 1:193
880 Citri et a. (1987) Nature 326:42-47
881 Cohen SM et al. (1989) Nature 338:432-434
882 Collier GE (1987) Genetic Maps vol. 4:374-391 (ed. SJ O'Brien), C S H Press
883 Cook JL et al. (1988) J Biol Chem 263:10858-10864
884 Cote S (1988) Genome 30, suppl. 1:194
885 Crews ST et al. (1988) Cell 52:143-151
886 Cronmiller C & TW Kline (1987) Cell 48:479-487
887 Cronmiller C et al. (1988) Genes & Develop 2:1666-1676
888 Das G et al. (1987) J Biol Chem 26:1187-1193
889 Davis MB & RJ MacIntyre (1988) Genetics 120:755-766
890 Dearolf CR et al.(1988) Dev Biol 129:169-178
891 Deng Y & TM Rizki (1988) Genome 30, suppl 1:192
892 DiBenedetto A et al. (1987) Dev Biol 119:242-251
893 DiNocera PP & G Casari (1987) Proc Natl Acad Sci USA 84:5843-5847
894 Dreesen TD et al. (1988) Mol Cell Biol 8:5206-5215
895 Driever W et al. (1987) Cell 54:83-93
896 Dubreuil et al. (1987) J Cell Biol 105:2095-2105
897 Edgar BA & PH O'Farrell (1989) Cell 57:177-187
898 Fechtel K et al. (1988) Genetics 120:465-474
899 Foster JL et al. (1988) J Biol Chem 263:1676-1681
900 Fouts D et al. (1988) Gene 63:261-275
901 Frei E et al. (1988) EMBO J 7:197-204
902 Freund JN & BP Jarry (1987) J Mol Biol 193:1-13
903 Friedman TB & AP Barker (1982) Insect Biochem 12:563-570
904 Frigerio G et al. (1986) Cell 47:735-746
905 Garofalo RS & OM Rosen (1988) Mol Cell Biol 8:1638-1647
906 Gaul U et al. (1987) Cell 50:639-64
907 Gaul U et al. (1989) Proc Natl Acad Sci USA 86:4599-4603

908 Georg EL et al. (1989) Mol Cell Biol 9:2957-2974
909 Gergen JP & BA Butler (1988) Genes & Develop 2:1179-1193
910 Gertler FB et al.(1989) Cell 58:103-113
911 Geyer PK & EA Fyrberg (1986) Mol Cell Biol 6:3388-3396
912 Goralski TJ et al. (1989) Cell 56:1011-1018
913 Green LL et al. (1988) Genome 30, suppl. 1:192
914 Grigliatti S et al. (1989) Dev Genet 10:33-41
915 Gubb D & A Garcia-Bellido (1982) J Emb Exp Morph 68:37-57
916 Gutierrez AG et al. (1989) Dev Genet 10:155-161
917 Hafen E et al. (1987) Science 236:55-63
918 Hall NA (1986) Biochem. Genet. 24:775-793
919 Hanke PD et al. (1987) J Biol Chem 262:17370-17373
920 Hanke PD & RV Storti (1988) Mol Cell Biol 8:3591-3602
921 Hartley DA et al. (1988) Cell 55:785-795
922 Hashimoto C et al. (1988) Cell 52:269-272
923 Haynes SR et al. (1989) Dev Biol 134:246-257
924 Henkemeyer MJ et al. (1987) Cell 51:821-828
925 Henkemeyer MJ et al. (1988) Mol Cell Biol 8:843-853
926 Hickey DA et al. (1988) Biochem. Genet. 26:757-768
927 Hiraizumi K & CC Laurie (1988) Biochem. Genet. 26:783-805
928 Hiromi Y & WJ Gehring (1987) Cell 50:963-974
929 Hirsh J et al. (1986) Mol Cell Biol 6:4548-4557
930 Hoey T et al. (1988) Mol Cell Biol 8:4598-4607
931 Hofmann A et al. (1987) Chromosoma 96:8-17
932 Hofmann A & G Korge (1987) Chromosoma 96:1-7
933 Hoogwerf AM et al. (1988) Genetics 118:665-670
934 Hoshizaki DK et al. (1987) Genet Res, Camb 49:111-119
935 Houpt DR et al. (1988) Genome 30:844-853
936 Howard K et al. (1988) Genes & Develop 2:1037-1046
937 Ish-Horowitz D & SM Pinchin (1987) Cell 5:405-415
938 Iverson LE et al. (1988) Proc Natl Acad Sci USA 85:5723-5727
939 Jackson FR et al. (1986) Nature 320:185-188
940 Jacob L et al. (1987) Cell 50:215-225
941 Jallon JM et al. (1988) Genet Res, Camb 51:17-22
942 James JM & GE Collier (1988) J Exp Zool 248:185-191
943 Jarry BP (1979) Mol Gen Genet 172:199-202
944 Johnson DH et al. (1987) Gene 59:77-86
945 Jokerst RS et al. (1989) Mol Gen Genet 215:266-275
946 Jones WK et al. (1988) Genetics 120:733-742
947 Kalderon D & GM Rubin (1988) Genes & Develop 2:1539-1556
948 Kalderon D & GM Rubin (1989) J Biol Chem 264:10738-10748
949 Kamb A et al. (1987) Cell 50:405-413
950 Katz F et al. (1988) EMBO J 7:3471-3477
951 Kay MA et al. (1988) Mol Gen Genet 213:354-358
952 Kelley MR et al. (1987) Cell 51:539-548
953 King RC et al. (1986) Dev Genet 7:1-20
954 Kirkland KC & JP Phillips (1987) Gene 61:415-419
955 Klambt C et al. (1989) Dev Biol 133:425-436
956 Kopczynski CC et al. (1988) Genes & Devlop 2:1723-1735
957 Kornber JS & SA Kaufman (1986) Chromosoma 94:285-216
958 Kral KG et al. (1986) Gene 45:131-137
959 Krause (sp?) HM et al. (1988) Genes & Develop 2:1021-1036
960 Kuhn R et al. (1988) EMBO J 7:447-454
961 Kwiatkowski J et al. (1989) Nuc Acid Res 17:1264
962 LaParco Y et al. (1986) Biol Cell 56:217-226
963 Laughton A et al. (1988) Development 104 suppl:75-83
964 Lebovitz RM et al. (1989) EMBO J 8:193-202
965 Lee H et al.(1988) Mol Cell Biol 8:4727-4735
966 Lehmann R et al. (1983) W Roux Arch Dev Biol 192:62-74
967 Lehmann R & C Nusslein-Volhard (1987) Dev Biol 119:402-417
968 Lehner CF & PH O'Farrell (1989) Cell 56:957-968
969 Leptin M et al. (1989) Cell 56:401-408
970 Lin H & MF Wolfner (1988) Genome 30, suppl. 1:193
971 Lin H & MF Wolfner (1989) Mol Gen Genet 215:257-265
972 Liu X et al. (1988) Genes & Develop 2:228-238
973 MacKay WJ & GC BEwley (1989) Genetics 122:643-652
974 MacKrell AJ et al. (1988) Proc Natl Acad Sci USA 85:2633-2637
975 Maki C et al. (1989) Gene 79:289-298
976 Manseau LJ et al. (1988) Genetics 119:407-420
977 Mansukhami A et al. (1988) Mol Cell Biol 8:615-623
978 Mariol M-C et al.(1987) Mol Cell Biol 7:3244-3251
979 Mark GE et al.(1987) Mol Cell Biol 7:2134-2140
980 McKeown M et al. (1987) Cell 48:489-499
981 Meisen N et al. (1988) Gene 74:457-464
982 Mismer D & GM Rubin (1989) Genetics 121:77-87
983 Mismer D et al. (1988) Genetics 120:173-180
984 Mlodzik M & WJ Gehring (1987) Cell 48:465-478
985 Monsma SA & MF Wolfner (1988) Genes & Develop 2:1063-1073
986 Montell C & GM Rubin(1988) Cell 52:77-772
987 Montell DJ & CS Goodman (1988) Cell 53:463-473
988 Montell C et al. (1987) J Neurosci 7:1558-1566
989 Mozer BA & IB Dawid (1989) Proc Natl Acad Sci USA 86:3738-3742
990 Mulligan PK et al. (1988) Mol Cell Biol 8:1481-1488
991 Munneke LR & GE Collier (1988) Biochem. Genet. 26:131-141
992 Namber U et al. (1988) Nature 336:489-492
993 Ng SC et al. (1989) Development 105:629-638
994 Nusslein-Volhard C et al. (1984) W Roux's Arch Dev Biol 193:267-282
995 Nichols R et al. (1988) J Biol Chem 263:12167-12170
996 Nishida Y et al. (1986) BBRC 141:474-481
997 Nishida Y et al. (1988) EMBO J 7:775-790
998 Oakeshott et al. (1987) Proc Natl Acad Sci U SA84:3359-3363
999 O'Donnell J et al. (1989) Genetics 121:272-280
1000 O'Donnell J et al. (1989) Genetics 121:272-280
1001 Oro AE et al. (1988) Nature 336:493-496
1002 Otto E et al.(1987) Mol Cell Biol 7:1710-1715
1003 Padgett RW et al.(1987) Nature 325:81-84
1004 Papazian DM et al. (1987) Science 237:749-753
1005 Parkhurst SM et al. (1988) Genes & Develop 2:1205-1215
1006 Patch NH et al. (1987) Cell 48:975-988
1007 Pauli D et al. (1988) J Mol Biol 200:47-53
1008 Payre F et al. (1988) Genome 30, suppl. 1:194
1009 Perrimon N (1988) Dev Biol 127:392-407
1010 Petruzelli L et al. (1986) Proc Natl Acad Sci USA 83:4710-4714
1011 Pirotta V et al. (1988) Genes & Develop 2:1839-1850
1012 Place, AR et al.(1987) Biochem. Genet.25:621-638
1013 Pollock JA & S Benzer (1988) Nature 333:779-782
1014 Pongs O et al. (1988) EMBO J 7:1087-1096
1015 Popodi E et al. (1988) Dev Biol 127:248-256
1016 Pribyl LJ et al. (1988) Dev Biol 127:45-53
1017 Price JV et al. (1989) Cell 56:1085-1092

1018 Priess A et al. (1988) EMBO J 7:3917-3927
1019 Provost NM et al.(1988) J Biol Chem 263:12070-12076
1020 Rabinow S et al. (1988) Dev Biol 126:294-303
1021 Rabinow S et al. (1988) Science 242:1570-1572
1022 Rafti F et al. (1988) Nuc Acids Res 16:4915-4926
1023 Rafti F et al. (1989) Nuc Acids Res 17:456
1024 Regan CL et al. (1988) Genome 30, suppl. 1:192
1025 Regulski M et al. (1987) EMBO J 6:767-777
1026 Reinke R et al. (1988) Cell 52:291-301
1027 Reinke R & SL Zipursky (1988) Cell 55:321-330
1028 Riggleman et al. (1989) Genes & Develop 3:96-113
1029 Rijsewik F et al. (1987) Cell 50:649-657
1030 Rosenberg UB et al. (1986) Nature 319:336-339
1031 Rosenthal A et al. (1987) EMBO J 6:433-441
1032 Rothberg JM et al.(1988) Cell 55:1047-1059
1033 Royden CS et al. (1987) Cell 51:165-173
1034 Rykowski MC et al. (1988) Cell 54:461-472
1035 Salkoff L et al. (1987) Science 237:744-749
1036 Salz HK et al. (1989) Genes & Develop 3:708-719
1037 Saunders RDC & M Bownes (1986) Mol Gen Genet 205:557-560
1038 Savant SS & GL Waring (1988) Dev Biol 135:43-52
1039 Saxton WM et al. (1988) Proc Natl Acad Sci USA 85:1109-1113
1040 Schaeffer E et al. (1989) Cell 57:403-412
1041 Schejter (sp?) ED & BZ Shilo (1989) Cell 56:1093-1104
1042 Scherer LJ et al. (1988) Dev Biol 130:786-788
1043 Schneider LE & PH Taghert (1988) Proc Natl Acad Sci USA 85:1991-1997
1044 Schneuwly S et al. (1989) Proc Natl Acad Sci USA 86:5390-5394
1045 Schulz RA et al. (1986) Proc Natl Acad Sci USA 83:9428-9432
1046 Schulz RA et al. (1989) Dev Biol 131:515-523
1047 Schulz RA et al. (1989) Genes & Develop 3:232-242
1048 Schwartz PE & WW Doane (1989) Biochem. Genet. 27:31-46
1049 Schwarz TL et al. (1988) Nature 331:137-142
1050 Seeger MA et al. (1988) Cell 55:589-600
1051 Seto NOL et al. (1987) Nuc Acid Res 15:10601
1052 Seto NOL et al. (1989) Gene 75:85-92
1053 Shieh B-H et al. (1989) Nature 338:67-70
1054 Smith VL et al. (1987) J Mol Biol 196:471-485
1055 Sole CK et al. (1987) Science 236:1094-1098
1056 Sondergaard L (1975) Hereditas 81:199-210
1057 Sondergaard L (1976) Hereditas 82:51-56
1058 Sondergaard L (1986) Hereditas 104:313-315
1059 Spana C et al. (1988) Genes & Develop 2:1414-1423
1060 Stellar H et al. (1987) Cell 50:1139-1153
1061 Stephenson EC et al. (1987) Dev Biol 124:1-8
1062 Stephenson EC et al. (1988) Genes & Develop 2:1655-1665
1063 Stewart R (1987) Science 238:692-694
1064 Stewart R. et al. (1988) Cell 55:487-495
1065 Struhl G et al. (1989) Cell 57:1259-1273
1066 Sun X-H et al. (1988) Mol Cell Biol 8:5200-5205
1067 Sun X-H et al. (1988) Genes & Develop 2:743-753
1068 Swaroop A et al. (1987) Proc Natl Acad Sci USA 84:6501-6505
1069 Takano T et al. (1989) Proc Natl Acad Sci USA 86:5000-5004
1070 Tempel BL et al. (1987) Science 237:770-775
1071 Templeton NS & SS Potter (1989) EMBO J 8:1887-1894
1072 Terpstra P & AB Geert (1988) J Mol Biol 202:663-665
1073 Thisse et al. (1988) EMBO J 7:2175-2183
1074 Thomas JB et al. (1988) Cell 52:133-141
1075 Thorig GEW et al. (1987) Biochem. Genet. 25:7-25
1076 Timpe LC & LY Jan (1987) J. Neurosci. 7:1307-1317
1077 Toffenetti J et al. (1987) J Cell Biol 104:19-28
1078 Tomlinson A et al. (1987) Cell 51:143-150
1079 Tomlinson A et al. (1988) Cell 55:771-784
1080 Uemura T et al. (1989) Cell 58:349-360
1081 Underwood EM & JA Lengyel (1988) Dev Genet 9:23-35
1082 Ursie D & B Ganetski (1988) Gene 68:267-274
1083 Uzvolgyi E et al. (1988) Proc Natl Acad Sci USA 85:3034-3038
1084 Van Vactor D et al. (1988) Cell 52:281-290
1085 Vassin H et al. (1987) EMBO J 11:3431-3440
1086 Vassin H & J Campos-Ortega (1987) Genetics 116:433-445
1087 Villares R & CV Cabrera (1987) Cell 50:415-424
1088 Vincent III WS et al.(1989) Genes & Develop 3:334-347
1089 Vinson CR et al. (1989) Nature 338:263-264
1090 von Kalm L et al. (1989) Proc Natl Acad Sci USA 86:5020-5024
1091 Wadsworth SC et al.(1988) Mol Cell Biol 8:778-785
1092 Wassenberg DR et al. (1988) J Biol Chem 262:10741-10747
1093 Weigel D et al. (1987) Mol Gen Genet 207:374-384
1094 Weigel D et al. (1989) Cell 57:645-658
1095 Whetten R et al. (1988) Genetics 120:475-484
1096 Whitfield WGF et al. (1989) Nature 338:337-340
1097 Williams JL et al. (1987) Mol Gen Genet 209:360-365
1098 Woods DF & PJ Bryant (1989) Dev Biol 135:43-52
1099 Wulczyn FG & R Kahmann (1987) Gene 51:129-137
1100 Yanicostas C et al. (1989) Mol Cell Biol 9:2526-2535
1101 Yang et al. (1988) Proc Natl Acad Sci USA 85:1864-1868
1102 Yarfitz S et al. (1988) Proc Natl Acad Sci USA 85:7134-7138
1103 Yedvobnick B et al. (1988) Genetics 118:483-497
1104 Yu et al. (1987) Proc. Natl. Acad. Sci. USA 84:784-788
1105 Zachar Z et al.(1987) Mol Cell Biol 7:2498-2505
1106 Zaker CS et al. (1987) J Neurosci 7:1550-1557
1107 Ziemer A et al. (1988) Genetics 119:63-74
1108 Zimmerman J et al. (1988) Genome 30, suppl. 1:193
1109 Zinn et al. (1988) Cell 53:577-587
1110 Zuker CS et al. (1988) Cell 55:75-82
1111 Matsumoda H et al. (1987) Proc Natl Acad Sci USA 84:95-989
1112 Maine EM et al. (1985) Cell 43:521-529

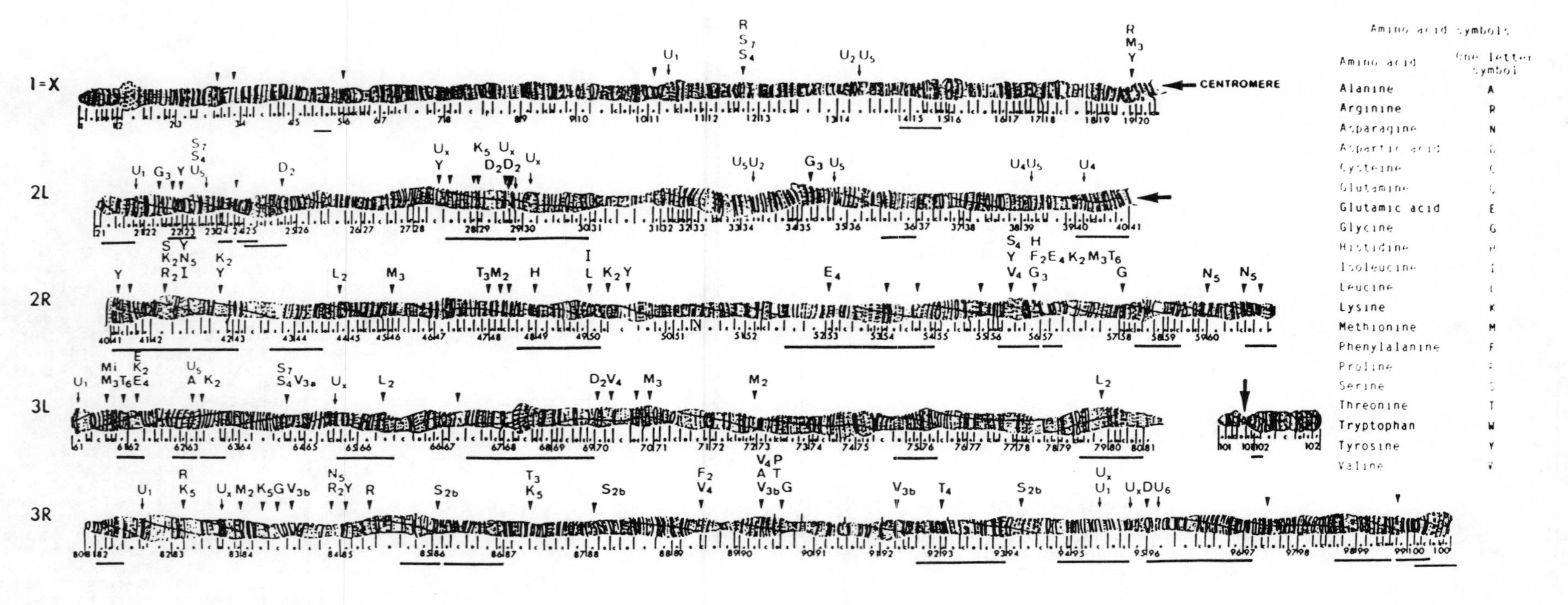

Amino acid symbols

Amino acid	One letter symbol
Alanine	A
Arginine	R
Asparagine	N
Aspartic acid	D
Cysteine	C
Glutamine	Q
Glutamic acid	E
Glycine	G
Histidine	H
Isoleucine	I
Leucine	L
Lysine	K
Methionine	M
Phenylalanine	F
Proline	P
Serine	S
Threonine	T
Tryptophan	W
Tyrosine	Y
Valine	V

Localization of tRNA and snRNA genes on _Drosophila melanogaster_ polytene salivary gland chromosomes. The tRNA genes without amino acid symbols have not been identified. U_X = unidentified snRNA. The bars delineate Minute deficiencies (42).

Drosophila melanogaster (fruitfly), 2 N = 8
Transfer RNA and snRNA in situ hybridization data

August 1989

Eric Kubli and Sylvia Schwendener
Zoolog. Inst. Univ. Zuerich-Irchel
Winterthurerstrasse 190
CH-8057 Zuerich
Switzerland

Chromosome	Labeled Region	Gene(s)	References
1	3D	tRNA	11
1	3F	tRNA	11
1	5F, 6A	tRNA	11
1	11A	tRNA	11
1	11B	U-1	2/32
1	12DE	tRNA-Ser-4, tRNA-Ser-7	3/11/15
		tRNA-Arg	28
1	14B	U-2, U-5	31/32
1	19F	tRNA-Met-3, tRNA-Tyr-1	17
		tRNA-Arg	28
2L	21E	U-1	2/32
2L	22BC	tRNA-Gly-3	15/33
2L	22DE	tRNA	11
2L	22F, 23A	tRNA-Tyr-1	9/17/23
2L	23D, 23DE	tRNA-Ser-4, tRNA-Ser-7, U-5	3/11/15/32
2L	24DE	tRNA	11
2L	25D	tRNA-Asp-2	34
2L	28C	tRNA-Tyr-1, snRNA	9/27/38
2L	28D	tRNA	11
2L	28F2, 29A	tRNA	11
2L	29A	tRNA-Lys-5	37
2L	29D	tRNA-Asp-2	11/15/34/39
2L	29DE	snRNA	27
2L	29E	tRNA-Asp-2	11/15/34/39
2L	29EF	snRNA	27
2L	30A	snRNA	27
2L	34AB	U-2, U-5	31/32
2L	35B, 35BC	tRNA-Gly-3	11/15/25
2L	35EF	U-5	32
2L	39B	U-4, U-5	32
2L	40AB	U-4	32
2R	41AB	tRNA-Tyr-1	9/17
2R	41CD	tRNA	11
2R	42A	tRNA-Asn-5, tRNA-Arg-2	9/11/12/15/21
		tRNA-Ile, tRNA-Lys-2,	35/39/40/41
		tRNA-Ser, tRNA-Tyr-1	7/30
2R	42E	tRNA-Lys-2, tRNA-Tyr-1	9/11/12/15/39
2R	44EF	tRNA-Leu-2	11/17
2R	46A1-2	tRNA-Met-3	15/39
2R	47F	tRNA-Thr-3	14/17
2R	48B5-7	tRNA-Met-2	11/15/39
2R	48CD	tRNA	11
2R	48F, 49AB	tRNA-His	8/11/35

Chromosome	Labeled Region	Gene(s)	References
2R	49F, 50AB	tRNA-Leu, tRNA-Ile	11/29/30
2R	50BC, 50C1-4	tRNA-Lys-2, tRNA-Tyr-1	11/12/17/39
2R	52F, 53A	tRNA-Glu-4	11/22
2R	54A	tRNA	11
2R	54D	tRNA	11
2R	55EF	tRNA	11
2R	56D	tRNA-Tyr-1	9/11
2R	56D3-7	tRNA-Val-4, tRNA-Ser-4(?)	11/14/15/39
2R	56E	tRNA-His	35
2R	56E-57D, 56EF	tRNA-Gly-3, tRNA-Phe-2(?) tRNA-Lys-2, tRNA-Met-3 tRNA-Glu-4, tRNA-Thr-6	5/11/14/15/16 19/22/25/39
2R	58AB	tRNA-Gly-GGA	11/17
2R	59F	tRNA-Asn-5	35
2R	60C	tRNA-Asn-5	35
2R	60E	tRNA	11
3L	61A	U-1	2/32
3L	61D1-2	tRNA-Met-3, tRNA Met$_i$	11/15/36
3L	61F	tRNA-Thr-6	14
3L	62A	tRNA-Glu, tRNA-Glu-4, tRNA-Lys-2	15/20/22
3L	63A	tRNA-Ala, U-5	11/32/35
3L	63B	tRNA-Lys-2	39
3L	64D	tRNA-Val-3a, tRNA-Ser-4+7	11/15/17/39
3L	65CD	snRNA	27
3L	66B5-8	tRNA-Leu-2	11/13/14/17
3L	67B	tRNA	11
3L	69F-70A	tRNA-Asp-2	8/11
3L	70BC	tRNA-Val-4	1/15/39
3L	70DE	tRNA	11
3L	70F1-2	tRNA-Met-3	15/39
3L	72F-73A	tRNA-Met-2	11/15/37/39
3L	79F	tRNA-Leu-2	11/17
3R	82E	U-1	2/32
3R	83AB	tRNA-Lys-5, tRNA-Arg	18/28
3R	83D	snRNA	27
3R	83F-84A	tRNA-Met-2	11/14/15/39
3R	84AB	tRNA-Lys-5	15/18
3R	84C	tRNA-Gly-GGA	17
3R	84D3-4	tRNA-Val-3b	10/11/15/24/ 26/39
3R	84EF	tRNA-Arg-2, tRNA-Asn-5	8/11/15/35/39
3R	85A	tRNA-Tyr-1	9/14/17/38
3R	85C	tRNA-Arg	18/28
3R	86A	tRNA-Ser-2b	11/14/17
3R	87BC	tRNA-Lys-5, tRNA-Thr-3	11/14/15/17
3R	88AB	tRNA-Ser-2b	11/14/17
3R	89BC	tRNA-Val-4, tRNA-Phe	1/11/39
3R	90BC	tRNA-Val-3b, tRNA-Pro tRNA-Thr	6/10/18/24/ 26/39
3R	90C	tRNA-Val-4, tRNA-Ala	1/11/35
3R	90DE	tRNA-Gly-GGA	11/17
3R	92B1-9	tRNA-Val-3b	10/11/15/39
3R	93A1-2	tRNA-Thr-4	11/14/17
3R	94A6-8	tRNA-Ser-2b	14/17
3R	95BC, 95C	snRNA, U-1	2/27/32

Chromosome	Labeled Region	Gene(s)	References
3R	95EF	snRNA	27
3R	95F-96A, 96A	tRNA-Asp, U-6	4/11/32
3R	97CD	tRNA	11
3R	99EF	tRNA	11

References:

1. Addison et al. 1982, J. Biol. Chem. 257, 670-673.
2. Alonso et al. 1984, J. Mol. Biol. 180, 825-836.
3. Cribbs et al. 1987, J. Mol. Biol. 197, 397-404.
4. Das et al. 1986, Nucl. Acids Res. 14, 7816-7816.
5. De Lotto et al. 1982, Molec. Gen. Genet. 188, 299-304.
6. DeLotto and Schedl, 1984, J. Mol. Biol. 179, 587-605.
7. Dingermann et al. 1982, J. Biol. Chem. 257, 14738-14744.
8. Dudler et al. 1980, Nucl. Acids. Res. 8, 2921-2938.
9. Dudler et al. 1981, Chomosoma 84, 49-60.
10. Dunn et al. 1979, J. Mol. Biol. 128, 277-287.
11. Elder et al. 1980, In: Soell-Schimmel-Abelson (eds.) "Transfer RNA", Vol. 2, pp. 317-323, CSH.
12. Gergen et al. 1981, J. Mol. Biol. 147, 475-499.
13. Glew et al. 1986, Gene 44, 307-314.
14. Hayashi et al. 1980a, Genetics 94, 42-43.
15. Hayashi et al. 1980b, Chromosoma 76, 65-84.
16. Hayashi et al. 1981, Chromosoma 82, 385-397.
17. Hayashi et al. 1982, Chromosoma 86, 279-292.
18. Hayashi, S., Tener, G., personal communication.
19. Hershey and Davidson 1980, Nucl. Acids Res. 8, 4899-4910.
20. Hosbach et al. 1980, Cell 21, 169-178.
21. Hovemann et al. 1980, Cell 19, 889-895.
22. Kubli and Schmidt 1978, Nucl. Acids Res. 5, 1465-1478.
23. Kubli et al. 1988, In: Tuite-Picard-Bolotin (eds.) "Genetics of Translation", NATO ASI Series H: Cell Biology, Vol. 14, 236-247
24. Leung et al. 1984, Gene 34, 207-217.
25. Meng et al. 1988, Nucl. Acids. Res 16, 7189.
26. Miller et al. 1981, Gene 15, 361-364.
27. Nag and Steffensen 1984, J. Cell. Biol. 99 (4 Part 2), 146A.
28. Newton et al. 1987, 12th International tRNA Workshop, Umea, Sweden.
29. Robinson and Davidson 1981, Cell 23, 251-259.
30. Saigo, K. 1986, Nucl. Acids Res. 14, 9526-9526.
31. Saluz, Schmidt, Alonso, Kubli, unpublished results.
32. Saluz et al. 1988, Nucl. Acids Res. 16, 3582.
33. Schedl and Donelson 1978, Biochim.Biophys.Acta 520, 539-554.
34. Schmidt et al. 1978, Molec. Gen. Genet. 164, 249-254.
35. Schmidt and Kubli 1980, Chromosoma 80, 277-282.
36. Sharp et al. 1981, Nucl. Acids Res. 9, 5867-5882.
37. Soell, D. personal communication.
38. Suter and Kubli 1988, Mol. Cell. Biol. 8, 3322-3331.
39. Tener et al. 1980, In: Soell-Schimmel-Abelson (eds.) "Transfer RNA", Vol. 2, pp. 295-307, CSH.
40. Yen et al. 1977, Cell 11, 763-777.
41. Yen and Davidson 1980, Cell 22, 137-148.
42. Zimm and Lindsley 1976, In: Fasman, G.D. (ed.) "Handbook of biochemistry and molecular biology", Vol.2, pp. 848-855, CRC.

CLONED GENES of Drosophila Melanogaster and Literature guide

Sept 1, 1989

John Merriam[1], Susan Adams[1], Geunbae Lee[2], David Krieger[3]

[1]Biology Department, [2]Computer Sciences Department,

[3]Graduate School of Library and Information Science

University of California, Los Angeles, CA 90024
FAX (213) 206-3987

The chromosomes are described in order by the locations of cloned genes and transformed inserts. Molecular summaries of the genes are appended when available. Transformed inserts are listed within [ ] symbols, followed by the markers used for selection. Preliminary information is included from the 30th U.S. and 11th European Drosophila Conferences.

Literature guide to cloned genes: under "key" a number designation describes the contents of the reference cited. 1 = the gene cloned; 2 = breakpoint identified in the clone region; 3 = transformed insert; 4 = chromosome walk; 5 = sequence determination; 6 = transcripts or transcript unit described; 7 = antibody raised and protein distribution described; 8 = promoter fusion with lacZ.

This publication is supported by grant R01 LM04896 from the National Library of Medicine. David Krieger is supported in part by a Graduate Research Fellowship of the National Science Foundation.

Locn	Genes-Name	Key	References
1A	telomeres	1	Cell34:85
1A	anonymous	1	Chrm89:206
1A	[P[rib7(1A)ry+/Sco;ry^506]];rDNA, ry+	3	GeneDev2:1745
1A	[WG1152];ry+	3	Bloomington Stock
1A	ARS element	5	DevB127:45
1A	ARS element	1	MGG197:342
1A	yellow	15	EMBOJ5:2657
1A	yellow	1	EMBOJ1:1185
1B	anonymous	1	Wolfner,Cornell
1B	scute	1	EMBOJ1:1185
1B	yellow	1	Eggleston,UWisc
1B	[10-2];neo-R, hsp 70, tropomyosin I	3	EMBOJ6:1375
1B	[27X-F];hsp28, ry+	3	MCB6:663
1B	[g71:1];sgs3+, ry+	3	Cell34:37
1B	P element insertions at yellow	6	NAR16:3039
1B	5'seq. upstream of yellow transcript	5	GeneDev1:996
1B	AS-C	56	Cell50:415
1B	yellow, achete	12	PNAS82:7369
1B	scute gamma: asense (ase) DNA binding	5	Modolell,Madrid
1B	elav	56	Science242:1570
1B	vnd (human APP)	156	PNAS86:2478
1B	anonymous	1	EMBOJ1:1185
1B	elav	6	DevB126:294
1B	su(s), m(1)Bld & flank. embryon. RNA	56	Genet122:625
1B	suppressor of sable su(s)	1	MCB6:1520

Locn	Genes-Name	Key	References
1BC	[26.1];hsp26, ry+	3	EMBOJ5:755
1C	[WG1131];ry	3	Bloomington Stock
1CD	[Bs2.71-2];chorion+, ry+	3	Cell34:47
1DE	su(wa)	16	MCB7:2498
1E	[Strain 23-2];ry+, D. teissieri 5S genes	3	Samson,CGM-CNRS
1EF	[23.3];hsp23, ry+	3	EMBOJ5:755
1F	[swallow-bb524];ry+, sww+	3	GeneDev2:1655
1F	[DA24-14];adh+,Ddc+	3	Cell34:37
1F	[AS410-R704.2, R702.1];ry+	3	Cell34:47
2A	[27P X/X-F];hsp28, ry+	3	MCB6:663
2A	[P[(w,ry)E]5];w, ry+	3	EMBOJ4:3489
2B	fs(1)Nas	4	Degelmann,Dusldrf
2B	[P(ry,hsp0-1)8];ry+, hsp	3	Laverty,UCBrk
2B	[Tf(1)Gr304-1];ry+	3	Meyerowitz,CalTec
2B	br, rbp, l(1)pp-1, early puff	46	Galceran,Madrid
2B	[g711:2];Sgs3+, ry+	3	Cell34:37
2B	armadillo (arm)	156	Weischaus,Prncton
2C	[AS369];ry	3	Bloomington Stock
2C	[GR792];w	3	Bloomington Stock
2C	[Pc[ry(delta0-1)]48];ry+	3	Laverty,UCBrk
2D	[WG1164];ry	3	Bloomington Stock
2D	polyhomeotic	46	Santamaria,Gifsur
2D	Pgd+	16	Lucchesi,ChaplHil
2DE	EGF like repeats	1	EMBOJ6:3431
2E	anonymous	1	EMBOJ1:1185
2E	fs(1)K10	56	GeneDev2:891
2EF	fs(1)pecanex,pcx1	1	LaBonne,CaseW
2F	Draf-1 essential gene	156	EMBOJ7:775
2F	Draf-1, proto oncogene	15	MCB7:2134
3A	[7-1;7-2];sgs4+, ry+	3	PNAS82:5055
3A	tko (prokaryotic rp S12 homology)	56	Cell51:165
3A	zeste	15	EMBOJ6:791
3A	zeste	56	MGG211:121
3B	anonymous	1	Cell28:165
3B	anonymous	1	MGG202:102
3B	[Tf(1)GR420-3];ry+	3	Meyerowitz,CalTec
3B	per	15	PNAS81:2142
3B	per, L(1)BA11	12	Cell38:701
3B	white	4	EMBOJ2:927
3BC	anonymous	1	Lengyel,UCLA
3C	3C-1 follicle cell specific expression	156	DevB124:441
3C	intracellular follicle cell, 2 genes	1	Waring,Marquette
3C	anonymous	1	BBA867:209
3C	distal to white	1	Goldberg,Cornell
3C	white -blood (Wbl)	15	EMBOJ5:3343
3C	white	1	Cell25:693
3C	irregular chiasmata C	1	Ramos,Freiburg
3C	Notch 5'regulatory region in interband	6	Cell54:461
3C	Notch	12	Cell38:701
3C	Notch	1	EMBOJ6:3431
3C	sequence of 8 Spl and Ax mutants	5	Cell51:538
3C	Notch, AxE2, spl missense mutants	56	EMBOJ6:3407
3C	Notch	15	MCB6:3094
3C	intermolt I RNA	1	Wolfner,Cornell
3C	notch-epidermal growth factors	1	Cell43:567
3C	Notch	4	PNAS80:1977
3C	Notch	12	PNAS81:2142
3C	Sgs-4	1	Korge,Berlin
3C	Sgs 4	1	Cell29:1041

Locn	Genes-Name	Key	References
3C	Sgs-4	12	NAR11:737
3D	intermolt I RNA	1	Wolfner,Cornell
3D	dunce (phosphodiesterase)	15	PNAS83:9313
3E	CA/GT Z DNA probe	1	EMBOJ6:1781
3F	[g71dx:2];Sgs3+, ry+	3	Cell34:37
3Z	anonymous	1	Wolfner,Cornell
4B	[cHB delta -59];hsp70-lacZ, ry+	3	Cell40:805
4B	[p[lac,ry+]A31];lac Z, ry+	3	Deve105:35
4B	mei-9	1	Mason,UCDavis
4B	norpA(Phospholipase C)	456	Cell54:723
4BC	anonymous	1	Cell28:165
4C	anonymous	1	Lengyel,UCLA
4C	unn	1	Jackson,Worcester
4C	[Strain 23-2];ry+, D. teissiere 5S genes	3	Samson,CGM-CNRS
4D	[R405.1, AS338, AS339];ry+	3	Cell34:47
4E	[AS757];ry	3	Bloomington Stock
4E	ovo	4	EMBOJ8:1549
4F	[GR875];w	3	Bloomington Stock
4F	anonymous	1	Wolfner,Cornell
4F	late IV RNA	1	Wolfner,Cornell
5AB	[P-l(2)gl-24E];lgl+,neoR	3	Cell50:215
5AB	anonymous	1	Wolfner,Cornell
5C	actin (cytoplasmic)	156	GeneDev1:1161
5C	actin	1	Cell19:365
5C	actin	1	GeneDev1:1161
5C	[HSAd010];hsp28A, 1.4 Kb adenovirus DNA	3	Science218:348
5D	[AS304];ry	3	Bloomington Stock
5D	ribosomal protein 7/8	1	MCB4:2643
5E	swallow	16	DevB124:1
5E	swallow (maternal)	156	GeneDev2:1655
5EF	anonymous	1	Wolfner,Cornell
5F	maternal restricted TU	16	Stephenson,Rochst
6AB	[P[(w,ry)H]4];w, ry+	3	EMBOJ4:3489
6C	[AS314];ry	3	Bloomington Stock
6E	l(1)ogre (optic ganglion reduced)	45	Kankel,Yale
6F	Sxl (sex lethal)	46	GeneDev3:708
6F	[S6.9-2];chorion+,lac+,ry+	3	Cell34:47
6F	cm mutants from unstable chromosome	1	Sheen,Minnesota
6F	Sxl; RNP; sex differential splicing	56	Cell55:1037
6F	Sex-lethal (sxl)	1	Cell43:521
7A	l(1)inR1	1	Cell43:521
7A	[C2.1];esc, white	3	EMBOJ4:979
7B	cut	1	PNAS80:1678
7B	cut H60X	56	Nature333:629
7B	cut	4	Cell42:869
7C	defective chorion-1 (dec-1)	16	GeneDev2:341
7C	[swallow-bb421];ry+,sww+	3	GeneDev2:1655
7C	ribosomal protein (Rps14)	156	MCB8:4314
7D	CA/GT Z DNA probe	1	EMBOJ6:1781
7D	anonymous	1	BBA867:209
7D	anonymous	1	Lengyel,UCLA
7D	[R403.1];ry+	3	Cell34:47
7D	l(1)mys(beta subunit integrin:PSantigen)	1	Cell56:401
7D	[ch41];sev+, neo+	3	EMBOJ(1989)
7D	fs(1)h (female sterile (1) homeotic)	56	DevB134:246
7D	l(1)mys (integrin B)	5	PNAS85:2633
7F	ovarian tumor (otu)	45	NAR17:3304
7F	neuroglian (nrg) (contactin)	1	Goodman,Berkeley
7F	otu	16	MCB8:1481

Locn	Genes-Name	Key	References
7Z	anonymous	1	Wolfner,Cornell
8A	orthodenticle (otd) Hbox	15	Amalric,Toulouse
8BC	[tAP-25,3.2];Adh+	3	Posakony,SanDiego
8D	lozenge	1	Cell21:729
8E	[cHB delta-59];hsp70-lacz,ry+	3	Cell40:805
8F	Yolk proteins 1 and 2	1	Cell21:729
8F	yp2 yolk protein	5	JMB164:481
9A	YP1	6	MCB8:4756
9A	[H(har1)M11];ry+	3	EMBOJ8:211
9A	yolk polypeptide genes Yp1 & Yp2	15	NAR15:67
9AD	[R404.2];ry+	3	Cell34:47
9B	[SB2.1-5];chorion+, ry+	3	Cell34:47
9B	[tAP-24B,3.2];Adh+	3	Posakony,SanDiego
9B	[unn];ry+	3	Chovnick,UConn
9C	[S6.9-9];chorion+,lacZ+, ry+	3	Cell34:47
9D	[P(ry,hsp0-1)22];ry+, hsp	3	Laverty,UCBrk
9D	Sgs7,Sgs3 hsp70-B galactosidase fusion	1	Chrm93:461
9D	[Bg9.61 (19Kb)];ry+, hsp70-lacZ	3	Chrm93:461
9E	[HSH32];hsp70-h, neo R	3	Cell51:405
9E	Sgs7,Sgs3 hsp70-B galactosidase fusion	1	Chrm93:461
9E	[R701.1, AS406];ry+	3	Cell34:47
9E	[cHB lambda -23];hsp70-lacZ, ry+	3	Cell40:805
9E	raspberry, gua 1-pur1- ras complex	12	Nash,UAlberta
9F	mit(1)21 cytokinenin defective	1	Gatti,Roma
10A	sevenless is tyrosine kinase (c-ros)	57	GeneDev2:620
10A	vermilion: tryptophan oxygenase	1	Searles,NIEHS
10A	[GR856];w	3	Bloomington Stock
10A	sevenless (tyrosine kinase)	56	Cell54:299
10A	sevenless(sev)	4	Cell49:281
10A	vermilion (tryp oxygenase)	15	MGG202:102
10B	dishevelled (dsh), hopscotch	16	Marsh,Irvine
10B	glutamine synthetase GSII	1	MGG204:208
10BC	[tAP-20,3.2];Adh+	3	Posakony,SanDiego
10C	[EcoR1-15kb];sev+, neo+	3	Hafen,Zurich
10C	RpII215	15	MCB6:3312
10C	RNA polymerase II largest subunit	1	Cell31:585
10C	near m-dy complex	1	Jackson,Worcester
10D	[AR4-038];w	3	Science229:558
10E	casein kinase II beta subunit	156	MCB7:3409
10EF	[sev+TK100];sev+, neo+	3	Hafen,Zurich
10EF	late V RNA	1	Wolfner,Cornell
10F	minor heat shock cDNA from Kc cells	1	MCB8:4727
11A	LSP-1 alpha	16	JMAppGen1:371
11A	LSP-1 alpha	1	JM&AG1:371
11A	LSP1 alpha	4	Cell23:441
11A	gastrulation defective	1	PNAS82:7369
11A	gastrulation defective	4	Mahowald,CaseWest
11A	Larval Serum Protein 1 (LSP1)-alpha	1	JMB189:1
11B	snRNAU1	1	NAR16:3582
11C	PRD gene5 (paired homology)	16	Cell47:735
11CD	[H(har1)G8-2];ry+	3	EMBOJ8:211
11Z	[1-31];Adh, ry+	3	DevB125:64
12	anonymous	1	BBA867:209
12	anonymous	1	Wolfner,Cornell
12A	[HSH33];hsp70-h, neo R	3	Cell51:405
12A	[tAP-17,4.8];Adh+	3	Posakony,SanDiego
12AD	Yp3 (yolk polypeptide)	15	NAR15:67
12B	[P[(w,ry)E]2];w, ry+	3	EMBOJ4:3489
12B	[WG1136];ry	3	Bloomington Stock

Locn	Genes-Name	Key	References
12B	mus(1)101, garnet, l(1)d.dg-4	4	Axton,Imp.College
12BC	[BL930];w,ry+	3	Bloomington Stock
12BC	yolk protein 3	1	Cell21:729
12BC	[AR4-032(X)];w	3	Science229:558
12BC	[SRS3.9-1];chorion+, ry+	3	Cell34:47
12C	[AS783];ry	3	Bloomington Stock
12D	[R301.2];ry+	3	Cell34:47
12DE	ser 7,4,4-7 tRNA, respectively	1	Chrm76:65
12E	[bcd+5];Adh+, bcd+	3	EMBOJ7:1749
12E	[E 7-10];Sgs-4	3	Korge,Berlin
12E	[HSAd006];hsp28A, 1.4 Kb adenovirus DNA	3	MCB6:4126
12E	[cHB delta-73];hsp70-lacZ, ry+	3	Cell40:805
12EF	[GR871];w,ry+	3	Bloomington Stock
12F	tRNA	1	Davidson,CalTech
12F	[26.2];hsp26, ry+	3	EMBOJ5:755
12F	tandem repeated 2L1 sequence	1	Genet107:611
13AC	[BS.27-5];Chorion+, ry+	3	Cell34:47
13CD	[SB2.1-6];chorion+, ry+	3	Cell34:47
13E	[18-4 YP1-Adh];ryt, Adh, YP1	3	DevGenet10:24
13EF	D-myb (proto-oncogene)	45	Cell41:449
13EF	D-myb	156	EMBOJ6:3085
13EF	[P[(w,ry)E]3g;P[(w,ry)G]4];w, ry+ (w-)	3	EMBOJ4:3489
13F	Gapdh-2	6	MCB8:5200
13F	G-protein beta subunit [transducin]	156	PNAS85:7134
13F	Gapdh-2 URS1,2	8	GeneDev2:743
13F	scalloped (sd)	1	Campbell,Storrs
13F	Glyceraldehyde-3-phosphate dehydrogenase	1	JBC260:4345
13F	[26.3];hsp26, ry+	3	EMBOJ5:755
14A	[21-1];Sgs4+, ry+	3	PNAS82:5055
14B	snRNAU5	1	NAR16:3582
14B	800kb has 85 SARs	4	Miassod,Marseille
14BC	anonymous	1	Wolfner,Cornell
14C	(Bj6) puff binding protein	16	Saumweber,Cologne
14C	mei-41	4	Boyd,UCDavis
14CD	Na+ channel (para)	15	PNAS86:2079
14D	anonymous	1	Lengyel,UCLA
15A	PS2 alpha subunit of integrin	156	Cell51:929
15AB	head specific RNA	1	DevB94:451
15B	ribosomal protein S18	1	MCB4:2643
15DE	[BS2.7-10];chorion+, ry+	3	Cell34:47
15F	forked	4	MCB6:1
15F	forked	1	Cell41:429
15F	[63-19B[Ddc(-208,-59)]];Adh+, Ddc+	3	GeneDev1:510
16B	anonymous	1	Cell21:729
16BC	[S6.9-11];ch, lacZ, ry+	3	JMB187:33
16C	[H1];ry+	3	EMBOJ5:583
16D	[unn];ry+	3	Chovnick,UConn
16E	[27 N/P-A];hsp28, ry+	3	MCB6:663
16EF	shaker	4	Cabrera,Madrid
16F	Shaker cDNAs	56	EMBOJ7:1087
16F	[27C X/X-A];hsp28, ry+	3	MCB6:663
16F	Shaker (K+ channel)	5	Science237:770
16F	anonymous	1	Wolfner,Cornell
17A	porcupine (porc)	1	Perrimon,Harvard
17A	mesoderm specific	16	Gene74:457
17A	[unn];	3	PNAS83:701
17AB	anonymous	1	Davidson,CalTech
17B	[AS316];ry	3	Bloomington Stock
17B	[23.3];hsp23, ry+	3	EMBOJ5:755

Locn	Genes-Name	Key	References
17C	fused is ser thr protein kinase	5	Isnard-Lamour,Gif
17C	[2-41];Adh, ry+	3	DevB125:64
17C	[tAP-5];Adh+	3	Cell34:59
17CE	fused at 17CD	46	MCB7:3244
17CE	fused	4	Kalfayan,ChaplHil
17DE	[B1-2];w+	3	Cell36:469
17E	homology w/ DCO kinase catalytic subunit	156	GeneDev2:1539
17E	homol.to GAP-43,neuronal growth protein	156	Devel05:629
17EF	EGF like repeats	1	EMBOJ6:3431
18A	[H(har1)M13];ry+	3	EMBOJ8:211
18A	[hsneo;Rh1(-2800/+67)-cat];neo, CAT	3	Genet116:565
18A	PRD gene 10 (paired homology)	16	Cell47:735
18A	[R704.3];ry+	3	Cell34:47
18CD	maternal restricted TU	16	Stephenson,Rochst
18D	G6PD	1	Gene35:91
18D	anonymous	1	Belote,UCSD
18D	[16-3];sgs4+, ry+	3	PNAS82:5055
18D	[BS2.7-3];chorion+, ry+	3	Cell34:47
18E	G6PD	1	JJG60:455
19A	[WG1168];ry	3	Bloomington Stock
19A	[bcd+8];Adh+, bcd+	3	EMBOJ7:1749
19A	[cHB delta-89];	3	Cell40:805
19B	CA/GT Z DNA probe	1	EMBOJ6:1781
19C	[WG1085];ry	3	Bloomington Stock
19E	[GR817];w	3	Bloomington Stock
19E	[+65];	3	Cell40:805
19E	[tAP-1];Adh+	3	Cell34:59
19E	unc, en-het junction	4	MGG213:63
19E	unc	1	Chrm89:218
19EF	collagen-like gene	1	DevB104:187
19F	small optic lobes(sol), sluggish A(slgA)	16	Barleben,Freiburg
19F	Arg rTNA locus	1	Chrm76:65
19F	[P[(w,ry)E]1];w, ry+	3	EMBOJ4:3489
19F	anonymous 19F1,2	4	MGG213:63
19F	anonymous 19F6	4	MGG213:63
20A	[2];Act88F+	3	Cell40:101
20A	[Adh hs20A];adh+, hsp70	3	Cell51:763
20AB	collagen-like gene	1	DevB104:187
20BC	[23.1];hsp23, ry+	3	EMBOJ5:755
20C	ribosomal gene (rib7)	1	GeneDev2:1745
20C	[GR870];w,ry	3	Bloomington Stock
20CD	[P(ry,HsAFP)];ry+, HsAFP	3	Walker,Ontario
20D	[AR4-024, BL928];w,ry+	3	Science229:558
20F	suppressor of forked	1	Fitch,Iowa State
20F	suppressor of forked (su-f)	1	O'Hare,London
20G	[651-A[Ddc(-208,-106)]];Adh+, Ddc+	3	GeneDev1:510
21A	lethal(2)giant larvae is tumor suppressr	4	EMBOJ4:1551
21A	l(2)gl lethal(2) giant larvae	156	Cell50:215
21A	ARS element	1	MGG197:342
21A	telomeres	1	Cell34:85
21A	lethal(2) giant larvae	4	EMBOJ4:1551
21A	telomere	1	Chrm89:206
21A	l(2)gl	1	MGG204:58
21B	[AS52(fs)];neo	3	Bloomington Stock
21B	[AS760(l)];ry	3	Bloomington Stock
21B	anonymous	1	Wolfner,Cornell
21B	[P[(w,ry)G]1];w, ry (w-)	3	EMBOJ4:3489
21BC	[63-39A[Ddc(-208,-59)]];Adh+,Ddc+	3	GeneDev1:510
21C	rp21C	15	NAR15:10064

Locn	Genes-Name	Key	References
21C	double sex cognate	1	Baker,Stanford
21C	[AS744(fs)];ry	3	Bloomington Stock
21C	[AS739(fs)];ry	3	Bloomington Stock
21CD	anonymous	4	Jones,Harvard
21D	cadherin transmembrane glycoprotein	1	Goodman,Berkeley
21D	cGMP dependent kinase	156	GeneDev2:1539
21D	LSP1 beta	1	Cell23:441
21D	LSP-1 beta	1	MGG199:357
21D	anonymous	1	Levis,Seattle
21D	U1 snRNA	15	NAR12:8835
21D	[P[(w,ry)F]4-2];w, ry	3	EMBOJ4:3489
21D	[R602.1];ry+	3	Cell34:47
21D	Larval Serum Protein 1 (LSP1)-beta	15	JMB189:1
21DE	LSP-1 beta	16	JMAppGen1:371
21DE	[tAP-10,4.8];Adh+	3	Posakony,SanDiego
21E	snRNAU1	1	NAR16:3582
21E	anonymous	1	EMBOJ6:3431
21E	nina A:cyclosporin a binding protein	156	Nature338:67
21F	anonymous	1	Wolfner,Cornell
22A	[H(har1)M14];ry+	3	EMBOJ8:211
22A	anonymous	1	Wolfner,Cornell
22A	[R604.1];ry+	3	Cell34:47
22AB	[63-10C[Ddc(-208,-59)]];Adh+, Ddc+	3	GeneDev1:510
22AC	4S RNA	1	BBA520:539
22B	anonymous	1	Cell28:165
22B	[w20.2,w20.10];w+	3	EMBOJ3:2077
22BC	anonymous	1	Wolfner,Cornell
22F	tRNA^tyr-Y22Fa,b	1	Kubli,Zurich
22F	dpp has tgf homology	56	Nature325:81
22F	tyr tRNA	1	Gelbart,Harvard
22F	decapentaplegic complex	4	CSHQB50:119
23A	[+65];hsp70,ry+	3	Cell40:805
23A	anonymous	4	Hoffmann,Madison
23AB	[T13.5X];snake+	3	Nature323:688
23B	[AS56(fs)];neo	3	Bloomington Stock
23B	[GR840];ry	3	Bloomington Stock
23BC	maternally expressed cDNA clones	16	DevB124:1
23BC	maternal restricted TU	16	Stephenson,Rochst
23D	snRNAU5	1	NAR16:3582
23D	[TnJW2];adh+, dpp+	3	GeneDev1:615
23E	[P[rib7(23E)ry+/Sco;ry^506]];rDNA, rosy+	3	GeneDev2:1745
23E	ser 7 tRNA	1	Chrm76:65
23E	odd skipped	4	Coulter,Princeton
24A	cGMP dependent kinase	156	GeneDev2:1539
24AB	0-1 hrs.	1	Lengyel,UCLA
24AB	[P[(w,ry)D]4];w, ry+	3	EMBOJ4:3489
24BC	Shaw(Sh homology)	156	Science243:943
24C	[sev-Met2242.218];sev+, neo+	3	Cell54:299
24C	anonymous	1	Cell28:165
24CD	[AR4-24];w	3	Science229:558
24D	(fat)cadherin transmembrane glycoprotein	1	Goodman,Berkeley
24D	[lacZ;ry at sloppy paired];ry+	3	Bellen,Basel
24E	[H(har1)M3];	3	EMBOJ8:211
24E	anonymous	1	EMBOJ6:3431
25	[26.4];hsp26, ry+	3	EMBOJ5:755
25A	bsg25A	1	Lengyel,UCLA
25A	[6-1];sgs4+, ry+	3	PNAS82:5055
25B	[alpha T3.21];LSP1 alpha	3	Glover,London
25BC	[WG1110];ry	3	Bloomington Stock
25BC	[lacZ;ry at collagen];ry+	3	Bellen,Basel
25BC	[M8a];ry+, lacZ	3	Devel04:245
25BC	anonymous	1	Cell28:165
25C	DCg1 (basement membrane IV collagen)	15	EJBiochem165:587
25C	Dcg1 (basement membrane IV collagen)	6	Devel02:369
25C	collagen-like gene	1	DevB104:187
25C	[P[(w,ry)D]1];w, ry	3	EMBOJ4:3489
25C	[R401.3];ry+	3	Cell34:47
25CD	[2-3];Adh, ry+	3	DevB125:64
25D	[AS756(1)];ry	3	Bloomington Stock
25D	blastoderm specific	1	EMBOJ6:3431
25D	blastoderm specific gene(bsg)	1	NAR15:2309
25D	anonymous	1	MacIntyre,Cornell
25F	[H(har1)M1];ry+	3	EMBOJ8:211
25F	[H(har1)M8];ry+	3	EMBOJ8:211
25F	mst 323, testis-specific tu	16	DevB119:242
25F	[27P X/X-B];hsp28, ry+	3	MCB6:663
25F	glycerol-3-phosphate dehydrogenase	15	vonKalm,Syracuse
26A	mst355a,mst355b(male specif. transcript)	156	GeneDev2:1063
26A	vitelline membrane gene cluster	56	DevB127:248
26A	26A-1 follicle cell specific expression	156	DevB124:441
26A	[H(har1)M15];ry+	3	EMBOJ8:211
26A	egg shell proteins, 4 genes	1	Waring,Marquette
26A	GDPH: glycerol-3-phosphate dehydrogenase	1	JBC261:11751
26A	vitelline	1	EMBOJ4:147
26A	position effect variegation (PEV)	12	MacIntyre,Cornell
26A	[23.5];hsp26, ry+	3	EMBOJ5:755
26A	beta galactosidase	1	MacIntyre,Cornell
26AB	Kr homol. DNA finger protein coding gene	15	Cell47:1025
26AB	anonymous	1	Stephenson,Rochst
26B	acetylcholine receptor subunit	1	Wadsworth,Worcest
26B	[D1 and D4];rp49+, ry+	3	Nature317:555
26B	H2.0 homeobox visceral mesoderm	156	EMBOJ7:2151
26CF	[57-12A[Ddc(-208,-140)]];Adh+, Ddc+	3	GeneDev1:510
26F	[8.0:4 (period gene DNA)];Adh, ry+	3	JNeurogen3:249
27A	[2-2 Icarus-neo];hsp70	3	MCB6:1640
27AC	[icarus-neo];	3	Steller,MIT
27C	Pupal cuticle protein	1	Cell44:33
27C	Gar transformylase,synthetase,AIR synth.	1	PNAS83:720
27C	[33-4];sgs4+, ry+	3	PNAS82:5055
27C	gart	1	Nash,Alberta
27D	anonymous	1	Davidson,CalTech
27F	anonymous	1	Wolfner,Cornell
28	int oncogene homologue	1	Kiss,Szeged
28A	head specific TU	1	DevB94:451
28A	[BS2.7-11];chorion, lacZ, ry+	3	Cell34:47
28A	Int-1 homologues	156	PNAS85:3034
28A	wingless is int-1 homologue	156	Cell50:649
28A	wingless	56	Cell50:659
28A	nina C	156	Cell52:757
28A	wingless(wg) (segment polarity)	12	EMBOJ6:1765
28C	src homologue (Dsrc28C)	156	MCB7:2119
28C	tRNA^tyr-Y28C	15	MCB8:3322
28C	head specific TU	1	DevB94:451
28C	urate oxidase	1	Gene45:131
28C	anonymous	1	Davidson,CalTech
28D	laminin B2	15	NAR16:7205
28D	[S-Ica];hsp70, neomycin resistance	3	MCB6:1640
28D	[E 7-1];Sgs-4	3	Korge,Berlin

Locn	Genes-Name	Key	References
28D	CDNA, Kc cells	1	PNAS82:7369
28EF	anonymous	1	EMBOJ6:3431
29A	heterochromatic specific	1	Genet116:s4
29A	C1A9 antigen - nuclear protein	15	MCB6:3862
29A	lys 5 tRNA locus	1	Chrm76:65
29B	[R308.1];ry+	3	Cell34:47
29B	CDNA, Kc cells	1	PNAS82:7369
29C	SRC homologous	1	Nature302:837
29D	[hsDfd-58];ry+, Dfd hsp70	3	Cell55:477
30A	[28C-B (lost?)];ry+,hsp28	3	MCB6:663
30A	[cHB delta-89];hsp70-lacZ, ry+	3	Cell40:805
30A	P6	4	MGG199:357
30B	numb	1	Cell58:349
30B	P6	16	JMAppGen1:371
30B	[AS125(lethal)];neo	3	Bloomington Stock
30B	[AS765];ry	3	Bloomington Stock
30B	anonymous	1	Davidson,CalTech
30C	[AS156(lethal)];neo	3	Bloomington Stock
30C	[BL908];w,ry	3	Bloomington Stock
30C	[23.5];hsp23, ry+	3	EMBOJ5:755
30C	[AN22];w,ry+	3	Science229:558
30C	[S6.9-3];chorion+,lac+,ry+	3	Cell34:47
30C	DCO catalyt. subunit cAMP depend. kinase	156	GeneDev2:1539
30D	[H(har1)E14-2];ry+	3	EMBOJ8:211
30D	[H(har1)G9-1];ry+	3	EMBOJ8:211
30DE	anonymous	1	Wolfner,Cornell
30EF	anonymous	1	Davidson,CalTech
30F	[C20Pgd8.9-2];Pgd+, ry+	3	Lucchesi,ChaplHl
31	[63-10C[Ddc(-208,-59)]];Adh+, Ddc+	3	GeneDev1:510
31A	anonymous	1	Lengyel,UCLA
31A	anonymous	1	Cell28:165
31B	[AS738(fs)];ry	3	Bloomington Stock
31B	[cHB delta-89];hsp70-lacZ, ry+	3	Cell40:805
31BD	trunk and 31C	1	Tucker,Rice
31C	anonymous	1	Wolfner,Cornell
31E	daughterless (da)	156	GeneDev2:1666
31F	anonymous	1	Wolfner,Cornell
32AB	head specific TU	1	DevB94:451
32BC	[BS2.7];chorion+,ry+	3	Cell34:47
32CD	myogenic cell RNA	1	DevB90:272
32CD	[cp70 delta B];hsp70-lacZ,ry+	3	Cell40:805
32CF	oocyte-specific TU	46	Graziani,Napoli
32DEF	hup, wdl, dal, abo	4	DevGenet10:33
32E	vitelline membrane protein (VMP)	156	DevGenet10:33
32E	[yp1-Adh-32E];neoR, yp-1, Adh	3	MGG210:153
32EF	vitelline membrane	1	EMBOJ4:147
32F	[g6:5];Sgs3+,ry+	3	Cell40:349
33	anonymous	1	BBA867:209
33A	[HSH12];hsp70-h, neo R	3	Cell51:405
33A	[lacZ;ry at spalt];ry+	3	Bellen,Basel
33A	spalt (sal)	56	EMBOJ7:197
33AB	extra sex combs	4	Struhl,Columbia
33B	anonymous	1	EMBOJ6:3091
33B	0-1,2-5,3-5 TU	16	Lengyel,UCLA
33B	anonymous	1	Wolfner,Cornell
33B	[26.3];hsp26, ry+	3	EMBOJ5:755
33BE	paired (prd) with H box	46	Nature321:493
33C	paired (prd) with H box	56	Cell47:735
33CD	UV and heat shock response	16	MCB6:4767

Locn	Genes-Name	Key	References
33DE	[line 3];rp21, ry+	3	MGG213:354
34A	snRNAU5	1	NAR16:3582
34A	[GR801];w	3	Bloomington Stock
34AB	anonymous	1	Stephenson,Rochst
34B	[H(har1)E8-2];ry+	3	EMBOJ8:211
34C	vitelline, Oregon R ovaries, cDNA	1	EMBOJ4:147
34D	[H(har1)M18-1];ry+	3	EMBOJ8:211
34D	[H(har1)M5-1];ry+	3	EMBOJ8:211
34D	[g7:4];Sgs3+,ry+	3	Cell40:349
34E	[BL919, P(w)11P];w,ry	3	Bloomington Stock
34EF	[P1/23];per(period), ry+	3	Nature326:390
34EF	[g711:1];Sgs3+,ry+	3	Cell40:349
34F	head specific TU	16	DevB94:451
35A	[15-1;15-2];sgs4+, ry+	3	PNAS82:5055
35A	Adh, outspread, no ocelli	45	JMB186:689
35A	Sco, noc	12	Genet119:647
35AB	anonymous	1	DevB86:438
35B	Adh	1	DevB114:194
35B	[AS54(fs)];neo	3	Bloomington Stock
35B	[AS57(fs)];neo	3	Bloomington Stock
35B	Adh promoters&larv.,adult enhancer	6	Nature337:279
35B	Adh, noc, osp, l(2)br22	1	JMB186:689
35B	no-ocelli	1	Nature316:81
35B	tRNA-gly (5 genes)	5	NAR16:7189
35B	alcohol dehydrogenase	1	PNAS77:5794
35B	crinkled locus	1	Chrm91:54
35BC	vasa	156	Cell55:577
35C	[AS382];ry	3	Bloomington Stock
35C	anonymous	1	Wolfner,Cornell
35D	[WG1033];ry	3	Bloomington Stock
35D	snail (sna)	456	Nature330:395
35DE	[14.6:21 (period gene DNA)];Adh, ry+	3	JNeurogen3:249
35DE	[S11.4-1];chorion+	3	Cell34:47
35EF	snRNAU5	1	NAR16:3582
36A	homology w/ DCO kinase catalytic subunit	156	GeneDev2:1539
36A	[H(har1)M7];ry+	3	EMBOJ8:211
36A	[H2-8];ry+, sgs-4, lcp	3	EMBOJ6:2249
36A	[26Z-36A];hsp26, lacZ, ry+	3	EMBOJ5:747
36A	[tAP-3];Adh+	3	Cell34:59
36B	myosin heavy chain	16	Gene74:457
36B	muscle specific gene	156	Rozek,CaseWestern
36B	[AS769];ry	3	Bloomington Stock
36B	walked from myosin heavy chain	1	Emerson,UVirg
36B	myosin heavy chain	1	PNAS83:1393
36BC	[Q21a];ry+, lacZ	3	Deve104:245
36C	[AS754];ry	3	Bloomington Stock
36C	dorsal	1	Nature311:262
36C	[tAP-8C,4.8];Adh+	3	Posakony,SanDiego
36C	dorsal (c-rel)	56	Science238:692
36D	anonymous	4	MCB1:475
36E	[WG1100];ry	3	Bloomington Stock
36E	[lacZ;ry at fasciclin III];ry+	3	Bellen,Basel
36E	fasciclin III	1	Cell48:975
36F	anonymous	1	Cell28:165
36F	male specific lethal-1 (msl-1)	46	Lucchesi,ChaplHil
37A	[56-3A[Ddc(-208,insert)]];Adh+, Ddc+	3	GeneDev1:510
37A	[26Z-37A];hsp26, lacZ, ry+	3	EMBOJ5:747
37A	[P[(w,ry)G]2];w, ry (w-)	3	EMBOJ4:3489
37B	ovarian cDNA 109	1	EMBOJ4:147

Locn	Genes-Name	Key	References
37B	Ddc	12	MCB1:475
37BC	[unn];Adh+ Hsp82	3	Viglianti,Harvard
37C	Ddc distal enhancer region	5	GeneDev3:676
37C	dDC and 3' flanking gene	15	Nature322:279
37C	Cc	1	NAR14:6169
37C	Ddc	15	EMBOJ5:3335
37C	Ddc (dopa decarboxylase)	15	EMBOJ5:2663
37C	Ddc and amd	15	Genet114:469
37C	amd (alpha methyldopa hypersensitive)	12	Genet114:453
37C	Cs gene	56	MGG209:290
37D	[H1-10];ry+, sgs-4, lcp	3	EMBOJ6:2249
37D	SD tandem duplication	46	Ganetzky,Madison
37E	Ref(2)P	1	Contamine,Gif sur
38A	[AS58(fs)];neo	3	Bloomington Stock
38A	[GR883];ry,w	3	Bloomington Stock
38A	E86 (optic lobe Hbox)	15	Jones,Yale
38A	anonymous	1	MCB1:475
38AB	[P6a];ry+, lacZ	3	Deve104:245
38B	[P(ry, HsAFP)2];ry+, HsAFP	3	Walker,Ontario
38B	[cHB delta-59];hHsp70-lacZ, ry+	3	Cell40:805
38BC	[tAP-19,4.8];Adh+	3	Posakony,SanDiego
38CD	[H(har1)15-1];ry+	3	EMBOJ8:211
38D	[C20Pgd7.4-D];Pgd+, ry+	3	Lucchesi,ChaplHil
38D	[1];Act88F	3	Cell40:101
38E	caudal (cad)	15	EMBOJ4:2961
38E	caudal(cad)	456	Cell48:465
38E	[unn];ry+	3	Chovnick,UConn
38E	caudal	1	Nature324:537
38EF	caudal (S67)	15	PNAS83:4809
39	[AS443(mf)];ry,Adh	3	Bloomington Stock
39B	snRNAU4,U5	1	NAR16:3582
39B	[28P-C (lost?)];ry+ hsp28	3	MCB6:663
39BC	[E 5-5];Sgs-4	3	Korge,Berlin
39BC	[S6.9-8];chorion+,lac+,ry+	3	Cell34:47
39C	[AS55(fs)];neo	3	Bloomington Stock
39CD	[H(har1)G8-1];ry+	3	EMBOJ8:211
39D	anonymous	1	BBA867:209
39D	H1	15	NAR14:5563
39DE	Histone H3	5	Genet122:87
39DE	Histone H1, H2A, H2B, H3, H4	5	NAR17:225
39DE	histone	1	CSHQB42:1047
39DE	[H5];ry+	3	EMBOJ5:583
39E	chromocenter	1	Wolfner,Cornell
39E	[B4];rp49+, ry+	3	Nature317:555
39EF	[AR4-2];w+,ry+	3	Cell36:469
39F	[hsp-per:9];hsp, per	3	Ewer,Brandeis
39F	[GR832];w,ry	3	Bloomington Stock
39F	anonymous	1	Wolfner,Cornell
40	chromocenter	1	Wolfner,Cornell
40A	teashirt (tsh) zn fingers	16	Kerridge,Marseill
40A	[AS53(fs)];neo	3	Bloomington Stock
40A	[BL909, BL924, GR805];w,ry+	3	Bloomington Stock
40A	[26Z-84D, 278-40A];hsp26, lacZ, ry+	3	EMBOJ5:747
40A	[R 30 b polo];ry-+, P-beta gal	3	Kerridge,Marseill
40AB	snRNAU4	1	NAR16:3582
41	Responder (Rsp) to SD	15	Cell54:179
41	[AS770];ry	3	Bloomington Stock
41	chromocenter	1	Wolfner,Cornell
41A	[BL915, BL921, GR819];w,ry+	3	Bloomington Stock

Locn	Genes-Name	Key	References
42A	[pPA-1 abl];abl, Adh	3	Cell51:821
42A	cluster of asn, arg, lys, ile tRNAs	4	Cell22:137
42A	anonymous	1	Cell28:165
42A	[+411];hsp70, ry+	3	Cell40:805
42A	[R301.1];ry+	3	Cell34:47
42A	[tAP-13,4.8];Adh+	3	Posakony,SanDiego
42AB	[R303.1];ry+	3	Cell34:47
42BC	anonymous	1	JCB102:2076
42CD	[14.6:63 (period gene DNA)];Adh, ry+	3	JNeurogen3:249
42DE	[unn];ry+	3	Chovnick,UConn
42E	[AS323];ry	3	Bloomington Stock
42E	tRNA-lys-2	1	Cell34:47
42EF	anonymous	1	Wolfner,Cornell
42F	[28X-C (lost?)];	3	MCB6:663
42F	[S6.9-4];chorion+,lac+,ry+	3	Cell34:47
43	[R704.1];ry+	3	Cell34:47
43A	anonymous	1	Davidson,CalTech
43A	prickle-1, spiny leg-1, Draf-2	4	Gubb,Cambridge
43A	Draf-2 proto oncogene	1	MCB7:2134
43AB	head specific TU	16	DevB94:451
43BC	maternal restricted TU	16	Stephenson,Rochst
43C	[R304.1];ry+	3	Cell34:47
43CD	[yp1-Adh-43CD];neoR, yp-1, Adh	3	MGG210:153
43CD	[+411];hsp70, ry+	3	Cell40:805
43D	torso (receptor tyrosine kinase)	156	Nature338:478
43DE	maternal restricted TU	16	Stephenson,Rochst
43E	Gapdh-1	6	MCB8:5200
43E	Glyceraldehyde-3-phosphate dehydrogenase	1	JBC260:4345
43E	[S6.9-7, AS317];ch, lacZ, ry+	3	JMB187:33
43E	[g7:7];Sgs3+,ry+	3	Cell40:349
43E	[p[lac, ry+]A2];lac Z, ry+	3	Deve105:35
43E	Glyceraldehyde 3 Phosphate Dehydrogenase	46	Warren,Canberra
44A	[act-line II];actin, beta-gal, ry+	3	DevB133:313
44C	anonymous	1	Chrm89:206
44C	[S6.9-7];ch, lacZ, ry	3	JMB187:33
44C	patched (ptc) has transmembrane domain	15	Ingham,Oxford
44CD	patched has 8-14 transmembr.segs.	156	Hooper,Boulder
44CD	LCP	4	Hooper,Boulder
44CD	[H(har1)G13-1];ry+	3	EMBOJ8:211
44CD	[H(har1)G21-1];ry+	3	EMBOJ8:211
44CD	head specific TU	16	Cell34:47
44D	[H(har1)M9];ry+	3	EMBOJ8:211
44D	larval cuticle protein	1	Cell25:165
44D	anonymous	1	Chrm89:206
44D	LCP1-4	1	AIBP(1986)119
44DE	Jonah to patched walk	456	Hooper,Boulder
44E	[R3.9-4, AS309];chorion+,ry+	3	Cell34:47
44EF	anonymous	1	Lengyel,UCLA
44F	[cp70ZT];hsp70-lacZ,ry+	3	Cell40:805
45A	anonymous	1	Cell28:165
45A	[S11.4-1, AS303];chorion+,ry+	3	Cell34:47
45AB	[F4];rp49+,ry+	3	Nature317:555
45B	0-3.5 hrs	1	Lengyel,UCLA
45C	homology w/ DCO kinase catlaytic subunit	156	GeneDev2:1539
45CD	[swallow-bb635];sww+	3	GeneDev2:1655
45D	[AS777];ry	3	Bloomington Stock
45D	[BL924];w,ry	3	Bloomington Stock
45D	[GR791];ry	3	Bloomington Stock
45D	anonymous	1	Cell28:165

Locn	Genes-Name	Key	References
45DE	[WG974];ry+	3	Bloomington Stock
45E	[DR-18];ry+,Ddc+	3	Cell34:37
45E	[cHB delta-73];hsp70-lacZ,ry+	3	Cell40:805
46B	maternal restricted TU	16	Stephenson,Rochst
46BC	[1-29];Adh, ry+	3	DevB125:64
46C	(eve) mutants	15	GeneDev2:1824
46C	eve reg.sequences (gap and pair rule)	8	Cell57:413
46C	[BL917, GR846];w,ry+	3	Bloomington Stock
46C	[TnJW3];adh+, dpp+	3	GeneDev1:615
46C	even-skipped (eve) with H box	156	Cell47:721
46C	eve H box	1	Science233:953
46C	[27X/A-2-A];hsp28,ry+	3	MCB6:663
46CD	[bb275];ry+, sww+	3	GeneDev2:1655
46D	[56-7A[Ddc(-208,insert)]];Adh+, Ddc+	3	GeneDev1:510
46DF	myogenic cell TU	16	DevB90:272
46E	[ch21];sev+, neo+	3	EMBOJ(1989)
46E	head specific TU	1	DevB94:451
47A	Goa-like protein	156	de Sousa,Seattle
47A	mst 325, testis specific tu	16	DevB119:242
47A	[A1-1];w+, ry+	3	DevB119:242
47A	[tAP-18,4.8];Adh+	3	Posakony,SanDiego
47AB	[P-1(2)gl-21E];lgl+,neoR	3	Cell50:215
47AC	[H(har1)E5-1];ry+	3	EMBOJ8:211
47C	[cHB delta-89];hsp70-lacZ,ry+	3	Cell40:805
47D	[GR806];w	3	Bloomington Stock
47D	[27 N/P-B];hsp28, ry+	3	MCB6:663
47D	[P[(w,ry)E]8];w, ry+	3	EMBOJ4:3489
47E	head specific TU	16	DevB94:451
47EF	[line 4];rp21, ry+	3	MGG213:354
47EF	[P1/14];per(period), ry+	3	Nature326:390
47F	anonymous	1	Cell40:37
47F	myogenic cell TU	16	DevB90:272
47F	[yp1-Adh-47F/48A];neoR, yp-1, Adh	3	MGG210:153
48A	invected (engrailed complex)	45	GeneDev1:19
48AB	engrailed (en)	5	Cell40:37
48AB	engrailed, invected	6	EMBOJ6:2803
48AB	[g7:3];Sgs3+,ry+	3	Cell40:349
48B	Met 2 tRNA	1	Chrm76:65
48B	[26Z-48B];hsp26, lacZ, ry+	3	EMBOJ5:747
48B	[tAP-6];Adh+	3	Cell34:59
48C	[AS408];ry	3	Bloomington Stock
48C	[AS59(fs)];neo	3	Bloomington Stock
48C	anonymous	1	Wolfner,Cornell
48D	F1, female enriched RNA (all stages)	156	PNAS82:5795
48D	[DR-9];ry+,Ddc+	3	Cell34:37
48E	anonymous	1	Wolfner,Cornell
48E	Deb-A, Deb-B, Deb-C	456	BBA867:209
48EF	[tAP-4];Adh+	3	Cell34:59
48F	his tRNA genes	1	NAR8:2921
48F	head specific TU	16	DevB94:451
48F	[13-2];neo-R, hsp70, tropomyosin I	3	EMBOJ6:1375
49A	calmodulin: additional 5' exon	6	Beckingham,Rice
49A	calmodulin	1	GeneDev1:1161
49A	Calmodulin protein	1	NAR15:3335
49B	[26Z-49B];hsp26, lacZ, ry+	3	EMBOJ5:747
49B	proximal breakpoint of aristapedioid	1	Adler,UVirg
49C	[BL915, P(w)24S];w	3	Bloomington Stock
49C	anonymous	1	Cell28:165
49CD	anonymous	1	Cell28:165

Locn	Genes-Name	Key	References
49CDE	[23.4];hsp23, ry+	3	EMBOJ5:755
49D	Bicaudal D (lethal allele vr22-P3)	1	Beckingham,Rice
49D	[-51];hsp70,ry+	3	Cell40:805
49D	[A1];rp49+,ry+	3	Nature317:555
49D	[AR4-042];w	3	Science229:558
49D	[S38M-5];s38+,M13+,ry+	3	Cell34:47
49DE	anonymous	1	Wolfner,Cornell
49DEF	vg (vestigial insert mutants)	5	MCB8:1489
49DEF	vestigial (vg)	12	EMBOJ7:1355
49E	Posterior Sex Comb- Suppressor 2 zest	456	Adler,UVirg
49EF	[S3.8-6];chorion+,ry+	3	Cell34:47
49F	muscle specific Troponin C(ca++)	1	JMB156:449
49F	[tAP-9,4.8];Adh+	3	Posakony,SanDiego
50A	[P[(w,ry)F]1];w, ry	3	EMBOJ4:3489
50AB	anonymous	1	EMBOJ6:3091
50AB	[H(har1)E5-1];ry+	3	EMBOJ8:211
50AB	cluster of leu and ile tRNAs	1	Cell23:251
50B	anonymous	1	DevB94:451
50B	[R306.1, S6.9-6];ry+,lac+,chorion+	3	Cell34:47
50BC	cluster of lys tRNAs	1	JMB147:475
50C	{GR794];ry	3	Bloomington Stock
50C	mastermind (mam)	1	EMBOJ6:3431
50C	anonymous	4	BBA867:209
50C	anonymous	1	Cell28:165
50C	[13C-1 hsp prom-P (Icarus)];hsp70	3	MCB6:1640
50CD	anonymous	1	Wolfner,Cornell
50D	[WG1038];ry	3	Bloomington Stock
50F	double sex cognate	1	Baker,Stanford
51A	anonymous	1	BBA867:209
51B	tra2 is RNA binding protein	156	Cell55:1025
51B	head specific TU	16	DevB94:451
51C	[H1-8];ry+, sgs-4, lcp	3	EMBOJ6:2249
51CD	maternal restricted TU	16	Stephenson,Rochst
51D	[AS785];ry	3	Bloomington Stock
51D	anonymous	1	Wolfner,Cornell
51DE	anonymous	1	Cell28:165
51E	maternal specific RNAs	16	Lengyel,UCLA
51F	paragonia (male) specific Tu	1	Schafer,Duseldorf
52A	[AS784];ry	3	Bloomington Stock
52A	[GR800];w	3	Bloomington Stock
52A	[DR-12];ry+,Ddc+	3	Cell34:37
52A	[tYPb];ry+, Yp1,Yp2	3	PNAS82:7000
52A	anonymous	1	Belote,Syracuse
52B	anonymous	1	Cell28:165
52B	[cHB delta-89];hsp70-lacZ,ry+	3	Cell40:805
52B	[tAP-21,3.2];Adh+	3	Posakony,SanDiego
52BC	[P(w)5P];w	3	Levis,Seattle
52C	[BL921];w,ry	3	Bloomington Stock
52CD	[60-15A[Ddc(-208,-83)]];Adh+, Ddc+	3	GeneDev1:510
52CD	[H3];ry+	3	EMBOJ5:583
52D	slit has EGF-like homology	156	Cell55:1047
52D	flight muscle TU, alpha GPO	16	Genet116:249
52D	not double sex cognate	1	Baker,Stanford
52D	[SRS3.9-1];chorion+,ry+	3	Baker,Stanford
52D	07, 0-6	1	Lengyel,UCLA
52DF	anonymous	1	Wolfner,Cornell
52E	2 epidermal segment genes	1	EMBOJ6:3431
52F	[AS333];ry	3	Bloomington Stock
52F	[S6.9-10];chorion+,lac+,ry+	3	Cell34:47

Locn	Genes-Name	Key	References
53A	kinesin heavy chain	7	Nature338:355
53A	kinesin heavy chain	5	Cell56:879
53A	kinesin heavy chain	1	PNAS85:1864
53A	[S38M-6];s38+,M13+,ry+	3	Cell34:47
53BC	[swallow-bb344];ry+,sww+	3	GeneDev2:1655
53BC	[E 7-3];Sgs-4	3	Korge,Berlin
53C	homology w/ DCO kinase catlaytic subunit	156	GeneDev2:1539
53C	[WG947];ry	3	Bloomington Stock
53CD	amy pseudogene	1	Genet110:313
53CD	ribosomal protein A1	15	NAR15:987
53DE	[23.5];hsp23, ry+	3	EMBOJ5:755
53E	[AS771];ry	3	Bloomington Stock
53E	protein kinase C (PKC)	15	EMBOJ6:433
53E	[R3,9-1];chorion+,ry+	3	Cell34:47
53E	protein kinase C eye (dPKC 53E eye)	156	Cell57:403
53EF	[SB2.1-1];chorion+,ry+	3	Cell34:47
53F	anon.	1	Davidson,CalTech
53F	[278-53F];hsp26, lacZ, ry+	3	EMBOJ5:747
53F	[g4:1];Sgs3+,ry+	3	Cell40:349
54A	[AS767];ry	3	Bloomington Stock
54A	[2];Act88F+	3	Cell40:101
54A	amylase duplication	1	Genet110:313
54AB	[P1/30];per(period), ry+	3	Nature326:390
54B	[HSH21];hsp70-h, neo R	3	Cell51:405
54B	[K10;p(cos9)];K10+, crn+, pcx+, kz+	3	Cell40:827
54B	alpha amylase upstream neg regulat. site	5	NAR15:7184
54C	anonymous	1	BBA867:209
54C	[Adh,hs54c];Adh+,hsp70	3	Cell51:763
54D	PRD gene 11 (paired homology)	16	Cell47:735
54E	minor heat shock cDNA	1	MCB8:4727
54F	poly A binding protein (PABP)	16	Amalric,Toulouse
54F	anonymous	1	Wolfner,Cornell
54F	anonymous	1	Stephenson,Rochst
54Z	[23.2];hsp23, ry+	3	EMBOJ5:755
55A	[AS742(fs)];ry	3	Bloomington Stock
55A	staufen	16	Nusslein-Volhard
55BC	anonymous	1	EMBOJ5:583
55E	maternally expressed cDNA clones	16	DevB124:1
55F	[AS745(fs)];ry	3	Bloomington Stock
55F	maternal restricted TU	1	Cell34:47
56AB	[H2];ry+	3	EMBOJ5:583
56C	beta tubulin	1	DevB104:187
56D	[R3.9-6];chorion+,ry+	3	Cell34:47
56D	alpha tubulin	1	JMB156:449
56E	smooth (sm)	1	zur Lage,Edinbur
56E	[H(har1)G18-2];	3	EMBOJ8:211
56E	[H(har1)M22];ry+	3	EMBOJ8:211
56E	[P-1(2)gl-21C];lgl+,neoR	3	Cell50:215
56EF	anonymous	1	Wolfner,Cornell
56F	5S RNA	5	MCB8:1266
56F	tRNA Gly	1	Davidson,CalTech
56F	5S RNA	1	MGG188:299
56F	[DR-15, DR-5];ry+,Ddc+	3	Cell34:37
56F	[Q266];ry+, lacZ	3	Devel04:245
57	[HB4/Sc1];LSP1 beta	3	Glover,London
57A	[p[9-4A Ddc5]];Adh+, Ddc+	3	GeneDev1:510
57A	[alpha T3.21];LSP1 alpha	3	Glover,London
57AB	[27P X/X-A;27P X/X-E];hsp28, ry+	3	MCB6:663
57AC	actin 57A	16	Gene74:457

Locn	Genes-Name	Key	References
57B	[K59a];ry+, lacZ	3	Devel04:245
57B	actin	1	GeneDev1:1161
57B	(w26) homeo box protein	16	Gehring,Basel
57B	[27S-B];hsp28, ry+	3	MCB6:663
57B	[P[(w,ry)E]6;P[(w,ry)F]4-1];w, ry+	3	EMBOJ4:3489
57B	exuperantia (exu)	46	Hazelrigg,Utah
57B	H box homol.	1	McGinnis,Yale
57C	head specific TU	1	DevB94:451
57C	punch	4	O'Donnell,CarnMel
57C	tudor	1	Boswell,Boulder
57D	paragonia (male) specific Tu	16	Schafer,Duseldorf
57E	[GR852];w	3	Bloomington Stock
57F	EGF receptor mutants	126	Cell56:1085
57F	EGF receptor homolog (DER)	15	Cell46:1091
57F	c-erbB, EGF receptor protein	1	Cell40:599
57F	[unn];ry+	3	Chovnick,UConn
58A	[26.4];hsp26, ry+	3	EMBOJ5:755
58AB	ets-2 (erythroblastosis virus homolog)	156	DevB127:45
58AD	[R8a];ry+, lacZ	3	Devel04:245
58C	en-like homeobox	1	Gustafson,UCSF
58D	[BL909];w,ry	3	Bloomington Stock
58D	[27 X/A-1-B];hsp 28, ry+	3	MCB6:663
58EF	[R3.9-5];chorion+,ry+	3	Cell34:47
58F	anonymous	1	Wolfner,Cornell
58F	[cp70 delta B];hsp70-lacZ,ry+	3	Cell40:805
58F	[tAP-7A,4.8];Adh+	3	Posakony,SanDiego
59A	cyclin B	16	Nature338:337
59B	[A3-1];w+,ry+	3	Cell34:47
59C	twist	16	Thisse,Strasbourg
59C	[tAP-7A,4.8];Adh+	3	Posakony,SanDiego
59D	[27C X/X-B];hsp28, ry+	3	MCB6:663
59E	[S11.4-11];chorion+	3	Cell34:47
59EF	[DP1241];ry	3	Bloomington Stock
59EF	[WG1000];ry	3	Bloomington Stock
59E	anonymous	1	Belote,UCSD
60A	G protein (stimulatory, alpha:Gsa)	156	PNAS86:4321
60A	mesoderm specific	16	Gene74:457
60A	maternal restricted TU	1	Stephenson,Rochst
60A	anonymous	1	Wolfner,Cornell
60A	[R302.1];ry+	3	Cell34:47
60AB	[g1];Sgs3+,ry+	3	Cell40:349
60B	beta 3 tubulin locus	1	Chrm89:206
60B	[BS2.7-4];chorion+,ry+	3	Cell34:47
60BC	[H(har1)E5-4];ry+	3	EMBOJ8:211
60BC	maternal restricted TU	1	Stephenson,Rochst
60BC	[27P X/X-B];hsp28, ry+	3	MCB6:663
60C	[AS740, AS741(fs)];ry+	3	Bloomington Stock
60C	Na channel?	1	Science237:744
60C	beta tubulin locus	1	DevB104:187
60C	[C2];rp49+,ry+	3	Nature317:555
60C	[unn];ry+	3	Chovnick,UConn
60C	beta tubulin locus	1	JMB156:449
60C	[GR807, GR836];w	3	Bloomington Stock
60D	[GR884];w	3	Bloomington Stock
60DE	voltage sensitive Na channel	5	NAR15:8569
60DE	putative voltage sensitive Na channel	16	Science237:744
60E	DSC-Na+ channel	15	PNAS86:2079
60E	[S38Z-1];s38,lacZ,ry	3	Cell34:47
60E	[tAP-15A,4.8];Adh+	3	Posakony,SanDiego

Locn	Genes-Name	Key	References
60E	Distalless (Dll),previously Brista	456	Nature338:432
60E	zipper(zip)	56	EMBOJ7:1115
60E	zipper-gooseberry region	46	EMBOJ6:2793
60E	gooseberry (gsb)	1	Cell47:1033
60EF	myosin heavy chain(cytoplasmic)	456	EMBOJ8:913
60F	ARS element	1	MGG197:342
60F	telomeres	1	Cell34:85
60F	anonymous	1	Chrm89:206
60F	[S3.8-4];chorion+,ry+	3	Cell34:47
60F	tyrosine protein kinase related (dTKR)	56	GeneDev1:862
60F	BSH9, BSH4: gooseberry (prd homology)	56	GeneDev1:1247
60F	Kruppel	4	Nature313:27
61A	snRNAU1	1	NAR16:3582
61A	LSP-1 gamma	16	JMAppGen1:371
61A	[AS1];neo	3	Bloomington Stock
61A	[AS628(lethal)];ry	3	Bloomington Stock
61A	[hsneo;Rh1(-833/+67)-cat];neo, CAT	3	Genet116:565
61A	LSP1 gamma	1	Cell23:441
61A	LSP-1 gamma	1	MGG199:357
61A	telomeres	1	Cell34:85
61A	[23.2];hsp23, ry+	3	EMBOJ5:755
61A	[27C+G-B];hsp28, ry+	3	MCB6:663
61A	[Bg9.61];ry+, 19 Kb hsp70-lacZ fusion gene	3	Chrm93:461
61A	[S38Z-5];s38+,lacZ+,ry+	3	Cell34:47
61A	[tAP-2];Adh+	3	Cell34:59
61A	anonymous	1	Cell28:165
61A	Larval Serum Protein 1 (LSP1)-gamma	15	JMB189:1
61C	extra macrochaetae (emc), DNA binding	16	Modolell,Madrid
61C	[AS240-neo^R #1 lethal];neo^R	3	Science239:1121
61C	PRD gene2 (paired homology)	16	Cell47:735
61C	[Adh hs61C];Adh+,hsp70	3	Cell51:763
61D	extra macrochaete (emc)	156	Posakony,SanDiego
61D	[AS384];ry	3	Bloomington Stock
61D	[AS624(lethal)];ry	3	Bloomington Stock
61D	[bcd+];Adh+, bcd+	3	EMBOJ7:1749
61D	[AS241-neo^R #2, #3 lethal];neo^R	3	Science239:1121
61D	[sev-IL33.2];sev+, neo+	3	EMBOJ(1989)
61D	[28P-A (lost?)];ry+hsp28	3	MCB6:663
61D	[cp70 delta B];hsp70-lacZ,ry+	3	Cell40:805
61D	[tYPc];ry+, Yp1,Yp2	3	PNAS82:7000
61DE	[M32b];ry+, lacZ	3	Devel04:245
61E	[SRS3.9-4];chorion+,ry+	3	Cell34:47
61E	lysozyme genes (8)	1	Hultmark,Stockhlm
61F	double sex cognate	1	Baker,Stanford
61F	[P(ry,HsAFP)3];ry+, HsAFP	3	Walker,Ontario
61F	[L53b];ry+, lacZ	3	Devel04:245
62A	[AS243-neo^R #4, #5, #6 lethal];neo^R	3	Science239:1121
62A	cluster of glu tRNAs	1	Cell21:169
62A	anonymous	1	Wolfner,Cornell
62A	tRNA locus	1	Davidson,CalTech
62A	[P[(w,ry)D]2];w, ry+	3	EMBOJ4:3489
62A	[WG946];ry	3	Bloomington Stock
62AB	anonymous	1	Wolfner,Cornell
62AB	[tAP-27];Adh+	3	Posakony,SanDiego
62B	[AS72(ms)];neo	3	Bloomington Stock
62B	[AS246-neo^R #7 lethal];neo^R	3	Science239:1121
62B	ras 3	15	Cell37:1027
62B	ras3	1	GeneDev2:567
62B	adenine phosphoribosyltransferase (Aprt)	5	MCB9:2220
62B	adenine phosphoribosyltransferase (Aprt)	1	Gene59:77
62C	H-1 histone variant	15	Raff,Harvard
62C	[P-1(2)gl-21F];lgl+,neoR	3	Cell50:215
62CD	myogenic cell TU	16	DevB90:272
62D	anonymous	1	Wolfner,Cornell
62E	[P-1(2)gl-23A];lgl+,neoR	3	Cell50:215
62E	ribosomal protein L12 locus	1	MCB4:2643
62F	[AS247-neo^R #8 lethal];neo^R	3	Science239:1121
63A	Shab(Sh homology)	156	Science243:943
63A	snRNAU5	1	NAR16:3582
63AB	[HB4/Sc2];LSP1 beta	3	Glover,London
63AC	myogenic cell TU	16	DevB90:272
63B	hsp 83	1	MCB8:4727
63BC	hsp 82	15	JMB188:499
63BC	hsp 83	1	Cell18:1359
63BC	[28C-A (lost?)];ry+ hsp28	3	MCB6:663
63C	[AS2];neo	3	Bloomington Stock
63C	[AS774];ry	3	Bloomington Stock
63C	[AS248-neo^R #9, #10 lethal];neo^R	3	Science239:1121
63C	[sev-IL2.1];sev+, neo+	3	Hafen,Zurich
63C	[S38Z-6];s38+,lacZ+,ry+	3	Cell34:47
63C	[tYPg];ry+, Yp1,Yp2	3	PNAS82:7000
63CD	[unn];	3	PNAS83:701
63D	[GR862];w	3	Bloomington Stock
63E	20-hydroxyecdysone inducible	1	JBC261:5575
63E	[23-3];sgs4+, ry+	3	PNAS82:5055
63F	polyubiquitin (previou.minor heat shock)	156	MCB8:4727
63F	Ubiquitin	15	BBA868:119
63F	anonymous	1	Wolfner,Cornell
63Z	anonymous	1	Goldstein,ArizSU
64A	[AS444(1)];ry	3	Bloomington Stock
64A	Glutamic Acid Decarboxylase (Gad) locus	45	Jackson,Worcester
64AB	[H(har1)G15-1];ry+	3	EMBOJ8:211
64B	IMP-L2 (Inducible membrane-bound polysm)	6	DevB129:439
64B	tip E	1	Kaekar,NY,Einstein
64B	nicotinic acetylcholine receptor subunit	156	MCB8:778
64B	ras(Dm ras 64B)	156	Gene51:129
64B	ras 2	15	Cell37:1027
64B	ras2	156	GeneDev2:567
64B	SRC homologous	1	Cell32:589
64B	20-hydroxyecdysone inducible	1	JBC261:5575
64B	anonymous	1	Lengyel,UCLA
64B	ras oncogene	15	MCB5:885
64B	src kinase domain	15	BBA867:144
64B	[P[(w,ry)F]3];w, ry	3	EMBOJ4:3489
64B	[hsp-per:1];hsp, per	3	Ewer,Brandeis
64BC	anonymous	1	Cell28:165
64C	RAS homologous	1	Cell32:589
64C	[R405];ry+	3	Cell34:47
64C	[tAP-7B,4.8];Adh+	3	Posakony,SanDiego
64CD	[2-35];Adh, ry+	3	DevB125:64
64DE	[AS464, AS466(mf)];ry+	3	Bloomington Stock
64D	[AS86];neo	3	Bloomington Stock
64F	defective dorsal discs (ddd)	1	Shearn,JohnsHopkins
64F	homology w/ DCO kinase catalytic subunit	156	GeneDev2:1539
64F	[sev+TK41.2];sev+, neo+	3	EMBOJ(1989)
64F	anonymous	1	Wolfner,Cornell
64F	anonymous	1	Davidson,CalTech
64F	anonymous	1	Cell28:165

Locn	Genes-Name	Key	References
64F	[SB2.1-3];chorion+, ry+	3	Cell34:47
64F	[cHB delta -73];hsp70-lacZ,ry+	3	Cell40:805
64F	PRD gene1 (paired homology)	16	Cell47:735
65A	[AS602(lethal)];ry+	3	Bloomington Stock
65A	[AS249-neo^R #11 lethal];neo^R	3	Science239:1121
65A	anonymous	1	Wolfner,Cornell
65A	[SB2.1-3];chorion+, ry+	3	Cell34:47
65AB	[w47.1 N8];w+	3	EMBOJ3:2077
65B	pale?	16	White,Brandeis
65B	tyrosine hydroxylase (TH)	1	Neuron2:1167
65B	[H2-1];ry+, sgs-4, lcp	3	EMBOJ6:2249
65B	20-hydroxyecdysone inducible	1	JBC261:5575
65BD	[1-6];Adh, ry+	3	DevB125:64
65C	G protein alpha subunit (DGa1)	156	JBC263:12070
65C	[AS623(1)];ry	3	Bloomington Stock
65C	anonymous	1	Wolfner,Cornell
65D	[H(har1)M21];ry+	3	EMBOJ8:211
65D	PRD gene3 (paired homology)	16	Cell47:735
65D	[278-65D];hsp26, lacZ, ry+	3	EMBOJ5:747
65D	[28A-B (lost?)];hsp28,ry+	3	MCB6:663
65D	[S38Z-7];s38, lacZ, ry+	3	JMB187:33
65E	[AS3];neo	3	Bloomington Stock
65E	[GR795];ry	3	Bloomington Stock
65F	[AS115];neo	3	Bloomington Stock
65F	[g711:3];sgs3+,ry+	3	Cell40:349
66A	tryptophan hydroxylase (TPH)	16	White,Brandeis
66A	[C20Pgd7.4-K];Pgd+, ry+	3	Lucchesi,ChaplHil
66A	[AS250-neo^R #12 lethal];neo^R	3	Science239:1121
66A	[28X-D (lost?)];ry+ hsp28	3	MCB6:663
66AB	[P-1(2)gl-21D];lgl+,neoR	3	Cell50:215
66B	[AS4];neo	3	Bloomington Stock
66B	leu tRNA	1	Gene44:307
66C	IMP-E1	16	DevB129:428
66C	[GR881];w	3	Bloomington Stock
66C	20-hydroxyecdysone inducible	1	JBC261:5575
66C	anonymous	1	Wolfner,Cornell
66CD	anonymous	1	Wolfner,Cornell
66D	507: eye specific	16	DevB94:451
66D	[AS5];neo	3	Bloomington Stock
66D	[AS251-neo^R #13 lethal];neo^R	3	Science239:1121
66D	mst 349, testis specif.+2 non sex-specif	16	DevB119:242
66D	[DR-17];ry+,Ddc+	3	Cell34:37
66D	anonymous	1	Ish-Horowicz,ICRF
66D	anonymous	1	EMBOJ1:1185
66D	[AS597(1)];ry	3	Bloomington Stock
66D	[AS605(1)];ry	3	Bloomington Stock
66D	chorion protein genes	4	Cell19:905
66D	gene s18-1, s15-1, s19-1	15	Chrm92:124
66D	hairy	16	CSHQB50:135
66DE	[AS252-neo^R #14 lethal];neo^R	3	Science239:1121
66E	[AS6];neo	3	Bloomington Stock
66E	[BL916];w,ry	3	Bloomington Stock
66E	[HSAd001];hsp28A, 1.4 Kb adenovirus DNA	3	MCB6:4126
66E	[Tf(3L)Ga6.0-1];Adh+	3	Meyerowitz,CalTec
66EF	[26.2 delta];hsp26, ry+	3	EMBOJ5:755
66F	anonymous	1	Davidson,CalTech
67A	[AS253-neo^R #15 lethal];neo^R	3	Science239:1121
67A	anonymous	4	JMB156:449
67B	hsp26 promoter and consensus sequencer	5	EMBOJ7:2191

Locn	Genes-Name	Key	References
67B	[AS7];neo	3	Bloomington Stock
67B	[AS254-neo^R #16 lethal];neo^R	3	Science239:1121
67B	hsp23	1	EMBOJ5:755
67B	hsp 23	15	EMBOJ5:1667
67B	includes hsp 28,23,26	1	NAR8:4441
67B	anonymous	1	BBA867:209
67B	loci of hsp 22,23,26, and 28	1	PNAS77:5390
67B	hsp and flanking TU	16	DevB89:196
67B	hsp22, hsp23, hsp26, hsp27, 1, 2, 3	1	EMBOJ4:2949
67BC	[DP1234];ry+	3	Bloomington Stock
67BC	[27P X/X-D];hsp28, ry+	3	MCB6:663
67C	[AS8];neo	3	Bloomington Stock
67C	alpha tubulin	1	DevB104:187
67C	0-1 hrs.	1	Lengyel,UCLA
67C	alpha4 tubulin	1	PNAS83:8477
67D	[gl(67D)(8Kb)];sn(w)y, bw, Sgs7, Sgs3	3	Chrm93:461
67DE	anonymous	1	JCB102:2076
67E	[GR790];w	3	Bloomington Stock
67E	[27 C X/X-A];hsp28, ry+	3	MCB6:663
67F	polycombeotic	12	Phillips,UMelb
68A	{R7.7-1];chorion+,ry+	3	Cell34:47
68A	[unn];scarlet	3	Genet122:595
68A	Cu-Zn superoxide dismutase (SOD)	5	NAR15:10601
68A	Cu-Zn SOD	15	NAR15:5483
68AB	[AS111];neo	3	Bloomington Stock
68B	[AS626(lethal)];ry	3	Bloomington Stock
68BC	[P[rib7(68BC)ry+/Sco;ry^506]];rDNA, ry+	3	GeneDev2:1745
68C	[AS10, AS112, AS84, AS95];neo	3	Bloomington Stock
68C	[AS60(fs)];neo	3	Bloomington Stock
68C	[AS75(ms)];neo	3	Bloomington Stock
68C	glue gene cluster	15	PNAS83:8654
68C	rotated abdomen	1	Science239:1121
68C	intermolt IV, III, II RNA	1	Wolfner,Cornell
68C	[28X-A (lost?)];ry+, hsp28	3	MCB6:663
68C	[g7:5, g71:2];Sgs3+,ry+	3	Cell40:349
68C	anonymous	1	Cell28:165
68C	Sgs protein genes, Sgs 3,7,8	1	Cell28:165
68C	Sgs proteins	1	JMEv20:251
68C	[AS9];neo	3	Bloomington Stock
68CD	[WG956];ry+	3	Bloomington Stock
68CD	anonymous	1	Cell28:165
68D	[AS600, AS617(lethal)];ry+	3	Bloomington Stock
68D	[Xho-25];ry+, sgs-4, lcp	3	EMBOJ6:2249
68D	[S38Z-3];s38+,lacZ+,ry+	3	Cell34:47
68DE	cyclin A (l(3)reg11)	156	Cell56:957
68DE	[AS255-neo^R #17 lethal];neo^R	3	Science239:1121
68E	cyclin A	16	Nature338:337
68E	LSP-2	16	JMAppGen1:371
68E	[H(har1)E8-1];ry+	3	EMBOJ8:211
68E	LSP-2	1	MGG199:357
68E	[g6:4];Sgs3+,ry+	3	Cell40:349
68EF	anonymous	1	Wolfner,Cornell
68F	[AS11, AS12];neo	3	Bloomington Stock
68F	[AS256-neo^R #18 lethal];neo	3	Science239:1121
69	anonymous	1	BBA867:209
69A	Esterase 6 (natural populations)	5	PNAS86:1426
69AB	anonymous	1	EMBOJ6:3431
69B	[H(har1)E19-1];ry+	3	EMBOJ8:211
69C	[WG987];ry	3	Bloomington Stock

Locn	Genes-Name	Key	References
69C	fst 241 avary, germ-line specif. transc.	1	DevB119:242
69CD	[tAP-12,4.8];Adh+	3	Posakony,SanDiego
69CD	Hbox homology	1	McGinnis,Yale
69D	anonymous	1	Lengyel,UCLA
69F	myogenic cell TU	16	DevB90:272
70A	male accessory gland peptide	156	Cell54:291
70A	[R1b];ry+, lacZ	3	Deve104:245
70A	[WG1052];ry	3	Bloomington Stock
70A	[AS257-neo^R #19 lethal];neo^R	3	Science239:1121
70A	asp tRNA genes	1	NAR8:2921
70A	20-hydroxyecdysone inducible	1	JBC261:5575
70A	anonymous	1	Wolfner,Cornell
70AB	anonymous	1	Wolfner,Cornell
70AB	[S3.8-3];chorion+,ry+	3	Cell34:47
70BC	maternal restricted TU	16	Stephenson,Rochst
70BC	Val 4 tRNA locus	1	Chrm76:65
70C	[AS13];neo	3	Bloomington Stock
70C	[AS618(lethal)];ry	3	Bloomington Stock
70C	[Act-line I];actin, beta-gal, ry+	3	DevB133:313
70C	[GR879];w	3	Bloomington Stock
70C	[H(har1)M10];ry+	3	EMBOJ8:211
70C	[H(har1)M16];ry+	3	EMBOJ8:211
70C	[H(har1)M5-2];ry+	3	EMBOJ8:211
70C	anonymous	1	Wolfner,Cornell
70C	[S11.4-2];chorion+	3	Cell34:47
70C	Glued(Gl)	15	PNAS84:6501
70D	P1	16	JMAppGen1:371
70D	[AS83];neo	3	Bloomington Stock
70D	[26.3];hsp26, ry+	3	EMBOJ5:755
70D	P1	1	MGG199:357
70D	frizzled (fz)	456	Nature338:263
70DE	[m34a];ry+, lacZ	3	Deve104:245
70F	[WG1064];ry	3	Bloomington Stock
71	[AS302];ry	3	Bloomington Stock
71A	[AS14];neo	3	Bloomington Stock
71A	[C20Pgd8.9-8];Pgd+, ry+	3	Lucchesi,ChaplHil
71A	gastrula differential poly(A) RNA	1	DevB109:476
71AB	anonymous	1	Wolfner,Cornell
71AB	[-51];hsp70,ry+	3	Cell40:805
71AB	[26.3];hsp26, ry+	3	EMBOJ5:755
71B	non comp # 16 of beta tb 6 (nc16)	4	Fuller,Boulder
71C	[AS629(lethal)];ry+	3	Bloomington Stock
71C	[AS90];neo	3	Bloomington Stock
71C	[S38M-1];s38+,M13+,ry+	3	Cell34:47
71CD	gonadal (gdl), z600, Eip 28/29	56	GeneDev3:232
71CD	EIP 28/29 Kc polypeptide	156	JMB189:617
71DE	[P-1(2)gl-25A];lgl+,neoR	3	Cell50:215
71DE	late I TU	16	Wolfner,Cornell
71DE	late II,III TU	16	Wolfner,Cornell
71E	[AS15];neo	3	Bloomington Stock
71E	ecdysone induce puff:late/intermolt gene	1	JMB188:517
71F	[g7:2];Sgs3+,ry+	3	Cell40:349
71F	[tAP-11,4.8];Adh+	3	Posakony,SanDiego
71Z	[12-1];neo R, hsp 70, tropomyosin I	3	EMBOJ6:1375
72	[unn];	3	PNAS83:701
72A	homology w/ DCO kinase catalytic subunit	156	GeneDev2:1539
72B	[hsneo;Rh1(-833/+67)-cat4];neo, CAT	3	Genet116:565
72B	brahma	4	Kennison,Boulder
72BC	head specific TU	16	DevB94:451

Locn	Genes-Name	Key	References
72DE	anonymous	1	Davidson,CalTech
73A	transformer (tra)	15	Cell50:739
73A	[AS615(l)];ry	3	Bloomington Stock
73A	H box homol	1	McGinnis,Yale
73A	transformer	1	Belote,Syracuse
73A	double sex cognate, not transformer	1	Baker,Stanford
73A	anonymous	1	Belote,UCSD
73A	tra, Dash	4	EMBOJ5:3607
73A	scarlet (st)	1256	Genet122:595
73B	abl (tyrosine kinase)	56	MCB8:843
73B	[AS258-neo^R #20, #21 lethal];neo^R	3	Science239:1121
73B	Abelson SRC homologous	1	Cell35:393
73B	Dash, abl oncogene	1	EMBOJ4:2609
73B	[27S-D];hsp28, ry+	3	MCB6:663
73B	[Pc[ry(delta0-1)]2];ry+	3	Laverty,UCBrk
73B	abelson, proto-onvogene, st	4	Cell51:821
73BC	[swallow-bb154];ry+, sww+	3	GeneDev2:1655
73C	[AS260-neo^R #22 lethal];neo^R	3	Science239:1121
73C	cell adhesion transmembrane	16	Jimenez,Madrid
73D	minor heat shock locus	1	MCB8:4727
73D	Rh4, R7, opsin and larval photoreceptor	1	JNeurosci7:1558
73DEF	head specific TU	16	DevB94:451
73E	[GR866];w	3	Bloomington Stock
73E	[P(ry,HsAFP)6];ry+, HsAFP	3	Walker,Ontario
73F	[23UZ-1];lacZ,w+	3	Irvine,Stanford
73F	[AS106-bfs49];neo	3	Bloomington Stock
73F	[AS261-neo^R #23 lethal];neo^R	3	Science239:1121
74A	[H(har1)E14-1];ry+	3	EMBOJ8:211
74C	[AS603(lethal)];ry+	3	Bloomington Stock
74D	[AS98];neo	3	Bloomington Stock
74EF	E74	56	GeneDev3:782
74EF	eary ecdysone-responsive gene	456	Burtis,UCDavis
74F	[AS262-neo^R #24 lethal];neo^R	3	Science239:1121
75	anonymous	1	BBA867:209
75A	grd chloride channel protein	1	Betz,Heidelberg
75A	[AS104];neo	3	Bloomington Stock
75A	PRD gene8 (paired homology)	16	Cell47:735
75B	[AS16-bl, AS17-bl];neo	3	Bloomington Stock
75B	[AS788(lethal)];ry+	3	Bloomington Stock
75B	[C20Pgd7.4-C];Pgd+, ry+	3	Lucchesi,ChaplHil
75B	[AS263-neo^R #25 lethal];neo^R	3	Science239:1121
75C	[65-13A[Ddc(-208,-106)]];Adh+, Ddc+	3	GeneDev1:510
75C	[AS80];neo	3	Bloomington Stock
75C	[AS81];neo	3	Bloomington Stock
75C	[AS264-neo^R #26 lethal];neo^R	3	Science239:1121
75C	anonymous	1	Wolfner,Cornell
75C	mst(3)ag-15 paragonia transcript	16	Schafer,Duseldrf
75C	[23.5];hsp23, ry+	3	EMBOJ5:755
75C	[P[(w,ry)H]2-2];w, ry+	3	EMBOJ4:3489
75C	terminus (ter)/ Zn binding finger	56	DevB125:85
75C	[AS619(lethal)];ry+	3	Bloomington Stock
75CD	[R502.1];ry+	3	Cell34:47
75D	[AS18];neo	3	Bloomington Stock
75D	[AS62(fs)];neo	3	Bloomington Stock
75D	anonymous	1	Belote,UCSD
75D	[R706.1];ry+	3	Cell34:47
75DF	[Pw+run01];w+, runt	3	GeneDev2:1179
75EF	naked? H box homol.	1	McGinnis,Yale
75F	[AS19];neo	3	Bloomington Stock

Locn	Genes-Name	Key	References
76A	[AS20];neo	3	Bloomington Stock
76A	[AS431];ry+	3	Bloomington Stock
76A	[H(har1)E5-2];ry+	3	EMBOJ8:211
76A	anonymous	1	Wolfner,Cornell
76A	[A38M-4, AS431];s38+,M13+,ry+	3	Cell34:47
76A	AS265-neo^R #27 lethal];neo^R	3	Science239:1121
76AB	[P-1(2)gl-21A];lgl+,neoR	3	Cell50:215
76B	Shal(Sh homology)	1	Science243:943
76C	[AS456(1)];ry+	3	Bloomington Stock
76D	[AS627(lethal)];ry+	3	Bloomington Stock
76D	[WG958(lethal)];ry+	3	Bloomington Stock
76DE	maternal restricted transcript	1	Stephenson,URoch
76DE	anonymous	1	EMBOJ6:3431
76F	anonymous	1	Cell28:165
77A	[cHB delta-73];hsp70-lacZ, ry+	3	Cell40:805
77B	[hsneo;Rh19-833/+67)-cat 41];neo,CAT	3	Genet116:565
77B	[AS266-neo^R #28 lethal];neo^R	3	Science239:1121
77BC	[AS76(ms)];neo	3	Bloomington Stock
77BC	[WG1025(lethal)];ry+	3	Bloomington Stock
77BD	[61-25A[Ddc(-208,-38)]];	3	GeneDev1:510
77C	[AS21];neo	3	Bloomington Stock
77DE	[P(ry, HsAFP)1];ry+, HsAFP	3	Walker,Ontario
77E	[26.2];hsp26, ry+	3	EMBOJ5:755
77E	Knirps-related (knrl) steroid receptor	45	Nature336:493
77F	RI	156	GeneDev2:1539
78A	[35UZ-1];lacZ,W+	3	Irvine,Stanford
78A	[AS22];neo	3	Bloomington Stock
78B	[278-78B];hsp26, lacZ, ry+	3	EMBOJ5:747
78BC	[R603.1];ry+	3	Cell34:47
78C	[AS267-neo^R #29 lethal];neo^R	3	Science239:1121
78CD	[AS595(lethal)];ry+	3	Bloomington Stock
78CD	[P[(w,ry)H]1];w, ry+	3	EMBOJ4:3489
78D	[WG1021];ry	3	Bloomington Stock
78D	[cHB delta-89];hsp70-lacZ, ry+	3	Cell40:805
78D	polycomb	4	Zink,Heidelberg
78E	[26.2];hsp26, ry+	3	EMBOJ5:755
79B	actin	8	DevB133:313
79B	[AS23];neo	3	Bloomington Stock
79B	[AS772];ry	3	Bloomington Stock
79B	actin	1	GeneDev1:1161
79CD	AP endonuclease (mutagen induced repair)	15	Grabowski,Loyola
79D	[AS620(lethal)];ry+	3	Bloomington Stock
79D	[AS87(fs)];neo	3	Bloomington Stock
79D	[AS97];neo	3	Bloomington Stock
79D	[AS268-neo^R #30 lethal];neo^R	3	Science239:1121
79E	[TnJA3];adh+, dpp+	3	GeneDev1:615
79E	[+411];hsp70,ry+	3	Cell40:805
79E	anonymous	1	Goldstein,ArizSU
79E	[AS630(lethal)];ry+	3	Bloomington Stock
79EF	[H(har1)E5-3];ry+	3	EMBOJ8:211
79F	[AS613(lethal)];ry+	3	Bloomington Stock
79F	[AS787];ry+	3	Bloomington Stock
79F	[8-1];sgs4+, ry+	3	PNAS82:5055
80	chromocenter	1	Wolfner,Cornell
80A	casein kinase II alpha subunit	156	MCB7:3409
80A	[p[lac, ry+]A37];lac Z,ry+	3	Deve105:35
80A	[B1,F2];rp49+,ry+	3	Nature317:555
80A	[hsp-per:13];hsp, per	3	Ewer,Brandeis
80C	ribosomal protein-21 (rp21)	6	MGG213:354

Locn	Genes-Name	Key	References
80C	Kc cells	1	PNAS82:7369
80F	chromocenter	1	Wolfner,Cornell
81	chromocenter	1	Wolfner,Cornell
81F	[GR804, GR860];w	3	Bloomington Stock
81F	[WG993];ry	3	Bloomington Stock
82A	[P-1(2)gl-22A];lgl+,neoR	3	Cell50:215
82A	[TnJA2];adh+, dpp+	3	GeneDev1:615
82A	anonymous	1	JMB156:449
82A	[w47.4L];w+	3	EMBOJ3:2077
82AB	[23.4];hsp23, ry+	3	EMBOJ5:755
82B	[Adh hs82b];Adh+,hsp70	3	Cell51:763
82BC	[g4:3];Sgs3+,ry+	3	Cell40:349
82BC	[tAP-8B,4.8];Adh+	3	Posakony,SanDiego
82C	CA/GT Z DNA probe	1	EMBOJ6:1781
82C	[28X-E (lost?)];ry+ hsp28	3	MCB6:663
82C	[tYPa];ry+, Yp1, Yp2	3	PNAS82:7000
82C	[AS621(lethal)];ry+	3	Bloomington Stock
82D	[AS24];neo	3	Bloomington Stock
82E	snRNAU1	1	NAR16:3582
82E	[23UZ-2];lacZ,W+	3	Irvine,Stanford
82E	[AS103];neo	3	Bloomington Stock
82E	[AS608(lethal)];ry+	3	Bloomington Stock
82E	[AS-neo^R #31, #32 lethal];neo^R	3	Science239:1121
82E	2 SnRNA U1 pseudogenes (inverted)	15	NAR12:8835
82F	head specific TU	16	DevB94:451
82F	[5-1];sgs4+, ry+	3	PNAS82:5055
82F	[AR4-025, BL925];w,ry+	3	Science229:558
83A	[P-1(2)gl-21H];lgl+,necR	3	Cell50:215
83A	anonymous	1	Wolfner,Cornell
83A	[unn];ry+	3	Chovnick,UConn
83AB	anonymous	1	Wolfner,Cornell
83AB	Lys 5 tRNA locus	1	Chrm76:65
83B	[WG1020, WG1046];ry+	3	Bloomington Stock
83B	anonymous	1	Wolfner,Cornell
83B	[27S-E];hsp28, ry+	3	MCB6:663
83B	[AS25];neo	3	Bloomington Stock
83BC	[SB2.1-2];chorion+,ry+	3	Cell34:47
83C	[neo^R #33 lethal];neo^R	3	Science239:1121
83C	anonymous	1	Cell28:165
83C	[AS611(lethal)];ry+	3	Bloomington Stock
83CD	maternal restricted TU	16	Stephenson,URoch
83CD	[23.1];hsp23, ry+	3	EMBOJ5:755
83D	[AS73(ms)];neo	3	Bloomington Stock
83F	anonymous	1	Wolfner,Cornell
83F	[R3.9-3,9/ S38M-2];ry+, s38+, M13+	3	Cell34:47
84A	bicoid(bcd) (maternal)	56	EMBOJ7:1749
84A	labial (lab)	1567	EMBOJ7:2569
84A	[AS637(lethal)];ry+	3	Bloomington Stock
84A	bicoid(bcd), PRD gene6 (paired homology)	16	Cell47:735
84A	bcd local. signal lies 3'end transcript	56	Nature336:595
84A	zen	1	Doyle,ColumbiaU
84A	labial (lab), (Hbox)	4567	GeneDev3:399
84A	labial (F90-2)	1	PNAS83:4809
84A	deformed (Dfd)	15	EMBOJ6:767
84A	Antp	1	EMBOJ5:733
84A	[AS599(lethal)];ry+	3	Bloomington Stock
84AB	[P-1(2)gl-22B];lgl+,neoR	3	Cell50:215
84AB	Lys 5 tRNA Ylocus	1	Chrm76:65
84B	control elements of ftz	1	Cell50:963

Locn	Genes-Name	Key	References
84B	alpha tubulin	1	DevB104:187
84B	[27P X/X-E];hsp28, ry+	3	MCB6:663
84B	S60 (zen)	15	Nature323:76
84B	Antp, Scr, Dfd	12	EMBOJ4:3757
84B	Antp (P2)	6	GeneDev2:1615
84B	Antennapedia	1	MCB6:4667
84B	amalgam(ama)	156	Cell55:589
84B	antennapedia	5	MCB6:4676
84B	delta tubulin	1	JMB156:449
84B	alpha1 tubulin	1	PNAS83:8477
84B	alpha tubulin	4	GeneDev2:477
84BC	ovarian cDNA 96	1	EMBOJ4:147
84BC	anonymous	1	Wolfner,Cornell
84BC	[S3.8-1];chorion+,ry+	3	Cell51:763
84BD	Male germ cell specific tu	16	Schafer,Duseldorf
84C	[hsPneosev-15];hsneo	3	Science236:55
84C	[unn];ry+	3	Chovnick,UConn
84C	stk,hat,rue:l(3)84Cb,ted:Ccl(3)84Ce,Gld	4	EMBOJ5:2939
84C	Gld, ted, YYRRbox (CTGA tandem reports)	5	NAR16:3375
84C	glucose dehydrogen., eclosion genes A-D	1	EMBOJ5:2939
84D	[neo^R #34 lethal];neo^R	3	Science239:1121
84D	alpha tubulin locus	1	DevB104:187
84D	tRNA	1	Davidson,CalTech
84D	anonymous	1	Cell28:165
84D	Val 3b tRNA locus	1	Chrm76:65
84D	[28X-B (lost?)];ry+, hsp28	3	MCB6:663
84D	[unn];hsp26-lacZ,ry+	3	Cell40:805
84D	responsible for dominant phenotype(rfd)	1	EMBOJ6:201
84D	overlaps Val 3b tRNA, rotund	4	Kerridge,Marseill
84D	delta tubulin	1	JMB156:449
84D	alpha3 tubulin	15	PNAS83:8477
84DE	[AS604(lethal)];ry+	3	Bloomington Stock
84DE	[P-1(2)gl-21G];lgl+,neoR	3	Cell50:215
84DE	[AS272-neo^R #35 lethal];neo^R	3	Science239:1121
84DF	[14-1];neo-R, tropomyosin I, hsp 70	3	EMBOJ6:1375
84E	[WG1173];ry	3	Bloomington Stock
84E	20-hydroxyecdysone inducible	1	JBC261:5575
84E	[cHB delta-194, g6:2];hsp70-lacZ,ry+	3	Cell40:805
84E	[g6:2];Sgs3+,ry+	3	Cell40:349
84E	doublesex (dsx)	56	Cell56:997
84E	anonymous	1	EMBOJ2:2027
84F	[AS609, AS614 (lethal)];ry+	3	Bloomington Stock
84F	[GR802, GR812];w	3	Bloomington Stock
84F	cluster of arg and asn tRNAs	1	NAR8:2921
84F	maternal restricted TU	16	Stephenson,Rochst
84F	[1];Act88F+	3	Cell40:101
84F	[A3];rp49+,ry+	3	Nature317:555
84F	[g71:3];Sgs3+,ry+	3	Cell40:349
85	anonymous	1	BBA867:209
85	anonymous	1	Wolfner,Cornell
85A	belle, helicase homology	16	Fuller,Boulder
85A	tRNA^tyr-Y85Aa-c, tRNA^tyr-Y85Ade	15	MCB8:3322
85A	[AS26, AS102];neo	3	Bloomington Stock
85A	[H(har1)M12];ry+	3	EMBOJ8:211
85A	[WG1122];ry	3	Bloomington Stock
85A	hunchback	1	Posakony,SanDiego
85A	Dhod	4	Rawls,Kentucky
85A	tRNA locus	1	Davidson,CalTech
85A	[27 X/A-1-A];hsp28, ry+	3	MCB6:663

Locn	Genes-Name	Key	References
85A	[R309.1];ry+	3	Cell34:47
85A	[p[lac, ry+]A18];lac Z, ry+	3	Deve105:35
85A	hb upstream bind. sites for bcd protein	5	Nature337:138
85A	hunchback	45	Nature327:383
85A	[AS63(msfs)];neo	3	Bloomington Stock
85B	[814,CH8];sry+,ry+	3	Nature317:555
85AB	hb upstream enhancers to lacZ	8	Nature340:363
85B	[WG950];ry	3	Bloomington Stock
85C	[24-1];sgs4+, ry+	3	PNAS82:5055
85C	Elongation factor 1 alpha, F1 (Ef-1,F1)	15	NAR16:3175
85C	[AS781];ry	3	Bloomington Stock
85C	[AS94];neo	3	Bloomington Stock
85C	[AS273-neo^R #36 lethal];neo^R	3	Science239:1121
85C	tRNA	1	Davidson,CalTech
85C	anonymous	1	MacIntyre,Cornell
85C	Arg tRNA	1	Chrm76:65
85C	[AS107];neo	3	Bloomington Stock
85CD	PRD gene 7 (paired homology)	16	Cell47:735
85D	ras (Dm ras 85D)	156	Gene51:129
85D	[lacZ;ry at Dras1];ry+	3	Bellen,Basel
85D	[AS274-neo^R #37 lethal];neo^R	3	Science239:1121
85D	ras 1	15	Cell37:1027
85D	Dyps-clps homologue	1	TrGen3:69
85D	RAS homologous	1	Cell32:589
85D	alpha tubulin locus	1	JMB156:449
85D	anonymous	1	BBA867:209
85D	alpha 2 tubulin locus	1	Chrm89:206
85D	[BS2.7-7];chorion+,ry+	3	Cell34:47
85D	[P(ry,HsAFP)5];ry+, HsAFP	3	Walker,Ontario
85D	[P[(w,ry)H]3];w, ry	3	EMBOJ4:3489
85D	[g6:1];Sgs3+,ry+	3	Cell40:349
85D	beta tubulin locus	1	DevB104:187
85DE	double sex cognate	1	Baker,Stanford
85E	[WG1092];ry	3	Bloomington Stock
85E	l(3)c43	1	Cheney,Hopkins
85E	alpha tubulin	1	DevB104:187
85E	0-1 hr. TU	16	Lengyel,UCLA
85E	anonymous	1	Cell28:165
85E	[26.4];hsp26, ry+	3	EMBOJ5:755
85E	alpha tubulin locus	1	JMB156:449
85E	alpha2 tubulin	15	PNAS83:8477
85E	metallothionein	15	JBC260:1527
85E	metallothionein (MT)	15	Genet112:493
85E	metallothionein (Mtn)	125	PNAS83:6025
85F	[S17b];ry+, lacZ	3	Deve104:245
85F	[P[(w,ry)H]2-1];w, ry+	3	EMBOJ4:3489
85F	[tAP-15B,4.8];Adh+	3	Posakony,SanDiego
86	anonymous	1	BBA867:209
86	anonymous	1	Wolfner,Cornell
86A	[AS27, AS96];neo	3	Bloomington Stock
86A	[20-1];sgs4+, ry+	3	PNAS82:5055
86AB	[AS109];neo	3	Bloomington Stock
86B	[AS751];ry	3	Bloomington Stock
86B	[WG1088];ry	3	Bloomington Stock
86C	[AS28];neo	3	Bloomington Stock
86C	[P[(w,ry)F]2];w, ry+	3	EMBOJ4:3489
86D	[AS766];ry	3	Bloomington Stock
86D	[AS88];neo	3	Bloomington Stock
86D	[R311.1, AS329];ry+	3	Cell34:47

Locn	Genes-Name	Key	References
86D	[g7:6];Sgs3+,ry+	3	Cell40:349
86E	[S38Z-2];s38+,lacZ+,ry+	3	Cell34:47
86E	[AS114];neo	3	Bloomington Stock
86EF	[yp1-Adh-86EF];neoR, yp-1, Adh	3	MGG210:153
86F	[AS89, AS93];neo	3	Bloomington Stock
86F	[HSH22];hsp70-h, neo R	3	Cell51:405
86F	[AS275-neo^R #38 lethal];neo^R	3	Science239:1121
86F	[278-86F];hsp26, lacZ, ry+	3	EMBOJ5:747
87A	[AS29];neo	3	Bloomington Stock
87A	hsp 70 and flanking	1	Cell14:921
87A	hsp70, Sn cell DNA	1	NAR8:4441
87A	[R307.1];ry+	3	Cell34:47
87A	hsp70 subclone	1	MCB8:4727
87A	hsp70	1	Cell18:1359
87AB	anonymous	1	BBA867:209
87AB	[28A-C (lost?)];ry+, hsp28	3	MCB6:663
87AB	[P1/35];per(period), ry+	3	Nature326:390
87B	seven-up: nuclear hormone receptor	16	Mlodzik,Berkeley
87B	[P-1(2)gl-24A];lgl+,neoR	3	Cell50:215
87B	[yp1-Adh-87B];neoR, yp-1, Adh	3	MGG210:153
87B	phosphatase 1 (lethal e 211)	1	Axton,London
87C	[WG1048];ry	3	Bloomington Stock
87C	hsp70, Sn cell DNA	1	NAR8:4441
87C	hsp70 and flanking	1	Cell14:921
87C	hsp 70	1	Cell18:1359
87CD	[C1-1];w+	3	Cell36:469
87CF	anonymous	1	Wolfner,Cornell
87D	[AS30];neo	3	Bloomington Stock
87D	[H(har1)M4];	3	EMBOJ8:211
87D	hsp70	1	NAR8:4441
87D	acetylcholinesterase (Ace)	5	Spierer,Geneva
87D	snake(snk+)	15	Nature323:688
87D	rosy (Xdh)	5	Genet116:67
87D	rosy Ace	1	JMB168:35
87DE	single minded(sim)	126	Cell52:133
87DE	sim	5	Cell52:143
87DE	[AS276-neo^R #39 lethal];neo^R	3	Science239:1121
87DE	rosy and Ace	4	JMB168:17
87DE	mesA,B,G9,S12,rosy,snake,hsc2,pic,...m32	4	JMB190:255
87DE	4 scaffold attachment regions(SARs)	1	JMB190:255
87E	Su-var(3)7, other transcripts	5	Gausz,Szeged
87E	[AS108];neo	3	Bloomington Stock
87E	[AS31];neo	3	Bloomington Stock
87E	[AS277-neo^R #40 lethal];neo^R	3	Science239:1121
87E	actin	1	GeneDev1:1161
87E	[unn];ry+	3	Chovnick,UConn
87E	AChE 16, 55 kD subunits	6	FEBS238:333
87E	Ace (proposed structural gene for AChE)	15	EMBOJ5:2949
87F	mst(3)gl-9	56	EMBOJ7:447
87F	mst(3)gl-9	16	MGG202:219
87F	[DRI-15];ry+,Ddc+	3	Cell34:37
87F	[R308.2, R404.1];ry+	3	Cell34:47
87F88	[AS32];neo	3	Bloomington Stock
88	anonymous	1	BBA867:209
88A	[AS33];neo	3	Bloomington Stock
88A	l(3)k43, imaginal disk and chorion amp.	1	Kelley,Texas
88A	empty spiracles (ems),homeobox homology	16	Gehring,Basel
88A	[cHB delta-23];hsp70-lacZ,ry+	3	Cell40:805
88A	[tYPe];ry+, Yp1,Yp2	3	PNAS82:7000

Locn	Genes-Name	Key	References
88A	empty spiracles? H box homol.	1	McGinnis,Yale
88AB	RNA pol II 140kd subcent (Rp11-140)	1	Greenleaf,NIEHS
88AB	[WG1091];ry+	3	Bloomington Stock
88B	trithorax (trx), (Rgbx)	156	PNAS86:3738
88B	[AS74(ms)];neo	3	Bloomington Stock
88B	minor heat shock cDNA	1	MCB8:4727
88BC	SU(HW) supprssor of Hairy Wing	156	GeneDev2:1205
88C	[AS34];neo	3	Bloomington Stock
88C	[AS278-neo^R #41 lethal];neo^R	3	Science239:1121
88C	anonymous	1	Cell28:165
88C	[S6.9-5];chorion+,lac+,ry+	3	Cell34:47
88D	[[GR841, GR876];w	3	Bloomington Stock
88D	[AS70, AS71, AS74(ms)];neo	3	Bloomington Stock
88D	[neo^R #42 lethal];neo^R	3	Science239:1121
88D	double sex cognate	1	Baker,Stanford
88E	[AS100];neo	3	Bloomington Stock
88E	[AS116];neo	3	Bloomington Stock
88E	[AS35];neo	3	Bloomington Stock
88E	[AS279-neo^R #43 lethal];neo^R	3	Science239:1121
88E	CA/GT Z DNA probe	1	EMBOJ6:1781
88E	hsp70	1	NAR8:4441
88E	[BS2.7-9, R401.2, AS331];chorion+,ry+	3	Cell34:47
88E	actin,tropomyosin,+ myofibril proteins	1	Cell37:469
88F	[AS36];neo	3	Bloomington Stock
88F	[AS428(lethal)];ry+	3	Bloomington Stock
88	tropomyosin	1	Davidson,CalTech
88F	tropomyosins 1 & 2	15	MCB6:1965
88F	muscle specific tropomysin	1	Cell41:57
88F	actin	1	DevB90:272
88F	[S6.9-1];chorion+,lac+,ry+	3	Cell34:47
88F	[cp70 delta B];hsp70-lacZ,ry+	3	Cell40:805
88F	easter is serine protease	156	Cell56:391
88F	3 tropomyosin loci	1	JMB156:449
88F	Tropomyosin 2	16	Gene74:457
89A	[AS598(lethal)];ry+	3	Bloomington Stock
89A	[DP1232];ry	3	Bloomington Stock
89A	[AS281-neo^R #44 lethal];neo^R	3	Science239:1121
89A	[+204];hsp70,ry+	3	Cell40:805
89A	[B1-1, GR818, GR821];w+	3	Cell36:469
89A	[Q39b];ry-+, P-beta gal	3	Kerridge,Marseill
89B	[65-16C[Ddc(-208,-83)]];Adh+, Ddc+	3	GeneDev1:510
89B	[AS37];neo	3	Bloomington Stock
89B	[GR877];w	3	Bloomington Stock
89B	[AS282-neo^R #45, #46 lethal];neo^R	3	Science239:1121
89B	[sev+TK91];sev+, neo+	3	EMBOJ(1989)
89B	Val 4 Phe 2 tRNA	1	Chrm76:65
89B	[27P X/X-C];hsp28, ry+	3	MCB6:663
89B	[28-2];sgs4+, ry+	3	PNAS82:5055
89B	[BS2.7-6];chorion+,ry+	3	Cell34:47
89B	[g7:8];sgs3+,ry+	3	Cell40:349
89B	Sb	1	Fristrom,UCBrk
89BC	anonymous	1	Finerty,EmoryU
89BC	[26.3];hsp26, ry+	3	EMBOJ5:755
89C	[AS284-neo^R #47 lethal];neo^R	3	Science239:1121
89D	Fasciclin I	156	Cell53:557
89D	[line 2];rp21, ry+	3	MGG213:354
89DE	[AS61(fs)];neo	3	Bloomington Stock
89E	Ubx	56	GeneDev3:243
89E	[6.1UZ-4];lacZ,w+	3	Irvine,Stanford

Locn	Genes-Name	Key	References
89E	[AS38];neo	3	Bloomington Stock
89E	Tab (Transabdominal)	1	GeneDev1:111
89E	anonymous	1	Bender,Harvard
89E	bithorax complex	4	Bender,Harvard
89F	35kDa subunit of proteasome (Pros-35)	1	Kloetzel,Heidlbrg
90A	[11-3];neo-R, hsp 70, tropomyosin I	3	EMBOJ6:1375
90B	Sgs-5	1	JMB190:149
90BC	anonymous	1	JMB188:517
90BC	Val tRNA, val 4 tRNA, Cal 3b, pro tRNA	1	Chrm76:65
90BC	Intermolt V RNA, sgs	1	Wolfner,Cornell
90C	tRNA	1	Davidson,CalTech
90CD	[H(har1)G8-3];ry+	3	EMBOJ8:211
90CD	[S3.8-2];chorion+,ry+	3	Cell34:47
90D	[AS39];neo	3	Bloomington Stock
90DE	[35UZ-2];lacZ,W+	3	Irvine,Stanford
90E	[AS625(lethal)];ry+	3	Bloomington Stock
90E	[AS285-neo^R #48 lethal];neo^R	3	Science239:1121
90E	[P[(w,ry)D]3, GR829];w, ry+	3	EMBOJ4:3489
90E	adult abdominal fat body expression	1	Bownes,Edinburgh
90EF	[p[Dr 4- Ddc5]];Adh+,Ddc+	3	GeneDev1:510
90EF	[DR-1];ry+,Ddc+	3	Cell34:37
90F	H box homol	1	McGinnis,Yale
90F	[R15b];ry+, lacZ	3	Deve104:245
90Z	anonymous	15	Raff,Harvard
91A	[sev-MC48.13];sev+, neo+	3	Cell54:299
91A	[26.3];hsp26, ry+	3	EMBOJ5:755
91A	glass (gl) has zn fingers	16	Moses,Berkeley
91AB	[AS286-neo^R #49 lethal];neo^R	3	Science239:1121
91AB	[AR01];w,ry+	3	Science229:558
91B	[WG1151];ry	3	Bloomington Stock
91B	[cHB delta-194];hsp70-lacZ,ry+	3	Cell40:805
91BD	ChA (choline acetyltransferase)	15	PNAS83:4081
91C	homology w/ DCO kinase catalytic subunit	156	GeneDev2:1539
91C	[AS632(lethal)];ry+	3	Bloomington Stock
91C	anonymous	1	Davidson,CalTech
91C	[A2-1];w+,ry+	3	Cell36:469
91D	[AS40];neo	3	Bloomington Stock
91D	[AS287-neo^R #50 lethal];neo^R	3	Science239:1121
91D	anonymous	1	Cell28:165
91D	[g4:2];Sgs3+,ry+	3	Cell40:349
91D	Rh2, ocellus opsin	156	Cell44:705
91E	[AS41];neo	3	Bloomington Stock
91F	nanos	16	Lehman,MIT
91F	[AS42];neo	3	Bloomington Stock
91F	[S3.8-5];chorion+,ry+	3	Cell34:47
92	anonymous	1	BBA867:209
92A	[AS43];neo	3	Bloomington Stock
92A	[AS462(lethal)];ry+	3	Bloomington Stock
92A	[H(har1)G12-1];ry+	3	EMBOJ8:211
92A	[WG1010(lethal)];ry+	3	Bloomington Stock
92A	[AS288-neo^R #51 lethal];neo^R	3	Science239:1121
92A	anonymous	1	Cell28:165
92A	[R3.9-2];chorion+,ry+	3	Cell34:47
92A	Delta (Dl)	456	GeneDev2:1723
92A	Delta (Dl)	46	EMBOJ6:3431
92AB	[23.3, 26.2];hsp23, ry+	3	EMBOJ5:755
92B	[AS91, AS92];neo	3	Bloomington Stock
92B	[BL918, BL920];ry+,w	3	Bloomington Stock
92B	[GR814, GR848];ry+,w	3	Bloomington Stock

Locn	Genes-Name	Key	References
92B	[GR845];w	3	Bloomington Stock
92B	[AS289-neo^R #52, #53 lethal];neo^R	3	Science239:1121
92B	nina E upstream regulators	58	Genet116:565
92B	opsin	15	Cell40:851
92B	opsin (ninaE)	1	EMBOJ6:443
92B	rhodopsin (nina E)	15	Cell40:839
92BC	[B2-1];w+	3	Cell36:469
92BC	[H4];ry+	3	EMBOJ5:583
92C	[DP1252];ry	3	Bloomington Stock
92D	Rh3,R7 and larval photoreceptor opsin	15	EMBOJ6:443
92E	metallothionein (Mto)	15	Wegnez,Paris
92E	[HSH11];hsp70-h, neo R	3	Cell51:405
92E	anonymous	1	Wolfner,Cornell
92F	[AS634(lethal)];ry+	3	Bloomington Stock
92F	[BS2.7-2];chorion+,ry+	3	Cell34:47
93	[AS383];ry	3	Bloomington Stock
93A	[C20Pgd8.9-1];6Pgd+, ry+	3	Lucchesi,ChaplHil
93A	[P1/36];per(period), ry+	3	Nature326:390
93AB	[TnJW1];adh+, dpp+	3	GeneDev1:615
93AB	[AS328-R310.1];ry+	3	Cell34:47
93B	Na+K+ ATPase (Sodium pump) alphasubunit	156	EMBOJ8:193
93B	[P-1(2)gl-24B];lgl+,neoR	3	Cell50:215
93B	[ch50];sev+, neo+	3	EMBOJ(1989)
93B	[AS291-neo^R #54 lethal];neo^R	3	Science239:1121
93B	rudimentary-like (r-l)	4	Rawls,UKy
93B	[AS44];neo	3	Bloomington Stock
93B	[AS633(lethal)];ry+	3	Bloomington Stock
93B	[WG997(lethal)];ry+	3	Bloomington Stock
93CD	ebony, 93D heat shock locus	1	EMBOJ3:2499
93D	[AS612(lethal)];ry+	3	Bloomington Stock
93D	[AS635(lethal)];ry+	3	Bloomington Stock
93D	[GR878, GR882];w	3	Bloomington Stock
93D	[AS292-neo^R #55 lethal];neo^R	3	Science239:1121
93D	heat shock	15	MGG204:334
93D	heat-shock	1	Chrm93:17
93D	anonymous	1	Rawls,UKy
93D	[26Z-93D];hsp26, lacZ, ry+	3	EMBOJ5:747
93D	[SB2.1-4];chorion+,ry+	3	Cell34:47
93D	heat shock (3 transcripts)	1	PNAS83:1812
93D	[g5:1];Sgs3+,ry+	3	Cell40:349
93E	insulin like receptor (DILR)	15	BBRC141:474
93E	dIRH (Dros. insulin receptor homolog)	6	MCB8:1638
93E	DIR-insulin receptor homologue	16	PNAS83:4710
93E	related NK-3, NK-4 (Hbox sequences)	16	Webber,NIH
93E	PRD gene9 (paired homology)	16	Cell47:735
93E	NK-1, NK-2	1	Webber,NIH
93E	[AS596(lethal)];ry+	3	Bloomington Stock
93F	[AS759];ry+	3	Bloomington Stock
93F	[C20Pgd8.9-6];Pgd+, ry+	3	Lucchesi,ChaplHil
93F	[AS616(lethal)];ry+	3	Bloomington Stock
94A	anonymous	1	BBA867:209
94A	anonymous	1	Wolfner,Cornell
94B	[P[rib7(94B)ry+/Sco;ry^506]];rDNA, ry+	3	GeneDev2:1745
94B	[28P-D (lost?)];ry+,hsp28	3	MCB6:663
94D	[AS414];ry+	3	Bloomington Stock
94D	anonymous	1	Wolfner,Cornell
94D	[P[(w,ry)E]7];w, ry+	3	EMBOJ4:3489
94D	[23.2];hsp23, ry+	3	EMBOJ5:755
94DE	[AS293-neo^R #56, #57 lethal];neo^R	3	Science239:1121

Locn	Genes-Name	Key	References
94E	anonymous	1	Davidson,CalTech
94E	[AR4-020, BL929, GR868];w,ry+	3	Science229:558
94EF	[swallow-bb854];sww+,ry+	3	GeneDev2:1655
94F	[AS606(lethal)];ry+	3	Bloomington Stock
94F	oocyte TU	16	Cell14:921
95A	[H(har1)M2];ry+	3	EMBOJ8:211
95A	[WG1120];ry	3	Bloomington Stock
95A	HMG CoA reductase	156	MCB8:2713
95A	[R601.1];ry+	3	Cell34:47
95AB	en-like homeobox	1	Gustafson,UCSF
95AB	[27S-C];hsp28, ry+	3	MCB6:663
95B	[AS113];neo	3	Bloomington Stock
95B	anonymous	1	Cell28:165
95C	snRNAU1	1	NAR16:3582
95C	[AS45];neo	3	Bloomington Stock
95C	[hsneo;Rh1(-833/+67)-cat20];neo, CAT	3	Genet116:565
95C	blastoderm-differential poly(A) TU	16	DevB109:476
95C	U1 snRNA	15	NAR12:8835
95C	[D1];rp49+,ry+	3	Nature317:555
95D	hsp68	1	Cell18:1359
95D	[27S-A];hsp28, ry+	3	MCB6:663
95D	[BS2.7-8];chorion+,ry+	3	Cell34:47
95D	[unn];ry+	3	Chovnick,UConn
95D	L-glutamate dehydrogenase (Gdh)	1	Cristos,Heraklion
95DE	[60-3A[Ddc(-208,-83)]];Adh+,Ddc+	3	GeneDev1:510
95E	mst 316, access. gland trans.+2 others	1	DevB119:242
95EF	mst 345a, mst 345b, 2 testis-spec.trans.	1	DevB119:242
95F	[AS46, AS105];neo	3	Bloomington Stock
95F	[BL926, GR873];w,ry+	3	Bloomington Stock
95F	crumbs (crb)	1	EMBOJ6:3431
95F	[28A-A (lost?)];ry+,hsp28	3	MCB6:663
95F	crumbs (crb) EGF like domains	5	Knust,Koln
96	[DP1230];ry	3	Bloomington Stock
96	[DR-2];ry+,Ddc+	3	Cell34:37
96A	snRNAU6	1	NAR16:3582
96A	nicotinic acetylcholine receptor, nAChR	45	EMBOJ7:611
96A	[AS401];ry	3	Bloomington Stock
96A	anonymous	1	Wolfner,Cornell
96AB	[P[(w,ry)G]3];w, ry+	3	EMBOJ4:3489
96B	[neo^R #58 lethal];neo^R	3	Science239:1121
96B	[S3.8-1];chorion+, ry+	3	Cell51:763
96B	[S38M-3];s38+,M13+,ry+	3	Cell34:47
96D	[AS47];neo	3	Bloomington Stock
96D	anonymous	1	Cell28:165
96EF	[P-l(2)gl-24D];lgl+,neoR	3	Cell50:215
96F	[AS48];neo	3	Bloomington Stock
96F	E(spl), myc homology	56	EMBOJ8:203
96F	E(spl), gro	4	EMBOJ6:4113
96F	E(spl), transducin homology	456	Cell55:785
96F	anonymous	1	Wolfner,Cornell
96F	anonymous	1	Wolfner,Cornell
97A	anonymous	1	Davidson,CalTech
97A	[28N];hsp28, ry+	3	MCB6:663
97A	[tAP-16,4.8];Adh+	3	Posakony,SanDiego
97AB	[+65];hsp70,ry+	3	Cell40:805
97AB	[28-term (lost?)];ry+,hsp28	3	MCB6:663
97B	[P[(w,ry)F]4-3];w, ry+	3	EMBOJ4:3489
97C	[H(har1)M17];ry+	3	EMBOJ8:211
97C	anonymous	1	Wolfner,Cornell

Locn	Genes-Name	Key	References
97CD	histone H2A variant (H2AvD)	15	NAR16:7487
97CD	[AS295-neo^R #59 lethal];neo^R	3	Science239:1121
97D	[P-l(2)gl-25B];lgl+,neoR	3	Cell50:215
97D	[lacZ;ry at Toll];ry+	3	Bellen,Basel
97D	[L2a];ry-+, P-beta gal	3	Kerridge,Marseill
97D	Toll	156	Cell52:269
97D	rough	15	Nature334:151
97D	rough (Hbox)	156	Cell55:771
97EF	[AS296-neo^R #60 lethal];neo^R	3	Science239:1121
97EF	beta tubulin	1	DevB104:187
97EF	[P1/25];per(period), ry+	3	Nature326:390
97F	[AS432];ry	3	Bloomington Stock
97F	[AS49];neo	3	Bloomington Stock
97F	beta tubulin	1	JMB156:449
97F	[unn];ry+	3	Chovnick,UConn
97F	lethal(3)malignant brain (l(3)mbt	1	Gateff,Mainz
98A	[pPA-1 abl];abl, Adh	3	Cell51:821
98A	[BS.27-1];chorion+,ry+	3	Cell34:47
98AB	rpL1 (ribosomal protein gene)	156	NAR16:4915
98B	[cHB delta-89];hsp70-lacZ,ry+	3	Cell40:805
98C	[H(har1)M6];ry+	3	EMBOJ8:211
98C	mst(3)gl12,13 testis-specific	16	Schafer,Duseldrof
98C	[AS402-R602.1, R705.1];ry+	3	Cell34:47
98CE	mst 336a,336b,plus non sex-specif.transc	1	DevB119:242
98D	forked head(fkh), distal of fkh (dfk)TU	4567	Cell57:645
98D	anonymous	1	EMBOJ6:3431
98DE	Hrb98DE (p9),(ss nuc.acid.bind. protein)	16	PNAS84:1819
98DE	oogenesis specific	1	Tucker,Rice
98E	[P-l(2)gl-21B];lgl+,neoR	3	Cell50:215
98E	Maternal restricted TU	16	Stephenson,Rochst
98E	anonymous	1	EMBOJ6:3431
98E	[23.2];hsp23, ry+	3	EMBOJ5:755
98EF	sry H (serendipity H1) "fingers"	156	MCB8:4459
98F	myosin alkali light chain (MLCALK)	6	DevB121:263
98F	protein kinase C (dPKC 98F)	156	Cell57:403
98F	[AS110];neo	3	Bloomington Stock
98F	anonymous	1	BBA867:209
98F	0-2.5 hrs. Tu	16	Lengyel,UCLA
98F	myosin alkali light chain	1	MCB4:956
98F	doa: dosage sensitive suppressor	1	Rabinow,Harvard
98F	maternal or yema Tu, yema gene region	46	DevB122:153
98F	string (stg)	156	Cell57:177
99A	[AS610, AS631(lethal)];ry+	3	Bloomington Stock
99A	[H(har1)M19];ry+	3	EMBOJ8:211
99A	[H(har1)M20];ry+	3	EMBOJ8:211
99A	[AS297-neo^R #61,#62 lethal@string];neo^R	3	Science239:1121
99A	[g6:3];Sgs+,ry+	3	Cell40:349
99AB	[DA24-44A[Ddc(-208)]];Adh+, Ddc+	3	GeneDev1:510
99B	anonymous	56	Jacq,Marseilles
99B	Homeo box	1	EMBOJ2:2027
99B	[delta 2-3, GR869];ry+	3	Cell44:219
99B	[WG963(l)];ry	3	Bloomington Stock
99C	[6.1UZ-3];lacZ,neoR	3	Irvine,Stanford
99C	Head specific RNA	1	DevB94:451
99C	acid phosphatase	12	MacIntyre,Cornell
99C	transient receptor potential	1	IOVS26:243
99CF	anonymous	1	Wolfner,Cornell
99D	janus A,B	156	MCB9:2526
99D	blastoderm-specific poly(A) TU sry	16	DevB109:476

Locn	Genes-Name	Key	References
99D	serendipity alpha, beta, delta TU	16	JMB186:149
99D	ribosomal protein, Minute	1	Nature285:674
99D	[SRS3.9-3];chorion+,ry+	3	Cell34:47
99E	anonymous	1	Wolfner,Cornell
99E	myosin light chain-2	15	Pardue,MIT
99E	myosin light chain	1	JMB156:449
99E	myosin light chain 2	1	DevB109:476
99E	[B 25];Sgs-4	3	Korge,Berlin
99E	myosin light chain-2 (MLC-2)	6	Genet122:139
99E	blastoderm differential poly(A) TU	16	DevB109:476
99E	cecropin gene cluster	1	Hultmark,Stockhlm
99EF	myosin light chain 2	16	Gene74:457
99F	[AS50];neo+	3	Bloomington Stock
99F	[AS601(1)];ry	3	Bloomington Stock
99F	[H(har1)M18-2];ry+	3	EMBOJ8:211
99F	[AS299-neo^R #63 lethal];neo^R	3	Science239:1121
99F	anonymous	1	Wolfner,Cornell
100	anonymous	1	BBA867:209
100A	cAMP dependent kinase	156	GeneDev2:1539
100A	[AS51];neo+	3	Bloomington Stock
100A	[AS622(lethal)];ry+	3	Bloomington Stock
100A	[AS82];neo+	3	Bloomington Stock
100AB	tailless (tll)	4	Pignoni,UCLA
100AB	anonymous	1	Goldstein,ArizSU
100B	head pecific TU	16	DevB94:451
100B	Ag24B10	15	PNAS82:1855
100B	anonymous	1	Davidson,CalTech
100B	anonymous	1	Cell28:165
100B	Chaoptin (chp), (Ag24B10)	56	Cell52:291
100C	double sex cognate	1	Baker,Stanford
100C	anonymous	1	Cell28:165
100CD	(k-pn), awd is NM23 homologue	5	Shearn,Hopkins
100CD	awd(abnormal wing discus), (Kpn)	1	DevB129:169
100D	tramtrack (DNA binding protein)	16	Harrison,Cambrige
100D	[C20Pgd7.4-A];Pgd+, ry+	3	Lucchesi,ChaplHil
100D	[act-line II];actin, beta-gal, ry+	3	DevB133:313
100D	anonymous	1	Cell28:165
100D	[unn];ry+	3	Chovnick,UConn
100E	F2, pupal mTU in both males and females	156	PNAS82:5795
100EF	[H(har1)G18-1];	3	EMBOJ8:211
100EF	205K MAP gene	1	JCB102:2076
100F	microtubule prot. & repetitive telomere	1	Levis,Seattle
100F	(LA9) DNA binding protein	1	Pradel,Marseille
100F	[WG1155];ry+	3	Bloomington Stock
100F	telomeres	1	Cell34:85
100F	anonymous	1	Chrm89:206
100F	[A4-4];w+,ry+	3	Cell36:469
101F	[line 1];rp21, ry+	3	MGG213:354
102C	homology to ribosomal spacer	1	DevB119:242
102C	anonymous	1	Cell28:165
102CD	anonymous	1	Cell25:693
102EF	anonymous	1	Davidson,CalTech
102F	ARS element	1	MGG197:342
102F	telomeres	1	Cell34:85
Het	simple sequence 1.672/cc satellite	15	NAR15:10061
Het	repeat element 1360, intercal het	15	Chrm97:247

Gene map of the Yellow Fever Mosquito
(Aedes (Stegomyia) aegypti) (2N=6)

August, 1989

Leonard E. Munstermann; Vector Biology Laboratory;
University of Notre Dame; Notre Dame, IN 46556, USA

Primary reference: J. Hered. 70:291-296 (1979).

Genetic loci of *Aedes aegypti*

No.	Gene symbol	Phenotypic feature	Linkage group	References
1.	Acp-2	Acid-phosphatase-2	I	43-45
2.	blp	black-palp	III	2,13,24,40,43,64
3.	blt	black-tarsus	III	3,13,18,22-24,31-36,39,41-43, 53,56,70,73,76,79-80,82
4.	bpd	black-pedicel	II	10,13,40
5.	Bt	Black-tergite	II	30,54
6.	bw	brown-eye	II	49,80
7.	bz	bronze-cuticle,sterility	I	2-5,13-14,18-19,40,75
8.	bz^2	bronze-cuticle,fertile	I	61
9.	co	compressed-antenna	III	13,24,64
10.	cr	cream-eye	II	43-45
11.	D	Distorter,sex ratio	I	13,21,46-48,50,52-53,79-82
12.	Dl	Dieldrin-resistance	II	13,15,25-26,29
13.	DDT(2)	DDT-resistance	II	10,13,25,29,31-32,34
14.	DDT(3)	DDT-resistance	III	24,31-32,34,76
15.	$DDT(3)^{py}$	Pyrethroid-resistance	III	31-33
16.	ds	dark-scutum	II	2,13,25
17.	Est-4	Esterase-4	I	35-36
18.	Est-5	Esterase-5	I	16
19.	$Est\text{-}5^{Mal}$	Malathion resistance	I	16
20.	Est-6	Esterase-6	II	15,17,22,35-36,43-45, 50,60,68
21.	$Est\text{-}6^{Mal}$	Malathion resistance	II	17
22.	f^m	filaria-susceptibility, *Brugia malaya*	I	13,37,59,67,83
23.	f^{m2}	fil.-susc., *Foleyella flexicauda*	I	67
24.	fr	fil.-susc., *Dirofilaria repens*	I	11,37
25.	fi	fil.-susc., *Dirofilaria immitis*	I	37,62-63
26.	fz	fuzzy-scale	III	2,13,24,40,43,64
27.	G	Gold-mesonotum	II	8,13
28.	gr	grey-body	I	13
29.	Gpd-1	Glycerol-3-phosphate-dehydrogenase	II	43-45,66
30.	Gpi-1	Glucosephosphate-isomerase-1	III	22,43-45,50
31.	h	halteres	II	8,13,25
32.	Hk-2	Hexokinase-2	III	64
33.	Hk-3	Hexokinase-3	III	64
34.	Hk-4	Hexokinase-4	III	6,9,44,64
35.	Ib	Intercalary-band	I	47
36.	Idh-1	Isocitrate-dehydrogenase-1	I	75

Linkage Map of *Aedes aegypti*

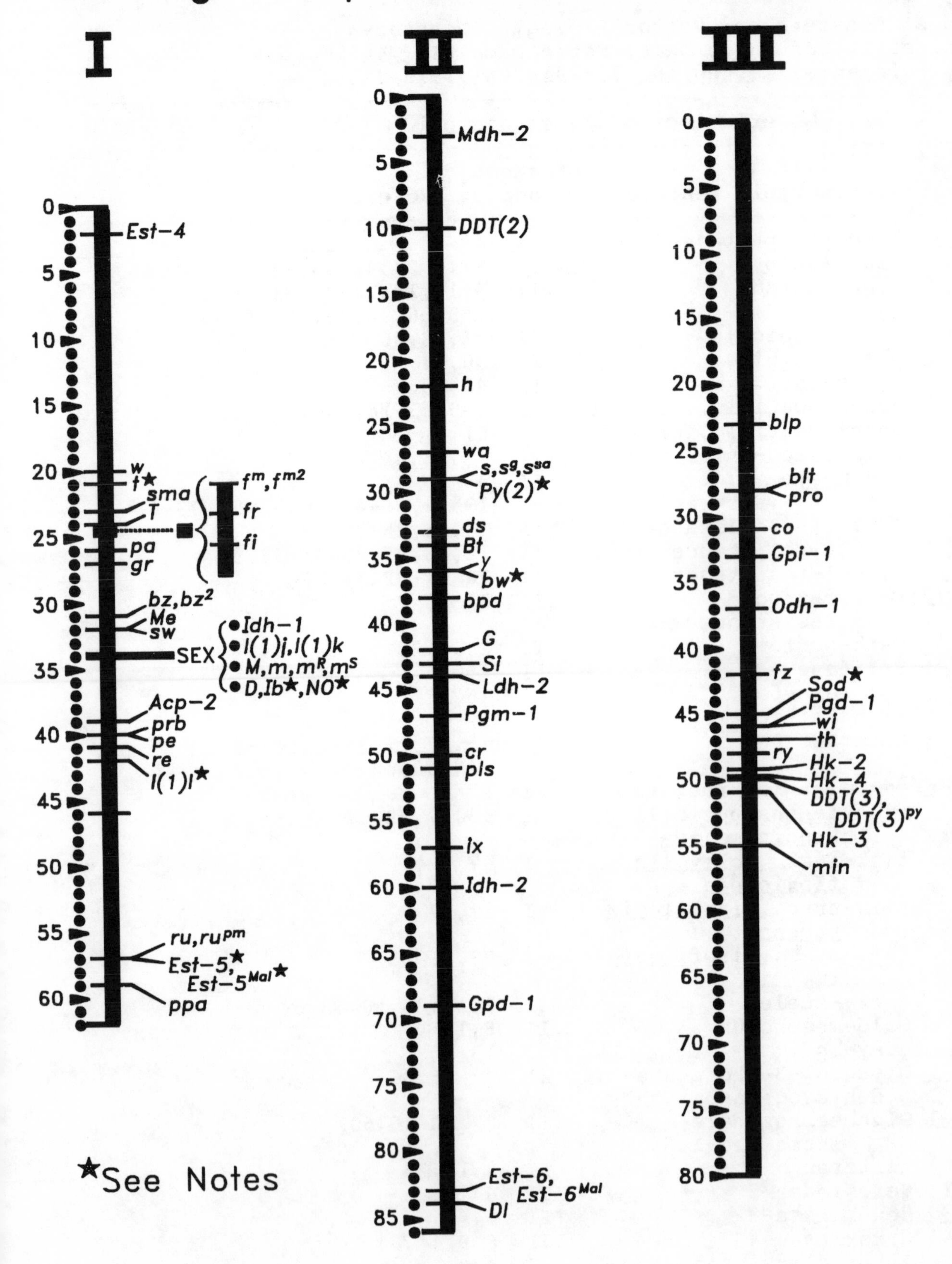

No.	Gene symbol	Phenotypic feature	Linkage group	References
37.	Idh-2	Isocitrate-dehydrogenase-2	II	9,22,43-45,50,66
38.	ix	intersex	II	13
39.	l(1)j	lethal	I	75
40.	l(1)k	lethal	I	75,78
41.	l(1)l	lethal	I	78
42.	Ldh-2	Lactate-dehydrogenase-2	II	22,43,45
43.	m, M	Sex	I	1-5,9,11,13-14,19,22,35-37, 39-53,56-57,61-62,65,67,69, 75,78-82
44.	m^R	resistant-Distorter	I	79,82
45.	m^S	susceptible-Distorter	I	47-48,52,79-82
46.	Mdh-2	Malate-dehydrogenase-2	II	43-45
47.	Me	Malic-enzyme	I	44,50,65
48.	min	miniature-body	III	2,13,24,40,43,64,73
49.	NO	Nucleolar-Organizer	III	3,18,27,39,73
50.	Odh-1	Octanol-dehydrogenase-1	III	16,43-45,50
51.	pa	pale-abdomen	I	13
52.	pe	palp-extended	I	19,57
53.	Pgd-1	Phosphogluconate-dehydrogenase-1	III	22,43-45,50
54.	Pgm-1	Phosphoglucomutase-1	II	7,9,43-45,50
55.	pls	plasmodium-susceptibility	II	13,26
56.	ppa	palp-antenna	I	55
57.	prb	proboscipedia	I	19,57
58.	pro	prolapsis (peritrofica)	III	45,58
59.	Py(2)	Pyrethroid-resistance	II	74
60.	re	red-eye	I	2-3,5,13,16,20,22,31-33,35-38, 40,42-43,48-49,51,53,55-57, 61-62,66,70,79-82
61.	ru	rust-eye	I	2,5,13,38,40,53,55-56
62.	ru^{pm}	plum-eye	I	45
63.	ry	rosy-eye	III	45
64.	s	spot-abdomen	II	2,3,8,10,13,15,18,22-23,25, 29,31-33,35-36,40-43,60,68, 74,76,79-80,82
65.	s^g	white-abdomen	II	71
66.	s^{sa}	sable-scale	II	45
67.	Si	Silver-mesonotum	II	13,18,26,40-41,43,66,70
68.	sma	small-antenna	I	14,57
69.	Sod	Superoxide-dismutase	III	6,9
70.	sw	short-wing	I	51,69
71.	t	tolerance-Distorter	I	48,82
72.	T	Terminalia	I	13,20
73.	th	tarsi-hooked	III	13
74.	w	white-eye	I	1-5,13-14,18,40-42,49,51, 53,55,67,76
75.	wa	wart-palp	II	13,25
76.	wi	withered-leg	III	13
77.	y	yellow-larva	II	7-8,13,16,25,29,35-36,43,49, 60,68,76-77

Notes: Aedes aegypti linkage maps

(1) Details concerning construction of the map model (see figure) are found in Munstermann and Craig (42) and based on the chiasmata frequencies of Ved Brat and Rai (68). The total map length (average number chiasmata/nucleus x 50 = 212 units) was partitioned to each linkage group by relative chromosome lengths. Relative map positions of loci and known linkage distances between them were centered on the expected maximum map length for each linkage group. Interlocus distances were rounded to the nearest unit unless interlocus comparisons were sufficiently accurate to warrant more detail, i.e, the filarial susceptibility complex (ref.37) and the hexokinase loci (ref.64).

(2) STARRED LOCI:
   A. Mapped with two point crosses only:
      --l(1)l (with respect to M)
      --bw (with respect to s)
      --Py(2) (with respect to s)
      --Sod (with respect to Hk-4).
      --t (with respect to re).
   B. An approximated position for a physical chromosomal marker
      --Ib, NO. For the first time the NO was demonstrated by in situ hybridization to be sex-linked (Kumar & Rai (27)), not on LG III (inferred from the LG III secondary constriction).

(3) Gene symbols and mutant names, wherever possible, were based on the conventions that Lindsley and Grell (28) followed for Drosophila melanogaster. Enzyme nomenclature followed the recommendations of the Commission on Biochemical Nomenclature (12).

(4) The first linkage map for Aedes aegypti was published by Craig and Hickey (13) 20 years ago. The present summary is an updated version of the 1979 map (42) and of the versions published in GENETIC MAPS vol.2, 3, and 4. Literature references previous to 1966 can be found in Craig and Hickey (13).

(5) This work was supported by NIH Grant AI-02753 to Prof. George B. Craig, Jr.

REFERENCES

1. BHALLA,S.C. 1968. Mosquito News 28:380-385.
2. BHALLA,S.C. 1971. Can.J.Genet.Cytol. 13:561-577.
3. BHALLA,S.C. 1973. Can.J.Genet.Cytol. 15:9-20.
4. BHALLA,S.C. & G.B.CRAIG,JR. 1967. J.Med.Entomol. 4:467-476.
5. BHALLA,S.C. & G.B.CRAIG,JR. 1970. Can.J.Genet.Cytol. 12:425-435.
6. BULLINI,L. & M.COLUZZI. 1978. Parassitologia 20:7-21.
7. BULLINI,L.,et al. 1972. Biochem.Genet. 7:41-44.
8. CARMELINA FLORES,C. & C.E.MACHADO-ALLISON. 1976. Acta Cient. Venez. 27:314-316.
9. CIANCHI,R.,et al. 1978. Parassitologia 20:47-58.
10. COKER,W.Z. 1966. Ann.Trop.Med.Parasitol. 60:347-356.
11. COLUZZI,M. & G.CANCRINI. 1974. Parassitologia 16:239-256.
12. COMMISSION ON BIOCHEMICAL NOMENCLATURE. 1984. Enzyme Nomenclature. Academic Press,Orlando,FL (USA).
13. CRAIG,G.B.,JR. & W.A.HICKEY. 1967. Genetics of Insect Vectors of Disease. J.W.Wright & R.Pal(eds). Elsevier,New York. p.67-131.
14. DUNN,M.A. & G.B.CRAIG,JR. 1968. J.Hered. 59:131-140.
15. FIELD,W.N. & J.M.HITCHEN. 1981. J.Med.Entomol. 18:61-64.
16. FIELD,W.N. & J.M.HITCHEN. 1987. J.Med.Entomol. 24:512-514.
17. FIELD,W.N.,et al. 1984. J.Med.Entomol. 21:412-418.
18. HALLINAN,E.,et al. 1977. Proc.XV Internat.Congr.Entomol. J.S. Packer & D.White,eds. Entomol.Soc Am.,College Park,MD (USA). p.117-128.
19. HARTBERG,W.K. 1975. Mosquito News 35:34-41.
20. HARTBERG,W.K. & G.B.CRAIG,JR. 1973. Mosquito News 33:206-214.
21. HICKEY,W.A. & G.B.CRAIG,JR. 1966. Genetics 53:1177-1196.
22. HILBURN,L.R. & K.S.RAI. 1982. J.Hered. 73:59-63.
23. HITCHEN,J.M. & R.J.WOOD. 1974. Can.J.Genet.Cytol. 16: 177-182.
24. HITCHEN,J.M. & R.J.WOOD. 1975. Can.J.Genet.Cytol. 17: 311-322.
25. HITCHEN,J.M. & R.J.WOOD. 1975. Can.J.Genet.Cytol. 17: 543-551.
26. KILAMA,W.L. & G.B.CRAIG,JR. 1969. Ann.Trop.Med.Parasitol. 63:419-432.
27. KUMAR,A. & K.S.RAI. 1989. Personal Communication.
28. LINDSLEY,D.L. & E.H.GRELL. 1968. Genetic variations of Drosophila melanogaster. Carnegie Inst.Washington, No.627.

References, cont'd.

---

29. LOCKHART,W.L.,et al. 1970. Can.J.Genet.Cytol. 12:407-414.
30. MACHADO-ALLISON,C.E. 1971. Ph.D. Diss.,U.Notre Dame,IN (USA).
31. MALCOLM,C.A. 1983. Genetica 60:213-219.
32. MALCOLM,C.A. 1983. Genetica 60:221-229.
33. MALCOLM,C.A. & R.J.WOOD. 1982. Genetica 59:233-237.
34. MARGHAM,J.P. & R.J.WOOD. 1975. Heredity 34:53-59.
35. MARVDASHTI,R. 1980. Ph.D. Diss., U.Manchester (England).
36. MARVDASHTI,R. 1985. J.Am.Contr.Assoc. 1:423-424.
37. MATTHEWS,H.A. 1987. Ph.D. Diss., U.Liverpool (England).
38. McCLELLAND,G.A.H. 1966. Can.J.Genet.Cytol. 8:192-198.
39. McDONALD,P.T. & K.S.RAI. 1970. Genetics 66:475-485.
40. McGIVERN,J.J. & K.S.RAI. 1972. J.Hered. 63:247-255.
41. McGIVERN,J.J. & K.S.RAI. 1974. J.Hered. 65:71-77.
42. MOTARA,M.A. & K.S.RAI. 1980. J.Entomol.Soc.Sth Afr. 43:129-138.
43. MUNSTERMANN,L.E. 1979. Ph.D. Diss.,U.Notre Dame,IN (USA).
44. MUNSTERMANN,L.E. 1981. Appl.Genet.Cytol.Insect Systemat.Evol. M.W. Stock,ed. Univ. Idaho,Moscow,ID (USA). p. 129-140.
45. MUNSTERMANN,L.E. & G.B.CRAIG,JR. 1979. J.Hered. 70:291-296.
46. NEWTON,M.E.,et al. 1976. Genetica 46:297-318.
47. NEWTON. M.E.,et al. 1978. Genetica 48:137-143.
48. OUDA,N.A. 1979. Ph.D. Diss.,U.Manchester (England).
49. OUDA,N.A. & R.J.WOOD. 1983. Ann.Trop.Med.Parasitol. 77:211-218.
50. PASHLEY,D.P. & K.S.RAI. 1983. Biochem.Genet. 21:1195-1201.
51. PEARSON,A.M. 1980. Heredity 44:425-429.
52. PEARSON,A.M. & R.J.WOOD. 1980. Genetica 51:203-210.
53. PEARSON,A.M. & R.J.WOOD. 1980. Genetica 54:79-85.
54. PETERSEN,J.L. 1977. Ph.D. Diss.,U.Notre Dame,IN (USA).
55. PETERSEN,J.L.,et al. 1976. J.Hered. 67:71-78.
56. REES,A.T. & J.M.HITCHEN. 1983. Proc.Spring Meeting Br. Soc.Parasitol. Salford (England).
57. ROBERTS,J.R. & W.K.HARTBERG. 1979. Mosquito News 39:348-359.
58. RODRIGUEZ,D.J. & C.E.MACHADO-ALLISON. 1977. Acta Biol.Venez. 9:347-375.
59. RODRIGUEZ,P.H. 1985. J.Med.Entomol. 22:366-369.
60. SAUL,S.H.,et al. 1976. Ann.Entomol.Soc.Am. 69:73-79.
61. SULAIMAN,I. 1982. J.Med.Entomol. 19:706-709.
62. SULAIMAN,I. 1985. Trop.Biomed. 2:47-53.
63. SULAIMAN,I. & H.TOWNSON. 1980. Ann.Trop.Med.Parasitol. 74:635-646.
64. TABACHNICK,W.J. 1978. Biochem.Genet. 16:571-575.
65. TABACHNICK,W.J. & J.M.LICHTENFELS. 1978. Isozyme Bull. 11:53.
66. TABACHNICK,W.J. & G.P.WALLIS. 1984. Am.Soc.Trop.Med.Hyg.Abstr. 1984:170.
67. TERWEDOW,H.A. & G.B.CRAIG,JR. 1977. Exp.Parasitol. 41:272-282.
68. TREBATOWSKI,A.M. & G.B.CRAIG,JR. 1969. Biochem.Genet. 3:383-392.
69. UPPAL,D.K.,et al. 1976. Heredity 36:147-150.
70. VANDEHEY,R.C.,et al. 1979. Ann.Entomol.Soc.Am. 72:509-513.
71. VANDEHEY,R.C. & G.B.CRAIG,JR. 1962. Ann.Entomol.Soc.Am. 55:58-69.
72. VED BRAT,S. & K.S.RAI. 1973. The Nucleus 16:184-193.
73. VED BRAT,S. & K.S.RAI. 1973. Genetics 74: 283s-284s.
74. WALKER & R.J.WOOD. 1984. Unpublished observations.
75. WALLIS,G.P. & W.J.TABACHNICK. 1982. J.Hered. 73:291-294.
76. WOOD,R.J. 1967. Genet.Res. 10:219-228.
77. WOOD,R.J. 1970. Genet.Res. 16:37-47.
78. WOOD,R.J. 1976. Genetica 46:49-66.
79. WOOD,R.J. & M.E.NEWTON. 1982. Rec.Dev.Genet.Insect Dis.Vectors. W.W.M.Steiner,et al.(eds). Stipes,Champaign,IL (USA). p.130-152.
80. WOOD,R.J. & N.A.OUDA. 1987. Genetica 72:69-79.
81. WOOD,R.J.,et al. 1977. J.Med.Entomol. 14:461-464.
82. WOOD,R.J.,et al. 1981. Cytogenet.Genet.Vectors. R.Pal,et al. (eds). Elsevier,Amsterdam (The Netherlands). p. 169-177.
83. ZIELKE,E. 1973. Z.Tropenmed.Parasitol. 24:36-44.

Linkage Map of *Aedes triseriatus*

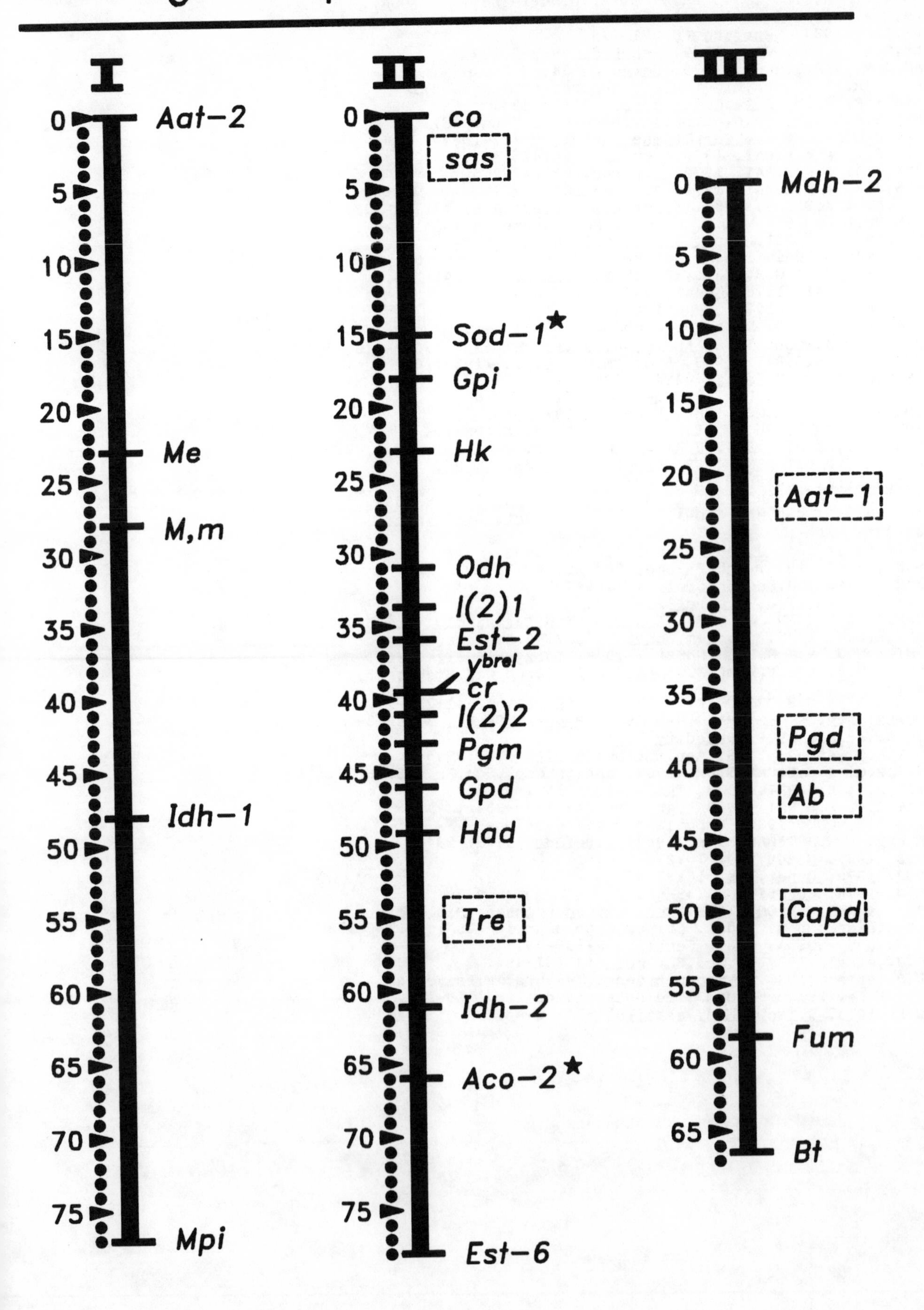

Gene Map of the Eastern North American Tree Hole Mosquito, *Aedes* (*Protomacleaya*) *triseriatus* (2N=6)

August, 1989

Leonard E. Munstermann; Vector Biology Laboratory
University of Notre Dame; Notre Dame, IN 46556, USA

Primary reference: Munstermann, L.E. 1981. In: Application of Genetics and Cytology in Insect Systematics and Evolution. M.W. Stock, ed. Forest, Wildl. & Range Expt. Sta., Univ. Idaho, Moscow, ID (USA). p.129-140.

Genetic loci of *Aedes triseriatus*

No.	Gene symbol	Phenotypic feature	Linkage group	References
1.	Aat-1	Aspartate-aminotransferase-1	III	2-3,5-6
2.	Aat-2	Aspartate-aminotransferase-2	I	2-3,5-6
3.	Ab	Abdominal-banding	III	5-6
4.	Aco-2	Aconitate-hydratase-2	II	2,4
5.	Bt	Banded-tarsi	III	5-6
6.	co	copper-thorax	II	6
7.	cr	cream-eye	II	3,5
8.	Est-2	Esterase-2	II	2,4
9.	Est-6	Esterase-6 (=Est-5)	II	2-3,5
10.	Fum	Fumarate-hydratase	III	2-3,5-6
11.	Gapd	Glyceraldehyde-3-phosphate-dehydrogenase	III	2,4
12.	Gpd	Glycerol-3-phosphate-dehydrogenase	II	2-3,5-6
13.	Gpi	Glucosephosphate-isomerase	II	2-3,5-6
14.	Had	Hydroxyacid-dehydrogenase (=Hydroxybutyrate-dehydrogenase)	II	1-3,5
15.	Hk	Hexokinase	II	1-3,5-6
16.	Idh-1	Isocitrate-dehydrogenase-1	I	2-3,5-6
17.	Idh-2	Isocitrate-dehydrogenase-2	II	2-3,5-6
18.	l(2)1	lethal	II	1
19.	l(2)2	lethal	II	1
20.	M, m	Sex	I	3,5-6
21.	Mdh-2	Malate-dehydrogenase-2	III	2-3,5-6
22.	Me	Malic-enzyme	I	2-3,5-6
23.	Mpi	Mannosephosphate-isomerase	I	2,4
24.	Odh	Octanol-dehydrogenase	II	1-3,5-6
25.	Pgd	6-Phosphogluconate-dehydrogenase	III	2-3,5
26.	Pgm	Phosphoglucomutase	II	2-3,5
27.	sas	silver-acrostichal-scale	II	6
28.	Sod-1	Superoxide-dismutase-1	II	2,4
29.	Tre	Trehalase	II	5-6
30.	y^{brel}	yellow-larvae(*Ae. brelandi*)	II	5-6

REFERENCES

1. MATTHEWS,T.C. & G.B.CRAIG,JR. 1989. J.Hered. 80:53-57.
2. MATTHEWS,T.C. & L.E.MUNSTERMANN. 1990. J.Hered. 81(1): in press.
3. MUNSTERMANN,L.E. 1981. In: Application of Genetics and Cytology in Insect Systematics and Evolution. M.W.Stock,ed. Forest,Wildl.& Range Expt.Sta.,Univ.Idaho,Moscow,ID (USA). p.129-140.
4. MUNSTERMANN,L.E. Unpublished data.
5. MUNSTERMANN,L.E.,et al. 1982. In: Recent Developements in the Genetics of Insect Disease Vectors. W.W.M.Steiner,et al.,eds. Stipes Publ.Co.,Champaign,IL (USA). p.433-453.
6. TAYLOR,D.B. 1982. Ph.D.Diss. U.Notre Dame,IN (USA).

Notes: *Aedes* (*Protomacleaya*) *triseriatus* and homologies with *Aedes*(*Stegomyia*) *aegypti*.

(1) For previous maps, most of the linkage relationships seen in Fig.1 were derived from interspecific crosses between the closely related species of the Triseriatus Group--*Ae.triseriatus*, *Ae.hendersoni* and *Ae.brelandi*. Mapping of the 21 enzyme loci was facilitated by the presence of both diagnostic alleles and alleles of high frequency difference between species pairs. Most of these loci have been remapped by intraspecific crosses with *Ae.triseriatus* (2).

(2) Loci for which the relative positions are determined or give inconsistent results are marked on the map in 2 ways:
  A. **Starred loci** are mapped with two-point crosses only:
    --**Sod** (with respect to **Odh**)
    --**Aco-2** (with respect to **Idh-2**)
  B. **Boxed loci** are those showing inconsistent results or for which there are only small sample sizes. The loci on Linkage Group III are positioned relative only to the end loci, *viz*., **Pgm** and **Fum**, but not to one another.

(3) More than 1,000 species have been described for the genus *Aedes*. All have 3 pairs of chromosomes with very similar metaphase morphologies. Polytene chromosome preparations are difficult to obtain in this genus. Comparisons of the linkage maps of *Aedes triseriatus* (subgenus *Protomacleaya*) and *Aedes aegypti* (subgenus *Stegomyia*) indicate at least one whole-arm, reciprocal translocation has occurred between the chromosomes corresponding to the second and third linkage groups (Fig.2). The crossed arrows in the figure show the presence of numerous inverted sequences as well. Apparently many more more chromosomal rearrangements than previously suspected have occurred in the evolutionary divergence of the subgenera in *Aedes* mosquitoes.

(4) This work was supported by NIH Grant AI-02753 to Prof.George B. Craig,Jr.

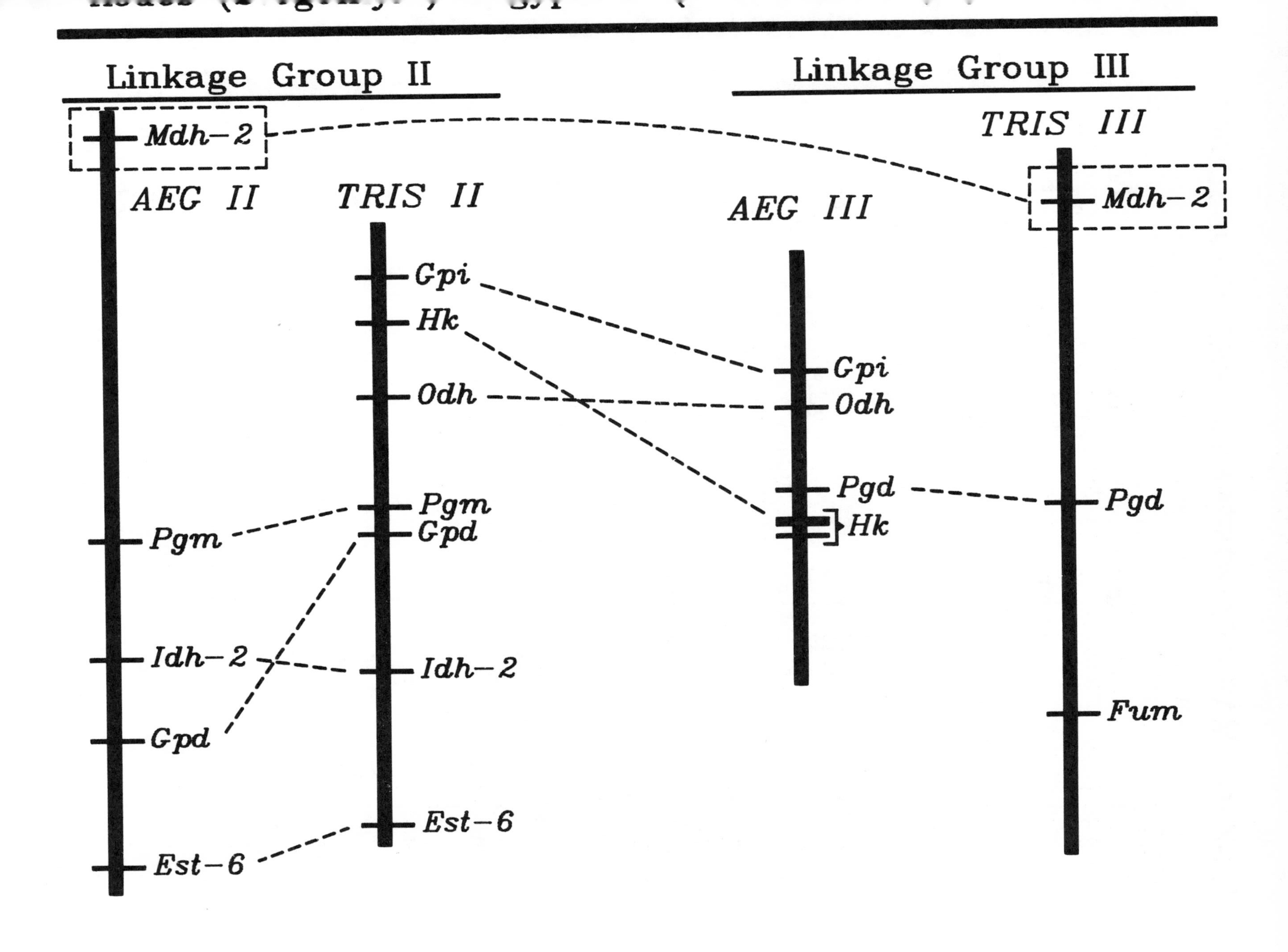
Linkage Group II
Linkage Group III
Mdh–2
AEG II
TRIS II
AEG III
TRIS III
Mdh–2
Gpi
Hk
Odh
Gpi
Odh
Pgd
Hk
Pgd
Pgm
Gpd
Pgm
Idh–2
Idh–2
Gpd
Fum
Est–6
Est–6

LINKAGE MAP OF *DROSOPHILA PSEUDOOBSCURA*

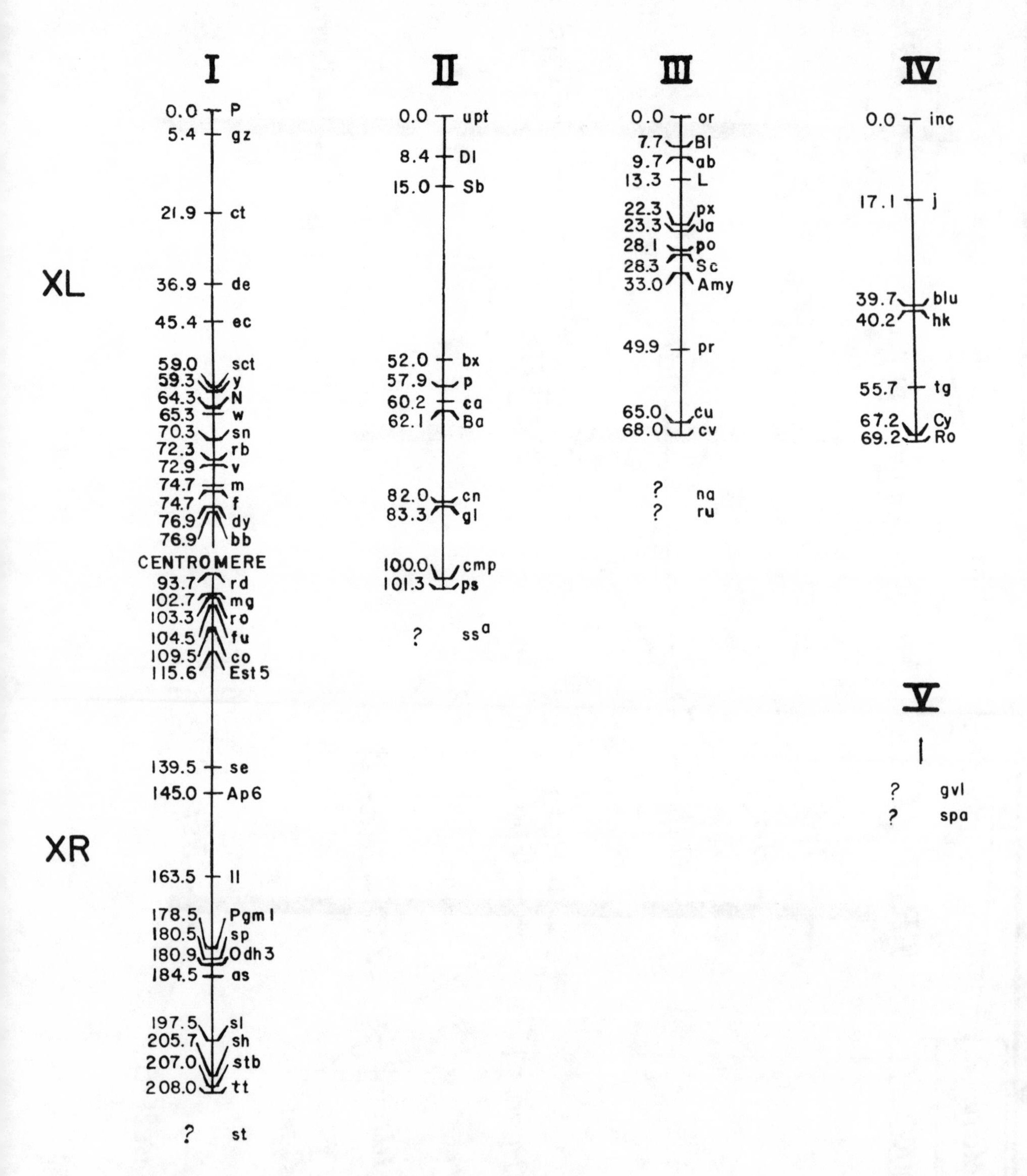

LINKAGE MAP OF THE FRUIT FLY *Drosophila pseudoobscura*

2N = 10

November 1986

Wyatt W. Anderson
Department of Genetics
University of Georgia
Athens, GA 30602

GENE SYMBOL	LOCUS	MUTANT EFFECT	MAP POSITION	REF
as	ascute	morphology	1-184.5	3,4,8
Ap6	*Adult acid phosphatase 6	adult enzyme	1-145.0	6
bb	bobbed	bristles, abdominal tergites	1-76.9	7,8
co	*compressed	eyes, morphology	1-109.5	3,4,7,8,10
ct	*cut	wings	1-21.9	3,4,8
de	dent	morphology	1-36.9	3,8
dy	dusky	wing color	1-76.9	3,7,8
ec	echinus	eyes, bristles, wings	1-45.4	8
Est5	*Esterase 5	enzyme	1-115.6	2,6
f	forked	bristles	1-74.7	5,7,8
fu	fused	wing veins	1-104.5	4,8
gz	glazed	eyes	1-5.4	4,8
ll	*lanceolate	wings	1-163.5	3,8
mg	magenta	eyes	1-102.7	3,8
m	miniature	morphology	1-74.7	3,4,8
N	Notch	wings	1-64.3	3,4,8
Odh3	*Octonol dehydrogenase 3	enzyme	1-180.9	6
Pgm1	*Phosphoglucomutase 1	enzyme	1-178.5	6
P	*Pointed	wings	1-0.0	3,4,8
rd	reduced	bristles	1-93.7	4,8
ro	rough	eyes	1-103.3	3,4,8
rb	ruby	eyes	1-72.3	8
st	scarlet	eyes	1-XR	8
sct	scutellar	bristles	1-59.0	3,4,8
se	*sepia	eyes	1-139.5	3,8
sh	*short	wing veins	1-205.7	3,4,7,8
sn	*singed	bristles	1-70.3	3,4,5,8
sl	slender	bristles	1-197.5	4,8
sp	*snapt	wing veins	1-180.5	3,8
stb	stubby	bristles	1-207.0	4,8
tt	*tilt	wings	1-208.0	3,8
v	*vermillion	eyes	1-72.9	3,4,8
w	*white	eyes	1-65.3	3,4,8
y	*yellow	body color	1-59.3	3,4,5,8
ss[a]	*aristapedia	aristae	2-?	3,8
Ba	*Bare	bristles	2-62.1	3,8,9
bx	*bithorax	morphology	2-52.0	3,8,9
cn	cinnabar	eyes	2-82.0	8
ca	claret	eyes	2-60.2	8
cmp	crumpled	wings, aristae	2-100.0	8
Dl	*Delta	wing veins	2-8.4	8
gl	*glass	eyes	2-83.3	3,8,9
ps	pauciseta	bristles	2-101.3	8,9
p	pink	eyes	2-57.9	3,8,9
Sb	Stubble	bristles	2-15.0	3,8
upt	*upturned	wings	2-0.0	8
ab	abrupt	wing veins	3-9.7	8,9
Amy	*Amylase	enzyme	3-33.0	11
Bl	*Blade	wing	3-7.7	8,9
cv	*crossveinless	wing vein	3-68.0	3,8,9
cu	curved	wings	3-65.0	8
L	*Lobe	eyes	3-13.3	3,8
Ja	Jagged	wings	3-23.3	8
na	narrow	wings	3-?	8
or	*orange	eyes	3-0.0	3,8,9
po	polychaete	bristles	3-28.1	8
pr	*purple	eyes	3-49.9	3,8,9
px	*plexus	wings	3-22.3	8
ru	rugose	eyes	3-?	8
Sc	*Scute	bristles	3-28.3	3,8,9
inc	*incomplete	wing veins	4-0.0	3,8,9
j	*jaunty	wings	4-17.1	3,8,9
blu	blunt	wings	4-39.7	8
hk	*hook	bristles	4-40.2	8
tg	tangled	wing veins	4-55.7	3,8,9
Cy	*Curly	wings	4-67.2	3,8,9
Ro	Rough	eye	4-69.2	8,9
gvl	grooveless	morphology	5-?	8
spa	sparkling	eye	5-?	1

*Indicates that the mutant is currently available from the National Drosophila Species Resource Center at Bowling Green State University.

The basic linkage map is that of Sturtevant and Tan (8). The distance between *co* and *se* has been increased by 23 map units (7), and the distance between *y* and *sn* has been increased by 3.8 map units (5). The designations of a few loci have been changed: Lobe (313.3) is used in place of Emarginate or Eyeless (3,8); cut (1-21.9) is used in place of beaded (3,4,9); and Delta (2-8.4) is used in place of Smoky (3,8,9).

REFERENCES

1. ANDERSON, W.W. and R.A. NORMAN. 1977. Dros. Inf. Serv. 52:11-12.
2. BECKENBACH, A.T. 1981. Dros. Inf. Serv. 56:23-24.
3. DONALD, H.P. 1936. J. Genet. 33:103-122.
4. LANCEFIELD, D.E. 1922. Genetics 7:335-384.
5. LEVINE, P.R. and E.L. LEVINE. 1954. Genetics 39:677-691.
6. PRAKASH, S. 1974. Genetics 77:795-804.
7. STURTEVANT, A.H. and E. NOVITSKI. 1941. Genetics 26:517-541.
8. STURTEVANT, A.H. and C.C. TAN. 1937. J. Genetics 34:415-432.
9. TAN, C.C. 1936. Genetics 21:796-807.
10. TAN, C.C. 1937. Proc. Natl. Acad. Sci. USA 23:351-356.
11. YARDLEY, D.G. 1974. Dros. Inf. Serv. 51:25.

LINKAGE GROUPS OF
Anopheles albimanus (Diptera: Culicidae)

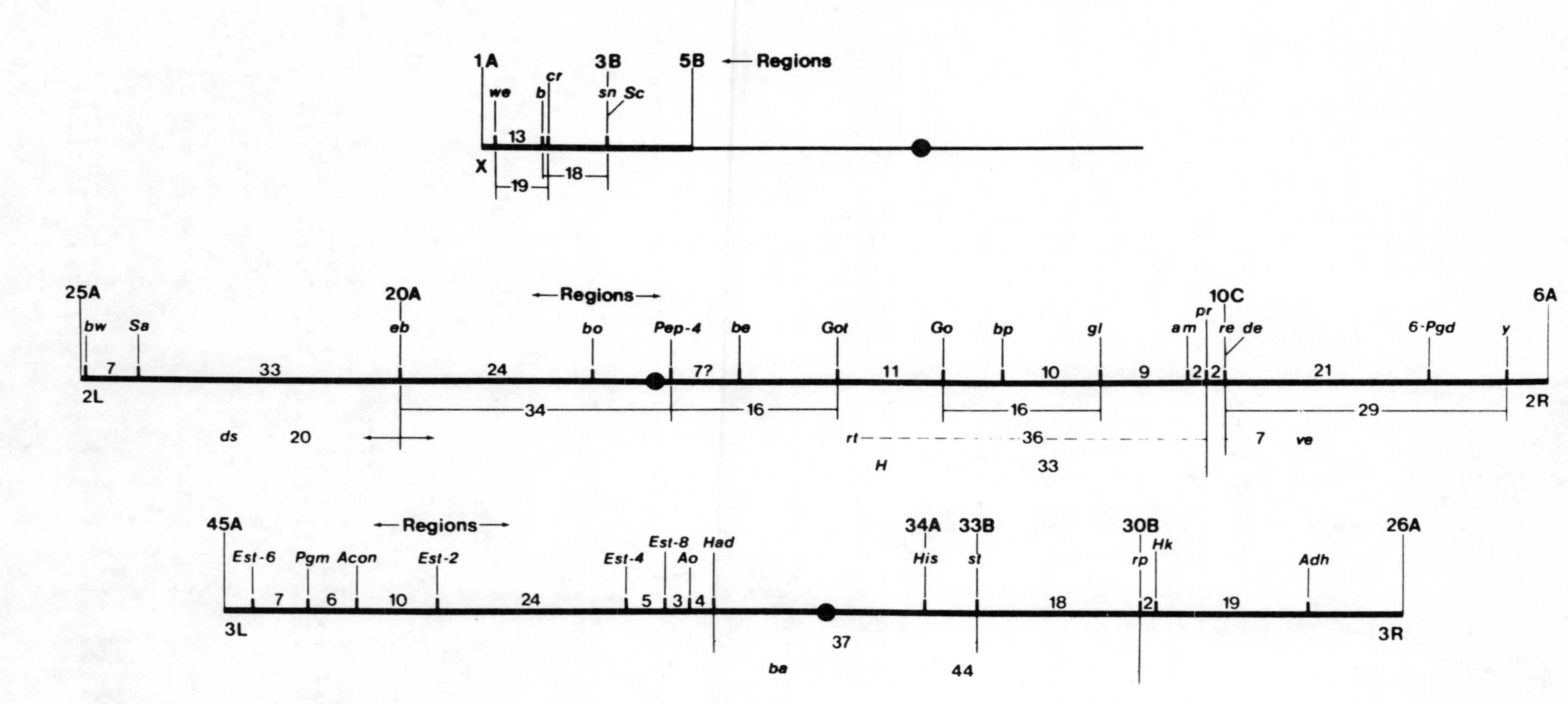

Linkage Map of the Mosquito (Anopheles albimanus) (2N=6)
August, 1989
Dr. Sudhir K. Narang
Dr. Jack A. Seawright
USDA, Agricultural Research Service
P. O. Box 14565
Gainesville, FL 32604

GENE SYMBOL	LOCUS DESCRIPTION	STAGE	INH	PEN	EXP	LG/CH	REF
cr	Curled, lateral abdominal hairs, lh(L,P)	L	Rec	F	U	X	21
we	White eye	L,P,A	Rec	F	U	X	21
sn	Snow eye	L,P,A	Rec	F	U	X	25
b	Bubble head, lh(female)	L,P	Rec	F	U	X	25,26
Sc	Scaleless palpi, lh (male larva)	A	Dom	F	V	X	28
D	Diminutive, associated with T(x;2R)8	A	Sd	F	U	X/2R	28
bw	Brown larva, apparent in adult male antenna	L,A	Rec	F	U	2L	27
eb	Ebony, lh	L	Sd	F	V	2L	1
ds	Diseased larva, lh(85-90% L,P)	L	Rec	F	U	2L	26
Sa	Sable, Lh(L,P)	L	Dom	F	V	2L	27
Dy	Dusky color	L	Sd	F	V	2L	28
bo	Bobbed anal setae, sterile females	L	Rec	F	U	2L	28
Tw	Twisted Tp(2R;3R)1	L	Dom	F	U	2R/3R	28
de	Dot eye, lh(L)	L	Rec	F	U	2R	26
bp	Bald palpi	A	Rec	F	V	2R	20
gl	Green larva	L,P	Rec	F	V	2R	18
re	Red eye	L,P,A	Rec	F	U	2R	22,23
am	Amber head capsule and saddle	L	Rec	F	U	2R	24
y	yellow	L	Rec	F	U	2R	27
ve	Vermillion eye	L,P,A	Rec	F	U	2R	22
pr^r	Propoxur resistance	L,P,A	Dom	F	U	2R	2
Pep-4	Peptidase-4	L,P,A	Cod	F	U	2R	13
Got-2	Glutamate oxaloacetate transaminase-2	L,P,A	Cod	F	U	2R	11
Go	Glucose oxidase	L,P,A	Cod	F	U	2R	10
6Pgd-2	6-Phosphogluconate dehydrogenase-2	L,A	Cod	F	U	2R	10
D	Diminutive	A	Sd	F	U	2R/X	28
be	Bent setae	L	Rec	F	U	2R	28

GENE SYMBOL	LOCUS DESCRIPTION	STAGE	INH	PEN	EXP	LG/CH	REF
H	Hooked (abdom. setae) lh	L	Sd	F	V	2R	28
sa	Short antenna in female, barren in male, female sterile	A	Rec	F	U	2R	28
ey	Eyeless, lh (most L)	L,P,A	Rec	F	U	2	28
ln	Lunate	L	Rec	F	U	2	26
rt	Retarded eye	L,P	Rec	F	U	2	28
K	Kinked palps	A	Dom	F	V	2	27
ba	Bald antenna, sex limited (males only)	A	Rec	I	V	3R	6,20
Adh	Alcohol dehydrogenase	L,P,A	Cod	F	U	3R	11
rp	Reduced palmate, ts	L	Rec	F	V(ts)	3R	19,23
Hk-1	Hexokinase-1	A	Cod	F	U	3R	4
st^{+}	Stripe	L,P	Dom	F	U	3R	16,23
st^{w}	White thorax, allele of st^{+}	L,P	Rec	F	U	3R	3
His	Histone gene repeat	L,P,A	-	-	-	3R	14
Tw	Twisted, associated with transposition Tp(2R;3R)1	L	Dom	F	U	3R/2R	28
Ao	Aldehyde oxidase	L,P,A	Cod	F	U	3L	9
Est-2	Esterase-2	L,P,A	Cod	F	U	3L	9,11,12
Est-4	Esterase-4	L,P,A	Cod	F	U	3L	5
Est-6	Esterase-6	L,P,A	Cod	F	U	3L	8,9,12
Est-8	Esterase-8	L,P,A	Cod	F	U	3L	5
Had	Hydroxy acid dehydroganase	L,P,A	Cod	F	U	3L	8
Acon-2	Aconitase-2	L,P,A	Cod	F	U	3L	12
Pgm	Phosphoglucomutase	L,P,A	Cod	F	U	3L	4
Dl	Dieldrin resistance	L,P,A	Dom	F	V	3	7
bl	Black larva, lh (L,P)	L	Rec	F	U	3	17
bar	Bar eye, lh (L,P)	L,P	Rec	F	U	Autosome	26
h	Hairy, lh(L)	L	Rec	F	U	Autosome	26
Acon-1	Aconitase-1	L,P,A	Cod	F	U	Autosome	15
Adk-2	Adenylate kinase-2	L,P,A	Cod	F	U	Autosome	15
a-Gpdh	a-Glycerophosphate dehydrogenase	L,P,A	Cod	F	U	Autosome	15
Got-1	Glutamate oxaloacetate transaminase-1	L,P,A	Cod	F	U	Autosome	15
Hk-2	Hexokinase-2	L,A	Cod	F	U	Autosome	15
Idh-1	Isocitrate dehydrogenase-1	L,P,A	Cod	F	U	Autosome	15
Mdh	Malate dehydrogenase	L,P,A	Cod	F	U	Autosome	15
Pgi	Phosphoglucose isomerase	L,P,A	Cod	F	U	Autosome	15
Pep-2	Peptidase-2	L,P,A	Cod	F	U	Autosome	15
Sodh-1	Sorbitol dehydrogenase-1	L,A	Cod	F	U	Autosome	15

Abbreviation used: L, larva; P, pupa; A, adult; INH, inheritance; PEN penetrance; EXP, expression; LG/CH, linkage group/chromosome arm (R=right arm, L=left arm); lh, lethal in homozygote condition; ts, temperature sensitive; Rec, recessive; Dom dominant; Sd, semidominant; Cod, Codominant; F, full; I, incomplete; U, uniform; V, variable.

The relative map lengths of linkage groups are proportional to the relative mitotic chromosome lengths. The cytological location of certain loci (eb, re, st and rp) on the polytene chromosomes were determined by deletion mapping (ref.23). Histone gene repeat was mapped to chromosome 3R by in-situ hybridization with cloned gene (ref.14). Approximately one-third of the mitotic chromosome is euchromatic (polytenized) and is represented as a thick line (region 1A-5B) in the map model. The remaining two-third is heterochromatic and is represented as a narrow line.

Generally crossing-over in male hybrids was 40-60% of that observed in female hybrids. Map distances derived from female F_1s were used in the map model. The distances among bw, sa and eb may not be acceptable estimates because these three mutants have similar, almost indistinguishable phenotypes and therefore only one-half of the recombinants in trihybrid crosses could be identified. Bw is probably located near the free end of 2L (ref.28) Tw (twisted) phenotype is associated with a transposition, Tp(2R,3R); of a region of 2R (break-points 7B, 13B) to 3R (region 28a). Loci marked with dotted lines were mapped by two-point crosses and precise gene order is not available. Double-arrows means that the locus may be either towards the distal or proximal end from the locus of reference.

Eye color mutants, we(X) and ve (2R) are epistatic to st^+ (3R) but re (2R) is not. we(X) pistatic to re and ve. ba(3R) is dominant in some genetic backgrounds.

References

1. BENEDICT, M.Q., et al. 1979. Can. J. Genet. Cytol. 21:193-200.
2. KAISER, P.E., et al. 1979. Can. J. Genet. Cytol. 21:201-211.
3. KAISER, P.E., et al. 1981. Mosq. News. 41:455-458.
4. NARANG, S., et al. 1981a. Mosq. News. 41:99-106.
5. NARANG, S., et al. 1981b. J. Hered. 72:157-160.
6. NARANG, S., et al. 1982. J. Med. Entomol. 19:195-197.
7. NARANG, S., and J.A. SEAWRIGHT, 1982. In Steiner et al. (eds.), Recent Developments in the Genetics of Insect Disease Vectors, A Symposium Proceedings. Stipes Publishing Co., Champaign, IL. 231-289.
8. NARANG, S., and J.A. SEAWRIGHT. 1983. Biochem. Genet. 21:885-893.
9. NARANG, S., and J.A. SEAWRIGHT. 1983. Biochem. Genet. 21:653-660.
10. NARANG, S., et al. 1983. Can. J. Genet. Cytol. 25:567-572.
11. NARANG, S., et al. 1984. Can. J. Genet. Cytol. 26:590-594.
12. NARANG, S., et al. 1986. Biochem. Genet. 25:67-77.
13. NARANG, S., et al. 1986. J. Heredity. 78:278-279.
14. NARANG, S., et al. 1985. Genetics. 110:s55.
15. NARANG, S., et al. 1989. Unpublished data.
16. RABBANI, M.G., and J.A. SEAWRIGHT. 1976. Ann. Entomol. Soc. Am. 69:266-268.
17. RABBANI, M.G., et al. 1976. Can. J. Gen. Cytol. 18:51-56.
18. SEAWRIGHT, J.A., et al. 1979a. Mosq. News. 39:55-58.
19. SEAWRIGHT, J.A., et al. 1979b. J. Hered. 70:321-324.
20. SEAWRIGHT, J.A., 1981. Mosq. News. 41:660-665.
21. SEAWRIGHT, J.A., et al. 1982. Can. J. Genet. Cytol. 24:661-665.
22. SEAWRIGHT, J.A., et al. 1982. Mosq. News. 42:590-593.
23. SEAWRIGHT, J.A., et al. 1984a. Mosq. News. 44:568-572.
24. SEAWRIGHT, J.A., et al. 1984b. Mosq. News. 78:177-181.
25. SEAWRIGHT, J.A., et al. 1985. Can. J. Genet. Cytol. 27:74-82.
26. SEAWRIGHT, J.A., and M.Q. BENEDICT. 1985. J. Am. Mosq. Control Assoc. 1:227-232.
27. SEAWRIGHT, J.A., et al. 1985. Ann. Entomol. Soc. Am 78:177-181.
28. SEAWRIGHT, J.A., et al. 1989. Unpublished data.

LINKAGE GROUPS OF *Anopheles quadrimaculatus* Species A (Diptera: Culicidae)

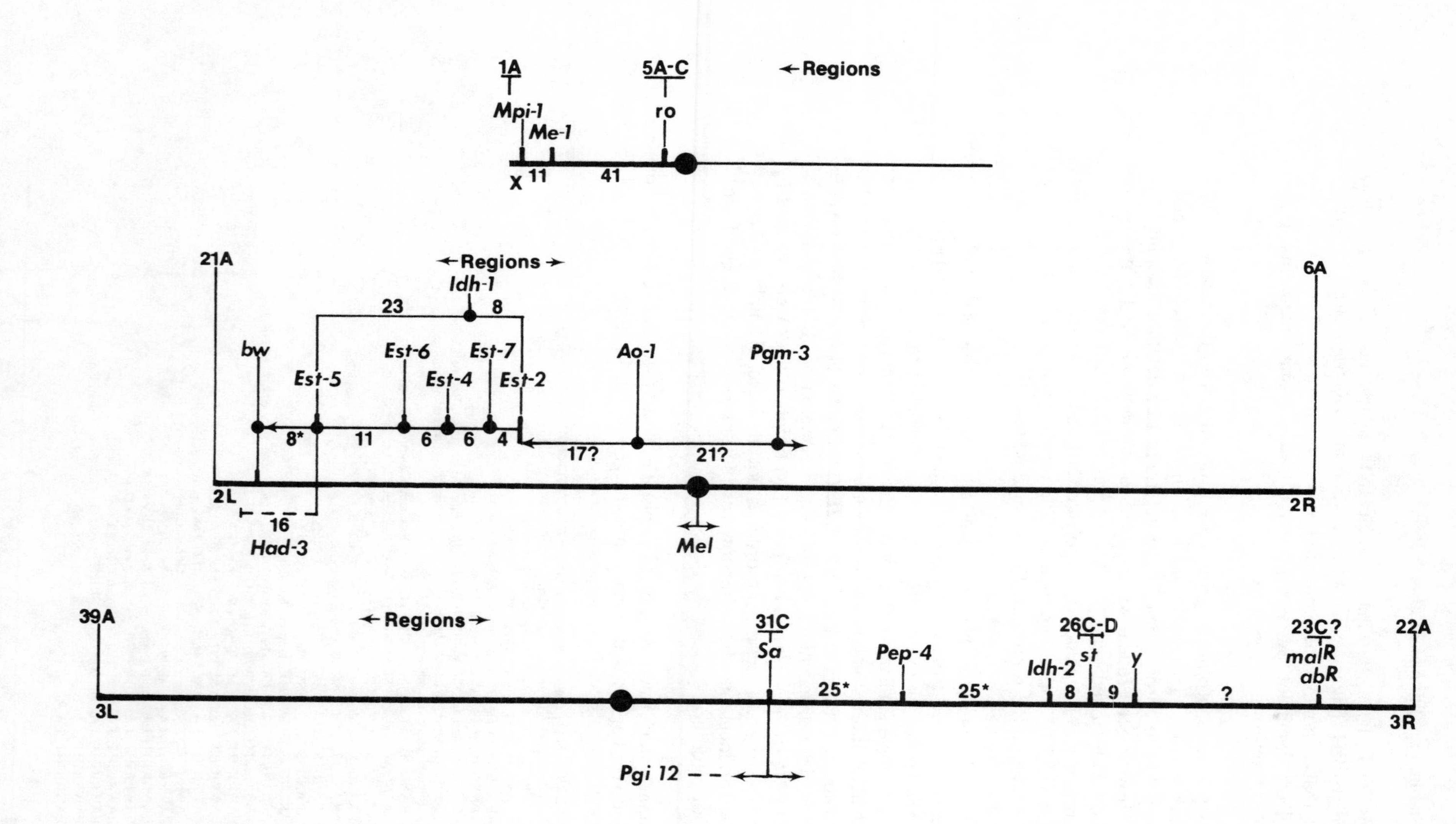

Linkage Map of the Mosquito (Anopheles quadrimaculatus species A) (2N=6)
August, 1989

Dr. Sudhir K. Narang
Dr. Jack A. Seawright
Ms. Sharon E. Mitchell
USDA, Agricultural Research Service
P. O. Box 14565
Gainesville, FL 32604

GENE SYMBOL	LOCUS DESCRIPTION	STAGE	INH	PEN	EXP	LG/CH	REF
ro	Rose eye	L,P,A	Rec	F	U	X	13
Me-1	Malic enzyme	L,P,A	Cod	F	U	X	9,16
Mpi-1	Mannose phosphate isomerase-1	A	Cod	F	U	X	9,16
Dl	Dieldrin resistance	A	ID	F	U	2	2,4
Mel	Melanotic, lh	L,P	Dom	F	V	2L	13
bw	Brown larva	L,P	Rec	F	V	2L	8,11,14
Idh-1	Isocitrate dehydrogenase-1	L,P,A	Cod	F	U	2L	6,9,15
Pgm-3	Phosphoglucomutase-3	A	Cod	F	U	2	15
Ao-1	Aldehyde oxidase-1	A	Cod	F	U	2	15
Est-2	Esterase-2	A	Cod	F	U	2	15
Est-4	Esterase-4	A	Cod	F	U	2	15
Est-5	Esterase-5	A	Cod	F	U	2	15
Est-6	Esterase-6	A	Cod	F	U	2	15
Est-7	Esterase-7	A	Cod	F	U	2	15
Had-3	Hydroxyacid dehydrogenase-3	L,P,A	Cod	F	U	2	15
Pgi	Phosphoglucose isomerase	L,P,A	Cod	F	U	3	16
Pep-4	Peptidase-4	L,P,A	Cod	F	U	3R	16
Idh-2	Isocitrate dehydrogenase-2	L,P,A	Cod	F	U	3R	6,16
St^{rd}	Red stripe	L,P	Dom	F	U	3R	12
Sa	Short antenna, lh	A	Dom	F	V	3R	13
St^{+}	Stripe	L,P	Dom	F	U	3R	3,8,11
mal^{R}	Malathion resistance	L,P,A	Dom	F	U	3R	7,18
ab^{R}	Abate resistance	L,P,A	Dom	F	U	3R	19
DDT^{R1}	DDT-resistance-1	L,P,A	Dom	F	U	2	1,4,18
DDT^{R2}	DDT-resistance-2	L,P,A	Dom	F	U	3R	4,18
y	Yellow larva	L,P	Rec	F	V	3R	6
Got-2	Glutamate oxaloacetate transaminase-2	A	Cod	F	U	autosome	15
Fdl	Fleur-de-leis	A	Dom	F	V	autosome	19
ba	Bald antenna, ssh sex-limited (males only) low expression in females	A	Rec	F	V	autosome	19
co	Collar, lh	P	Rec	-	V	autosome	19
bl	Black body, lh	L,P	Rec	F	U	autosome	17

GENE SYMBOL	LOCUS DESCRIPTION	STAGE	INH	PEN	EXP	LG/CH	REF
	Short palpi	A	Rec	I	V	autosome	8
	Long palpi	A	-	I	U V in males	autosome	8
	Wartoid palpi	A	-	-	U V in males	autosome	8
	Supernumerary mouth parts	A	Rec	-	V	autosome	8
	Wing Vein anastomosis	A	Rec	U	-	autosome	8
	Fused antennal segments (similar to Fdl mutant)	A	Dom	I	V	autosome	8
	Wart antennal section sex-limited (females only)	A	-	I	V	autosome	8

Notes: Anopheles quadrimaculatus species A linkage maps

(1) Abbreviation used: L, larva; P, pupa; A, adult; INH, inheritance; PEN penetrance; EXP, expression; LG/CH, linkage group/chromosome arm (R=right arm, L=left arm); lh, lethal in homozygote condition; Rec, recessive; Dom, dominant; ID, Incomplete dominance; F, full; I, incomplete, U, uniform; V, variable; ssh, semisterile in homozygotes.

(2) The relative map lengths of linkage groups are proportional to the relative mitotic chromosome lengths. The cytological location of certain loci (ro, st^+ and mal^R on the polytene chromosomes were determined by translocations (ref. 7, 11, and 13). Approximately l/3 of the mitotic chromosome (X) euchromatic (polytenized) and is represented as a thick line (region 1A-5C) in the map model. The remaining two-third is heterochromatic and is represented as a narrow line.

(3) Generally crossing-over in male hybrids was 40-60% of that observed in female hybrids. Map distances derived from female F_1s were used in the map model except those marked by * (from male hybrids). Loci marked with dotted lines were mapped by two-point crosses and precise gene order is not available. Double-arrows means that the locus may be either towards the distal or proximal end from the locus of reference.

(4) Eye color mutants, ro and brown body (bw) are epistatic to st^+ (3R).

REFERENCES

1. DAVIDSON, G. 1963. Bull. Wld. Hlth. Org. 29:177-184.
2. FRENCH, W.L. 1963. Thesis, University of Illinois, Urbana, Ill. 124 pp.
3. FRENCH, W.L. and J.B. KITZMILLER. 1962. Am. Zool. 177-184.
4. FRENCH, W.L. and J.B. KITZMILLER. 1964. Mosq. News 24:32-39.

6. KIM, S.S., et al. 1987a. J. Hered. 78:187-190.
7. KIM, S.S.,et al. 1987b. J. Am. Mosq. Control Assoc. 3:50-53.
8. KITZMILLER, J.B., and G.F. MASON. 1964. In J.W. Wright and R. Pal (eds.), Genetics of Insect Vectors of Disease. Elserier, Amsterdam. pp. 3-15.
9. LANZARO, G.C., et al. 1989. Unpublished.
10. MITCHELL, S.E. 1984. Master's Thesis, University of Florida, Gainesville, FL pp.
11. MITCHELL, S.E. and J.A. SEAWRIGHT. 1984a. J. Hered. 75:341-344.
12. MITCHELL, S.E. and J.A. SEAWRIGHT. 1984b. J. Hered. 75:421-422.
13. MITCHELL, S.E. and J.A. Seawright. 1989. J. Hered. 80:58-61.
14. NARANG, S., J.A. SEAWRIGHT. 1982. In Steiner et al. (eds.) Recent Developments in the Genetics of Insect Disease Vectors, A Symposium Proceedings. Stipes Publishing Co., Champaign, IL. pp. 231-289.
15. NARANG, S., et al., 1989a. J. Am. Mosq. Control Assoc. Submitted.
16. NARANG, S., et al., 1989b. J. Hered. Submitted.
17. SEAWRIGHT, J.A., and D.W. ANTHONY. 1972. Mosq. News. 32:47-50.
18. SEAWRIGHT, J.A., et al. 1986. In II Intl. Symp. Fruit Flies/Crete. pp 203-208.
19. SEAWRIGHT, J.A., et al. 1989. Unpublished.

Gene Map of the Parasitic Wasp *Nasonia vitripennis* (=*Mormoniella vitripennis*) 2N = 10

June, 1989

George B. Saul, 2nd
Department of Biology
Middlebury College
Middlebury, VT 05753

Linkage Group I

% Recombination	Mutant Gene: Symbol	Name	Phenotype
	rep	red-eyed pupae	slightly reddish eye color in +/R^{pe333} and +/R^{oy423} female pupae
2*			
	rdh 1	reddish	dark red eyes
11			
	rev 421	reverend	legs of pupae extend toward ventral midline
3			
	ga 251	garnet	red eyes
1*			
	hb 441	hunchback	thoracic segments compressed
3			
	R‡	R-complex	Many eye-color mutants; some also female-sterile or male-lethal
4			
	cur 321	current	red eyes
3*			
	cop 362	copper	frons copper-yellow
1*			
	cop 2†	copper	frons copper-yellow
1*			
	stp 211	stumpy	abdomen shortened
2			
	ga 351	garnet	red eyes
2			
	gl†	glass	eye facets poorly differentiated and reduced in number
1			
	pu	purple	dorsal thorax purple, frons blue
1*			
	ga 120†	garnet	red eyes
4*			
	cop 1	copper	frons copper-yellow
5			
	wa 362	white appendages	appendages (and entire body of young pupae) white, variable abnormalities of appendages
1*			
	stp 361	stumpy	abdomen shortened
1*			
	bk 362†	black	black eyes
17			
	vg	vestigial	rudimentary wings

Linkage Group II

% Recombination	Mutant Gene: Symbol	Name	Phenotype
	bl 108	blue	frons blue
15			
	rdh 5	reddish	red eyes
<1			
	cl 131†	cleft	ocellar region reduced; dorsal cleft between eyes; number of antennal segments reduced
<1			
	bl 106	blue	frons blue
<1			
	se 121†	small eyes	eyes small, fewer facets than normal
<1			
	mh 493	mahogany	dark red eyes
1			
	bl 109	blue	frons blue
25			
	unf 441†	unfolded	incomplete eclosion from pupal case in dorsal thorax; small mesothoracic wings

Linkage Group III

% Recombination	Mutant Gene: Symbol	Name	Phenotype
	st 5219	scarlet	bright red eyes
10			
	tl 627	tile	rust-red eyes
<1			
	bl 13	blue	frons blue
<1			
	bk 576	black	eyes slightly darker brown than wild type
3			
	bl 5101†	blue	frons blue
2			
	fx 331†	flexed	mesothorax duplicated, metathorax reduced; both pairs of wings of equal size; pupa flexed ventrally
9			
	cop 411†	copper	frons copper-yellow
37			
	bk 424	black	black eyes

Linkage Group IV

% Recombination	Mutant Gene		
	Symbol	Name	Phenotype
	bgs 532†	blue grass	frons green with blue glints
10			
	st 473	scarlet	bright red eyes
4			
	or 123	orange	light orange-red eyes
3			
	pu 416†	purple	frons deep blue or purple
1			
	vio 6	violet	frons deep blue or blue-purple

Linkage Group V

% Recombination	Mutant Gene		
	Symbol	Name	Phenotype
	ga 561	garnet	red eyes
10			
	pel 311	pellucid	gray-white eyes
12*			
	mod 306	modifier	changes red and scarlet eyes to yellow or orange
7*			
	st 318	scarlet	bright red eyes
13			
	mm(=bu) 251	mickey mouse (= bulgy)	eyes protuberant, dorsal head defective
5			
	pm 541	plum	frons blue or reddish-blue
6			
	sw 561†	short wings	small mesothoracic wings; metathoracic wings of pupae project out from body

*: Recombination frequency not measured directly: estimated from distances to common loci.

†: Mutant stock lost

‡: The R-complex is a group of loci which have not been shown to recombine. Mutations at four of the loci affect eye color, and at three other loci affect female fertility. An undetermined number of loci in the complex affect viability of males[3,5].

The five linkage groups of *Nasonia* have not been matched to specific chromosomes, so there *is no* evidence that the linkage groups represent all five chromosomes of the genome. In fact, unpublished data suggest that *bk 424* may show loose linkage with Group II mutants. Linkage *groups* II and III may therefore be part of a single group.

Most mutants of *Nasonia*, including many which have not been mapped, and a *variety of* wild type stocks, are maintained at Middlebury College by G.B. Saul. Collections are also maintained at Loyola College of Baltimore (G.W. Conner) and The University of Rochester (J.H. Werren).

REFERENCES

1. Altman, P.L. and D.S. Dittmer, ed. 1972. Biology Data Book, Vol. 1. Federation of American Societies for Experimental Biology, Washington, D.C., pp. 48-49.

2. Saul, G.B. and M. Kayhart 1956. Genetics 41:930-937.

3. Saul, G.B., et al. 1965 Genetics 52:1317-1327.

4. Saul, G.B., et al. 1967 Genetics 57:369-384.

5. Whiting, A.R. 1965 Adv. Genet. 13:341-358.

6. Whiting, A.R. 1967 Quart. Rev. Biol. 42:333-406.

7. Whiting, P.W. 1950 Genetics 35:699.

8. Whiting, P.W. 1955 Genetics 40:602.

INDEX